(De)Automating the Future

Historical Materialism Book Series

The Historical Materialism Book Series is a major publishing initiative of the radical left. The capitalist crisis of the twenty-first century has been met by a resurgence of interest in critical Marxist theory. At the same time, the publishing institutions committed to Marxism have contracted markedly since the high point of the 1970s. The Historical Materialism Book Series is dedicated to addressing this situation by making available important works of Marxist theory. The aim of the series is to publish important theoretical contributions as the basis for vigorous intellectual debate and exchange on the left.

The peer-reviewed series publishes original monographs, translated texts, and reprints of classics across the bounds of academic disciplinary agendas and across the divisions of the left. The series is particularly concerned to encourage the internationalization of Marxist debate and aims to translate significant studies from beyond the English-speaking world.

For a full list of titles in the Historical Materialism Book Series available in paperback from Haymarket Books, visit: www.haymarketbooks.org/series_collections/1-historical-materialism.

(De)Automating the Future

*Marxist Perspectives
on Capitalism and Technology*

Edited by
Johannes Fehrle
Marlon Lieber
J. Jesse Ramírez

Haymarket Books
Chicago, IL

First published in 2024 by Brill Academic Publishers, The Netherlands
© 2024 Koninklijke Brill NV, Leiden, The Netherlands

Published in paperback in 2025 by
Haymarket Books
P.O. Box 180165
Chicago, IL 60618
773-583-7884
www.haymarketbooks.org

ISBN: 979-8-88890-545-6

Distributed to the trade in the US through Consortium Book Sales and Distribution (www.cbsd.com) and internationally through Ingram Publisher Services International (www.ingramcontent.com).

This book was published with the generous support of Lannan Foundation, Wallace Action Fund, and the Marguerite Casey Foundation.

Special discounts are available for bulk purchases by organizations and institutions. Please call 773-583-7884 or email info@haymarketbooks.org for more information.

Cover art and design by David Mabb. Cover art is an adaption of *Luibov Popova Untitled Textile Design on William Morris Wallpaper*, screen print on wallpaper (2010).

Printed in the United States.

Library of Congress Cataloging-in-Publication data is available.

Contents

Notes on Contributors VII

Introduction: Marxism and the Technology Debate 1
   Johannes Fehrle, Marlon Lieber and J. Jesse Ramírez

PART 1
Histories of the (De)Automation Debate

1   Production and De-humanization: Herbert Marcuse, Günther Anders, and the Marxian Response to Automation   57
   Jason Dawsey

2   Nowhere to Go: Automation, Then and Now   83
   Jason E. Smith

3   Time to Automate: The Hidden Labour of Automation   111
   Christina Gratorp

PART 2
Key Concepts in the Automation Debate

4   Capitalism without Workers: On the Impossibility of Automation and Its Relation to the Question of Value   135
   Benjamin Ferschli

5   Deskilling: Automation and Alienation   158
   Amy Wendling

6   De-alienated Labour, Technology, and the Social Heart of Socialism   178
   Jeff Noonan

PART 3
Automation, Labour, and Resistance: Production, Distribution, Representation

7 The Transformation of the Retail Sector: Automation and the Warehouse 203
   Larry Liu

8 Automation along Global Supply Chains: How RFID-Systems Will Transform Work and Power Relations in the Supply Chain of Fast Fashion 226
   Steffen Reitz

9 End Meeting: A Workers' Inquiry into the Algorithmic University 256
   Robert Ovetz

10 Automation of Artistic Labour and Its Limits 274
   Jens Schröter

   Afterword: Stagnation, Circulation, and the Automated Abyss 288
   James Steinhoff, Atle Mikkola Kjøsen and Nick Dyer-Witheford

   Works Cited 311
   Index 347

Notes on Contributors

Jason Dawsey
is a historian in The National WWII Museum's Jenny Craig Institute for the Study of War and Democracy, where he researches prisoners of war, anti-Nazi resistance movements, and the Holocaust. Dawsey is the author of several articles and book chapters on the philosophical and political thought of the German-Jewish technology critic and anti-nuclear militant, Günther Anders, and co-edited (with Günter Bischof and Bernhard Fetz) *The Life and Work of Günther Anders: Émigré, Iconoclast, Philosopher, Man of Letters* (Studien Verlag, 2015). As part of his long interest in the history of socialism, he is currently working on a monograph about American Trotskyists during World War II.

Nick Dyer-Witheford
is a Professor in the Faculty of Information and Media Studies at the University of Western Ontario. He is, with Atle Mikkola Kjøsen and James Steinhoff, author of *Inhuman Power: Artificial Intelligence and the Future of Capitalism* (Pluto Press 2019) and, with Allessandra Mularoni, of *Cybernetic Circulation Complex: Big Tech and Planetary Crises* (Verso Books, forthcoming 2025).

Johannes Fehrle
is a member of the Biomaterialities Research Group at the Humboldt University of Berlin working on questions of political economy and ecology. He holds a Ph.D. in American literature and has published on a wide range of topics from literary, film, and media studies to political economy and ecology.

Benjamin Ferschli
is a doctoral research at the University of Oxford, researching work, employment, and technological change.

Christina Gratorp
is a PhD candidate in Environmental and Energy Systems Studies at Lund University prior to which she worked two decades as an embedded systems engineer.

Atle Mikkola Kjøsen
is Assistant Professor in the Faculty of Information and Media Studies at the University of Western Ontario. His current research lies at the intersections between Marxist political economy, media theory, retailing, and consumption, the latter with a focus on artificial intelligence and smart technology.

Marlon Lieber

is an assistant professor of American Studies at Goethe University Frankfurt. He is the author of *Reading Race Relationally: Embodied Dispositions and Social Structures in Colson Whitehead's Novels* (transcript, 2023) and the co-editor (with Dennis Büscher-Ulbrich) of a special issue of *Amerikastudien / American Studies* on "Marxism and the United States."

Larry Liu

is an assistant professor of sociology and anthropology at Morgan State University. His research interests are in the future of work, automation and technological change at the workplace. He has also published on political activism, universal basic income, and global political economy. His blogs are available at https://liamchingliu.wordpress.com/.

Jeff Noonan

is Professor of Philosophy at the University of Windsor. He is the author of *Critical Humanism and the Politics of Difference* (2003), *Democratic Society and Human Needs* (2006), *Materialist Ethics and Life-Value* (2012), *Embodiment and the Meaning of Life* (2018), *The Troubles with Democracy* (2019), and *Embodied Humanism* (2022) along with dozens of peer reviewed articles and book chapters. He has written for alternative and progressive websites in Canada and abroad and maintains an active blog at http://www.jeffnoonan.org/.

Robert Ovetz

is a senior lecturer in Political Science at San José State University. He is the author of *When Workers Shot Back: Class Conflict from 1877 to 1921* (Brill 2018 and Haymarket 2019), *We the Elite: Why the U.S. Constitution Serves the Few* (Pluto 2022), and a forthcoming book on NGOs and capitalism (Haymarket 2024). He is the editor of *Workers' Inquiry and Global Class Struggle: Strategy, Tactics, Objectives* (Pluto, 2020), co-editor with Kari Lydersen and Kevin Van Meter of *Real World Labor*, Vol. 4 (Dollars & Sense 2024), and an Associate Editor of and contributor to *The Routledge Handbook of the Gig Economy*, edited by Immanuel Ness (Routledge 2022). Robert is a labor writer for *Dollars & Sense* and *The Chief* magazine.

J. Jesse Ramírez

is a teacher and independent scholar. He is the author of three works of Marxist cultural studies: *Un-American Dreams: Apocalyptic Science Fiction, Disimagined Community, and Bad Hope in the American Century* (2022), *Rules of the Father in The Last of Us: Masculinity among the Ruins of Neoliberalism* (2022), and

Against Automation Mythologies: Business Science Fiction and the Ruse of the Robots (2020). Jesse has also coedited the collection *Work: The Labors of Language, Culture, and History in North America* (2021).

Steffen Reitz

is an independent researcher based in Frankfurt am Main. He has an M.A. in American studies, Latin American studies and Political Science and a teacher's degree. He's interested in media and technology studies, psychoanalysis and political economy. His research usually focuses on issues where these fields meet.

Jens Schröter

is Professor and chair of media studies at the University of Bonn and Co-Director of the vw-Main Grant "How is Artificial Intelligence Changing Science?" His recent publications include *Medien und Ökonomie* (Springer 2019), (with Christoph Ernst) *Media Futures: Theory and Aesthetics* (Palgrave 2021), and (with Andreas Sudmann et al., eds.) *Beyond Quantity: Research with Subsymbolic AI* (Transcript 2023). Visit www.medienkulturwissenschaft-bonn.de / www.theorie-der-medien.de / www.fanhsiu-kadesch.de.

Jason E. Smith

lives in Los Angeles, and is the author of *Smart Machines and Service Work: Automation in an Age of Stagnation* (Reaktion 2020). He frequently writes for the "Field Notes" section of the *Brooklyn Rail*, and is currently writing about the political economy of demographic decline.

James Steinhoff

is Assistant Professor in the School of Information and Communication Studies at University College Dublin. His research focuses on automation and the political economy of AI and data. He is the author of *Automation and Autonomy: Labour, Capital and Machines in the Artificial Intelligence Industry* (Palgrave 2021) and co-author of *Inhuman Power: Artificial Intelligence and the Future of Capitalism* (Pluto 2019).

Dr. Amy Wendling

is Professor of Philosophy and Associate Dean of Humanities and Fine Arts at Creighton University, a Jesuit University in Omaha, Nebraska, in the United States. She is the author of two books: *Karl Marx on technology and alienation* (Palgrave MacMillan, 2009 and in German with Dietz Verlag 2022) and *The ruling ideas: bourgeois political concepts* (Lexington 2012). She has written many articles about the implications of the capitalist social form on human life, most recently for the American Medical Association's Journal of Ethics, about how loneliness is bad for human health.

INTRODUCTION

Marxism and the Technology Debate

Johannes Fehrle, Marlon Lieber and J. Jesse Ramírez

When trying to wrap his head around the impact of steam and machinery, Karl Marx could not have imagined AI, the internet, or fully autonomous robots. Yet Marx's analysis of technology in texts like the much-discussed 'Fragment on Machines' in the *Grundrisse* still seems remarkably contemporary and clairvoyant today. When the 'Fragment' famously theorises how 'general social knowledge has become a direct force of production, and to what degree, hence, the conditions of the process of social life itself have come under the control of the general intellect',[1] we can relate this to open-source software and wikis. Likewise, Marx's description of the '*automatic system of machinery*' as the 'most complete, most adequate form' of machinery evokes in the twenty-first century images of robotic factories. Thus, it is hardly surprising that texts like the 'Fragment' have once again fallen on such fertile soil in the past decades.[2] The development of productive forces, which Marx describes as 'the material conditions to blow this foundation [of capital] sky-high', has inspired recent socialist, social democratic, and communist visions of automation by the likes of Paul Mason, Alex Williams and Nick Srnicek, Peter Frase, and Aaron Bastani.[3] The concept of the general intellect has been used by Mason to proclaim the imminent possibility of and, indeed, the *need* for a transition to 'postcapitalism', while Bastani has proposed a 'Fully Automated Luxury Communism'.[4]

Leftist automation discourse can build not only on the 'Fragment', but also on a rich foundation in Marx's and Engels's writing from the *Poverty of Philosophy* (1847) through the *Communist Manifesto* (1848) and the preface to the *Contribution to the Critique of Political Economy* (1859), which lay out the idea that productive forces will eventually come into such an irresolvable contradiction with the social relations that gave rise to them that they will explode the social order and give birth to a new one. This line of thinking, which sees capitalism's demise always around the next technological corner, remains strong

---

1 Marx 1993, p. 706.
2 Marx 1993, p. 692.
3 Marx 1993, p. 692; Mason 2016; Williams and Srnicek 2014; Srnicek and Williams 2015; Frase 2016; Bastani 2019.
4 Mason 2016; Bastani 2019.

today. Proponents usually invoke what Gerry Cohen has described as the idea of 'fettering',[5] which posits that the true potential of technology – to reduce necessary labour time to a minimum and unfold humanity's mastery over nature – is held back by capitalist social relations and the pursuit of surplus value. There are certainly passages in Marx's writing, not least in the 'Fragment', that seem to suggest that this is what Marx really thought.

There is, however, another strand in Marx's thinking on technology that is much darker. The 'Fragment' hints at another side of machinery: the same sentence developing the idea of an *'automatic system of machinery'* continues to describe 'workers themselves ... merely as its [the automaton's] conscious linkages'. Marx notes that 'the direct means of labour, is *superseded by a form posited by capital itself and corresponding to it*', and that the 'worker's activity, reduced to a mere abstraction of activity, is determined and regulated on all sides by the movement of the machinery'.[6] These less cheerful passages in the 'Fragment' and elsewhere have inspired a critical tradition in Marxist thought that sees technology as a materialisation of capitalism's logic of accumulation and of its abstract domination. Thus, these critics claim, technology cannot be easily separated from the antagonistic forces that give rise to it. In *Inhuman Power* (2019), Nick Dyer-Witheford, Atle Mikkola Kjøsen, and James Steinhoff offer a dystopian vision in which full automation does not abolish capitalism but supplies the technical means to literalize Marx's claim in *Capital* Volume I, that value is an 'automatic subject'. Moreover, Jason E. Smith has recently rediscovered radical auto-worker James Boggs's prophetic claim that automation creates a 'surplus people' no longer able to find steady employment. This might have debilitating consequences for the workers' movement since it exacerbates the fragmentation of those forced to rely on wage labour in order to have access to the means of life.

This volume gathers chapters that critically investigate automation's ambivalences from inter-disciplinary Marxist perspectives. The contributions raise questions about automation's affordances for postcapitalism, its transformation of manual and mental labour, and its role in the intensification of class antagonisms and exploitation. By examining the changed range of action of labourers *vis-à-vis* capital in workplaces that have become more automated in both production and distribution, the contributions re-examine old ways of resistance and suggest new ones. After all, capitalism only wants to automate the future in various ways; the point remains to change it.

---

5 Cohen 2000, pp. 326–41.
6 Marx 1993, pp. 692–3; second emphasis added.

Whereas many recent leftist analyses of technology take continuing automation for granted and project either mass unemployment or postcapitalist versions of the 'rise of the robots', this volume uses the figure of 'deautomating the future' to question such assumptions. It refocuses Marxist thought on core concepts and contexts and develops a more robust vocabulary for analysing and critiquing automation's past, present, and future. While they are unified by their shared Marxian frameworks, the contributions engage broadly with key problems in the automation debate from a range of thematic and disciplinary perspectives. They intervene in existing debates and add perspectives that have been mostly neglected or underdeveloped in previous contributions, such as the history of leftist automation discourse, racial capitalism, sectoral case studies, the need to resist automation to ensure pockets of non-alienated labour in a future socialist society, and the role of art and representation. In this way, the volume enters one of the central debates in current Marxist discourse, probing and developing its key assumptions, and widening the scope of our understanding of how capitalist technology and human resistance can shape our collective future.

1   Was Marx a Productive-Force Determinist?

It seems fitting for a volume on Marxist perspectives on capitalism and technology to provide at the outset some of the main answers given to the question of technology's role in maintaining capitalism and transitioning to socialism and communism from the late nineteenth to the early twenty-first century. After all, these provide the theoretical basis for current attempts to make sense of new technologies and their implications for workers' lives and the prospects for a world after capitalism. In the following, we will first turn to some of the main arguments in Marx's own thinking on technology and automation before tracing how the debate has unfolded. Finally, we provide an overview of the contributions to Marxist technology studies collected in this volume.

There are two broad strands in Marxist discourse regarding the significance of productive forces and relations of production. Both positions respond to an 'inconsistency', if not an outright contradiction, in Marx's writing: a tension between Marx's early theoretical model of 'productive-force determinism', which he was unwilling to abandon even as his historical findings from the 1860s onwards revealed that the modern use of machinery was a consequence rather than a cause of the transition from feudalism to capitalism, and his later prioritisation of social relations of production, for which the Economic

Manuscripts of 1861–63 serve as the watershed.[7] The history of Marxist thought on technology and automation moves around these two poles. Thus, we agree with Louis Althusser when he notes that it is possible to 'write a history of the Marxist worker's movement by considering the answer given to the following question: Within the unity of productive forces/relations of production, to which element should we assign *primacy*, theoretically and politically?'[8]

It is possible to discern a tendency to emphasise the power of the productive forces to influence the course of history, especially in texts written before the late 1850s. In the manuscripts usually called *The German Ideology*, a text often said to offer the first formulation of 'historical materialism [...] as an integral theory',[9] Marx and Engels write: '[m]en [*sic*; the original reads "die Menschen" (humans)] are the producers of their conceptions, ideas, etc., that is, real, active men, as they are conditioned by [*wie sie bedingt sind durch*] a definite development of their productive forces and of the intercourse corresponding to these'.[10] Formulations such as this may give readers the impression that the authors grant primacy to the forces of production in their analyses of history. Moreover, the manuscript includes several passages that theorise fettering. Marx and Engels write that there are '*social relations* corresponding to a definite stage of production', which can 'become fetters on the existing productive forces'.[11] Yet it would be too simplistic to equate productive forces with technology. There are other passages in which the notion of productive forces is clarified: 'a certain mode of production, or industrial stage, is always combined with a certain mode of co-operation, or social stage, *and this mode of co-operation is itself a "productive force"*'.[12] That is, the term 'productive forces' does not merely denote a set of machines, tools, and other instruments of labour – technology – but also includes the social relations human beings establish in and through their productive activity. Surveying the discussion about technological determinism in Marx in the mid-1980s, Donald MacKenzie concludes that '[t]he inclusion of labor power as a force of production thus admits conscious

---

7   Malm 2018a, pp. 176–7, and *passim*. On the role of technology in the transition from feudalism to capitalism, see also the historical studies by 'political Marxists' such as Robert Brenner (1977) or Ellen Meiksins Wood (2016, pp. 108–45). Søren Mau also argues that the 1861–3 manuscripts constitute 'the first thorough empirical and historical study of modern industrial production' (2021, p. 230, our translation).

8   Althusser quoted in Malm 2018a, p. 178.

9   Editors of Marx and Engels 1976a, p. xiii. On the editorial history of the 1845–6 manuscripts, see Carver and Blank 2014.

10 Marx and Engels 1976a, p. 36.

11 Marx and Engels 1976a, p. 231, original emphasis.

12 Marx and Engels 1976a, p. 43, emphasis added.

INTRODUCTION: MARXISM AND THE TECHNOLOGY DEBATE 5

human agency as a determinant of history: it is people, as much as or more than the machine, that make history'.[13] In short, despite the existence of passages that imply that the course of history is determined by the development of the forces of production, Marx avoids identifying the latter with technology by insisting that the way in which labour is socially organized is a productive force, too.

The manuscripts that became *The German Ideology* were never completed and left, as Marx famously put it, to the 'gnawing criticism of the mice'.[14] Yet one of the books whose publication Marx saw through, *The Poverty of Philosophy*, contains the statement that seems to most explicitly reveal his technological determinism: 'The hand-mill gives you society with the feudal lord; the steam-mill, society with the industrial capitalist'.[15] Here, Marx seems to put forward a 'universal-historical developmental hypothesis' according to which the forces of production account for the 'unity of history'.[16] But while Marx's quip is eminently quotable, *The Poverty of Philosophy* complicates what seems to be a unidirectional relationship between technology and society. On the one hand, we read that '[t]ill now the productive forces have been developed in virtue of this system of class antagonisms'.[17] This idea is further developed later in the text when Marx writes, in language anticipating chapter 15 of *Capital* Volume I, that 'strikes have regularly given rise to the invention and application of new machines. Machines were, it may be said, the weapon employed by the capitalists to quell the revolts of specialised labour'.[18] Here, social property relations and the class conflicts to which they give rise are the ground for technological innovation, not the other way around. On the other hand, Marx reiterates that the forces of production cannot be reduced to technology: 'Of all the instruments of production, the greatest productive power is the revolutionary class itself'.[19] Even if we cannot completely discount the possibility that *The Poverty of Philosophy* contains at least a 'soft determinism', as Robert Heilbroner has proposed,[20] Marx's attention to class relations as a factor informing the 'invention and application' of machines shows that he did not subscribe to a view of history in which an autonomous logic of technological development mechanically determines social relations even at this early stage.

---

13 MacKenzie 1984, p. 477.
14 Marx 1987b, p. 264.
15 Marx 1976, p. 166.
16 Elbe 2008, p. 9; our translations.
17 Marx 1976, p. 132.
18 Marx 1976, p. 207.
19 Marx 1976, p. 211.
20 Heilbroner 1967, p. 342.

Perhaps the long-running tendency to read Marx as a technological determinist – if a 'soft' one – can be explained by the fact that some of his most programmatic texts propound the fettering thesis. In *Manifesto of the Communist Party* and the 'Preface' to *Contribution to the Critique of Political Economy*, Marx repeatedly asserts that the end of the capitalist mode of production is imminent due to its internal contradictions. The vigorous, optimistic spirit of these passages may come at the expense of a rigorous conceptualisation of the relationships among the means of production, the forces of production, class struggle, social property relations, and historical tendencies. In the *Manifesto*, Marx and Engels claim that '[a]t a certain stage in the development' the 'means of production and exchange' within feudalism were 'no longer compatible' with feudal property relations and 'became so many fetters'. Consequently, these outdated property relations 'had to be burst asunder'.[21] Marx and Engels's equation of the productive forces with means of production and exchange may be read as suggesting that the former merely consist of a set of technologies (or 'means'), labour power notwithstanding. Later in the same text, when Marx and Engels begin to envision the rise of a revolutionary working-class movement, it is 'industry' or 'machinery' that seems to act as a subject, increasing the proletariat and 'equalis[ing]' the latter's 'interests and conditions of life'.[22] Thus, the 'development of Modern Industry' brings about the workers' 'revolutionary combination' in a self-undermining process that will 'inevitabl[y]' result in 'victory of the proletariat'.[23] Since the *Manifesto* does not elaborate on the conflict-ridden process whereby capitalists employ machinery as a weapon against striking workers, this passage can indeed suggest that technological development is the mechanism driving history.

The best-known version of the fettering thesis can be found in the 'Preface', where Marx writes:

> At a certain stage of development, the material productive forces of society come into conflict with the existing relations of production [...]. From forms of development of the productive forces these relations turn into their fetters. Then begins an era of social revolution.[24]

Nowhere in this brief text does Marx define the 'material productive forces', or, more precisely, whether labour power is part of them. This makes possible

---

21   Marx and Engels 1976b, p. 489.
22   Marx and Engels 1976b, p. 492.
23   Marx and Engels 1976b, p. 496.
24   Marx 1987b, p. 263.

the (mis)reading that identifies the forces of production exclusively with the instruments of production.

Ultimately, these programmatic claims that articulate the development of the forces of production with the promise of a coming revolution do not do justice to the more complex account that can be discovered in the main body of Marx's critique of political economy, both in its published version and in the many notebooks that Marx compiled in the years after 1857.

2   *Capital* and the Fraud of Automation[25]

Marx never used the term *automation*, which is an Americanism that first came into broad usage in the 1950s. The coining of *automation* is usually attributed to Delmar Harder, Vice President of Manufacturing at Ford Motor Company, although the word's novelty belies the fact that Harder was mainly referring to long-standing production processes. According to one origin story, Harder first used the word while in conversation with another Ford executive about improving parts-handling equipment and removing 'the delays often incurred due to the human element'.[26] As David Noble explains, when Harder called for more 'automation' at Ford, he 'simply meant an increase in the use of electro-mechanical, hydraulic, and pneumatic special-purpose production and parts-handling machinery which had been in existence for some time'.[27] Thus, while Marx never wrote about automation *per se*, he would have immediately understood that Harder was interested in implementing the techniques of the 'automatic workshop', which Marx first studied extensively for his critique of Proudhon in *The Poverty of Philosophy*. If Harder, in turn, had read the section of *Capital* Volume I, on the 'automatic system of machinery', he would have found a model of continuous production that basically describes his ambitions: 'As soon as a machine executes, without man's help, all the movements required to elaborate the raw material, and needs only supplementary assistance from the worker, we have an automatic system of machinery, capable of constant improvement in its details'.[28]

The improvements that Harder sought were not merely technical. Ford had always been anti-union, but the war years witnessed a dramatic spike in worker

---

25   The following sections reuse and update passages from Ramírez 2017.
26   Bean 1949, p. 389.
27   Noble 2011, p. 66.
28   Marx 1990a, p. 503.

resistance. Between 1941–1945, there were 773 strikes at Ford plants.[29] Automation was a solution not just to delays in production but to workers' control; not just to a general 'human element' but to the specific class power of organised labour. In this way, Ford continued a long tradition of using technology as an instrument of class struggle, a 'fix' whose history Marx's contemporaries had already noticed: 'According to Gaskell, the steam-engine was from the very first an antagonist to "human power", an antagonist that enabled the capitalist to tread underfoot the growing demands of the workers, which threatened to drive the infant factory system into crisis'.[30]

In Chapter 15 of *Capital*, on 'Machinery and Large-Scale Industry', Marx observes that 'it would be possible to write a whole history of the inventions made since 1830 for the sole purpose of providing capital with weapons against working class revolt'.[31] The part of that history that Marx tells in the chapter – a 'critical history of technology', as he intriguingly defines it in a footnote – hinges on what we call a process of 'unhanding'. According to Marx, the industrial revolution began not with new forms of motive energy such as steam, but rather with the transformation of the handy tool into a machine part: '[I]t is not labour, but the *instrument* of labour, that serves as the starting-point of the machine'.[32] Marx defines a machine as an appropriation of a worker's tool, which is consequently rendered *unhandy*: 'The machine, therefore, is a mechanism that, after being set in motion, performs with its tools the same operations as the worker formerly did with a similar tool'.[33] Only then does the human worker become a motive force that is functionally equivalent with wind, water, and steam – that is, until larger machine systems require motive forces more powerful than human muscle.

The concept of the unhandy helps to distinguish Marx's critical history of technology from Heideggerian phenomenologies that abstractly and ahistorically define tools as always already handy or 'ready-to-hand' (*zuhanden*).[34] The Heideggerian tool par excellence is the hammer. While a hammer is a rich example of preindustrial technical practice, it is useless for understanding unhandy tools like a steam hammer 'of such a weight that even Thor himself could not wield it'.[35] What Heidegger's phenomenology obscures is the

---

29 Noble 2011, p. 23.
30 Marx 1990a, pp. 562–3.
31 Marx 1990a, p. 536.
32 Marx 1990a, p. 500, n. 15; emphasis added.
33 Marx 1990a, p. 495.
34 Heidegger 1962, p. 98.
35 Marx 1990a, p. 507.

anti-phenomenological quality of capitalist machinery. Driven by the pursuit of surplus value, capitalist mechanisation radically curtails 'Dasein's' technical action. Harry Braverman influentially describes this process as the separation of 'conception' and 'execution'.[36] Workers who guide their tools according to their own knowledge and skills – that is, workers for whom tools are ready-to-hand – can exert control over the work process. Stripping that guiding role from workers by embedding capital's logic in machinery, whose functions can then be executed by deskilled machine tenders, remakes work and the working class in the interests of capital. Through what Marx calls the 'real subsumption of labour', the machine becomes 'capital's material mode of existence', 'the material foundation of the capitalist mode of production'.[37]

Apart from the controversies they sparked over the trajectory of deskilling, Braverman's contributions to Marxist thought can be faulted for relying too heavily on a mentalistic notion of skill. As Amy Wendling argues in her contribution to this volume, skill resides not only in workers' *ideas* or *imagination* ('conception') but also in their *perceptual-somatic* abilities. The anthropologist Tim Ingold provides a useful explanation of how working with a handy tool differs in this regard from working with unhandy machines. Ingold develops the following passage from Marx, which is liberally translated by Eden and Cedar Paul in a now obscure 1930 edition of *Capital*:

> ... the essential distinction, as [Marx] put it, lies 'between a man as a simple motor force and as a worker who actually handles tools' [*An vielem Handwerkszeug besitzt der Unterschied zwischen dem Menschen als bloßer Triebkraft und als Arbeiter mit dem eigentlichen Operateur eine sinnlich besonderte Existenz*].[38]

In Ingold's view, what Marx meant by the handling of tools is the worker's guidance, engaged attentiveness, continuous adjustment to the product and work environment, and awareness of the product's emerging form. Ingold also draws on the craftsmanship theorist David Pye's definition of skill as wilful *constraint* – not the application of force but its moderation, as when the dentist constrains the force of her drill using judgement, dexterity, and care.[39] Whether the tool is powered by human or nonhuman motive energy – which is to say, whether the tool is Heidegger's hammer or an electric drill – is of secondary

---

36 Braverman 1998, p. 79.
37 Marx 1990a, p. 554.
38 Ingold 2000, p. 301.
39 Pye 1979, pp. 50–1.

significance, for what determines skill is the unity of 'the technically effective gesture' and 'immediate sensory perception'.[40] Ingold's framework is a brilliant way to unpack what Marx only implies by 'the sensually distinct' difference between the worker as sheer motive power and the worker as operator proper, namely, the sensual, embodied, tacit intelligence expressed through the operator's hand and in reciprocity with the object and the work environment.

Thus, when conception and execution are split, it is not only the subsumption of conceptual knowledge but also the destruction and remaking of an embodied, sensorial manifold that constitutes capital's control of the work process. The unhanding process replaces the worker's careful constraint with mechanical constraint; when the tool is guided by machines, constraint is predetermined, automatic, constant, and invariable.[41] In Marx's account of the effects of large-scale industry, unhandy workers become 'living appendages' of a 'lifeless mechanism' that is dumb to changes in the work environment and in the product's emergent form.[42] Furthermore, such deskilled and cheapened work can now increasingly be done by exploited women and children and is spread out over a working day that the capitalist lengthens and intensifies in order to take full advantage of the machine's value before it wears down or becomes technologically obsolete. Workers who are replaced by machines become members of the 'surplus population', or 'a population which is superfluous to capital's average requirements'.[43]

In sum, Marx's account of capitalist automation in Chapter 15 foregrounds its destructiveness of work and workers (not to mention nature, which Marx addresses in the chapter's final section on agriculture). But what makes this destructiveness especially egregious is that large-scale industry could potentially shorten work time and reduce work's physical burden. After all, while the unhanding of tools impoverishes the worker's perceptual-somatic experience of work, it can also reduce the experience of hard physical labour in which the immediate and reciprocal relation with the object and the work environment is precisely the reason that such work is exhausting (and avoided by many who have the opportunity to do so). If you have ever used a jackhammer, for example, you know that you must carefully constrain the tool and dexterously adjust to changes, but this does not say much for jackhammering. If given the choice, most people prefer to unhand this tool as quickly as possible. Marx's central argument is not a nostalgic defence of handy labour but

---

40 Ingold 2000, p. 301.
41 Pye 1979, p. 55.
42 Marx 1990a, p. 548.
43 Marx 1990a, p. 782.

an indictment of capitalism that underscores the way this system can express technological potential only in inverted form: 'machinery in itself shortens the hours of labour', but 'when employed by capital it lengthens them'; it 'lightens labour, but when employed by capital it heightens its intensity'.[44]

As William Clare Roberts explains, Marx's goal in Chapter 15 is to demonstrate that capitalism is a 'thoroughgoing fraud, which will never deliver on the promise of less and more attractive labor'.[45] This promise is as old as Adam Smith, who claimed that every sensible person immediately understands 'how much labour is facilitated and abridged by the application of proper machinery', and as contemporary as Kevin Kelly, the cyber-utopian who thinks that once robots inevitably take our jobs, our natural response will be: 'Wow, now that robots are doing my old job, my new job is much more fun and pays more!'.[46] In Roberts's view, the real political potential that Marx sees in large-scale industry is that the breadth and depth of its destructiveness demonstrates to workers that capitalism's promises are hollow, and that they cannot return to handicraft production. This is the major sense in which capitalism's technological development serves as a material condition of postcapitalism: it is not merely the objective technical basis of post-scarcity and abundance, but a condition for the working class's realisation that the only way forward is a collective emancipation that appropriates and reorganises large-scale industry.

3 *Grundrisse*: Free Time

If there is a Marxist theory of revolutionary automation, it is not in *Capital*, but in the 1857–8 manuscripts that have come to be known as the *Grundrisse*. When Martin Nicolaus, the translator of the first complete English edition of the *Grundrisse*, introduced the 'unknown Marx' to readers of *New Left Review* in 1968, he described the text as 'the only work in which [Marx's] theory of capitalism from the origins to the breakdown was sketched out *in its entirety*'.[47] Nicolaus suggested that if the problem of capitalism's collapse had perplexed so many Marxists, it was not because Marx had no solution, but rather because so few Marxists had read the *Grundrisse*, a pre-*Capital* manuscript that was mostly unknown until the mid-1950s. In the published volumes of *Capital*, Marx never makes it all the way through his immanent critique of capital-

---

44   Marx 1990a, pp. 568–9.
45   Roberts 2017, p. 172.
46   Smith 1981, p. 19; Kelly 2012.
47   Nicolaus 1968, p. 43, emphasis added.

ism; the stalled project thus resembles 'a mystery novel which ends before the plot is unraveled'.[48] But in the *Grundrisse*, Nicolaus found the missing denouement. Here Marx 'envisages a capitalist productive apparatus more completely automated than that of any presently existing society' and shows how 'this economic organization must break down' at the very height of its productive powers.[49] Nicolaus concluded that the *Grundrisse* would reinvigorate Marxism because it proved that, far from being a relic of nineteenth-century thought, Marx's critique 'exposes even the most industrially advanced society at its roots'.[50]

Nicolaus's assessment of the *Grundrisse*'s significance was prescient. Perhaps more than any other of Marx's texts, the *Grundrisse* has become a fruitful source for developing Marxist theories of capitalism's crisis. One of the most influential readings originated in the Italian New Left. According to Mario Tronti, 'it was the *Grundrisse*, more than *Capital*, that seemed to offer theoretical weapons of sufficient analytical, stylistic and polemical novelty'.[51] The *operaismo* (workerist) movement considered the *Grundrisse* evidence that Marx was 'a proponent of [...] a catastrophic view of capitalism'.[52] In 1964, *Quaderni Rossi* published the most 'catastrophic' section of the text as 'Frammento sulle macchine'. Coincidentally, in 1964 quotations of the same section appeared in Marcuse's *One-Dimensional Man*. But while Marcuse approached the *Grundrisse* in the hypothetical subjunctive, contrasting what would happen *if* automation were fully realised with the actuality of its partial employment in capitalism, the Italian exegesis tended to be in the indicative, especially in the work of Antonio Negri. Negri argued that the radical transformation of production that Marx sketches in the 'Fragment' had already happened. As Negri and co-author Michael Hardt would later write in *Empire*, '[w]hat Marx saw as the future is our era'.[53]

In the 'Fragment', Marx claims that capitalism transforms production to suit its needs through the 'technological application of science', which in turn causes 'direct labour' to 'disappear as the determinant principle of production'.[54] The labourer becomes a 'watchman' and 'regulator' of machines who 'steps to the side of the production process instead of being its chief actor'.[55]

---

48   Nicolaus 1968, p. 54.
49   Nicolaus 1968, p. 57.
50   Nicolaus 1968, p. 61.
51   Tronti 2008, p. 232.
52   Tronti 2008, p. 232.
53   Hardt and Negri 2000, p. 364.
54   Marx 1993, pp. 699–700.
55   Marx 1993, p. 705.

Marx then makes a series of profound inferences from these changes. When 'labour in the direct form has ceased to be the great well-spring of wealth', he continues, 'labour time ceases and must cease to be its measure, and hence exchange value [must cease to be the measure] of use value'. In one of the few passages in which he directly refers to capitalism's demise, Marx concludes that 'production based on exchange value collapses'.[56] In other words, while *Capital* stresses the fraud of capitalism's development of large-scale industry, and depicts 'science, the gigantic natural forces, and the mass of social labour embodied in the system of machinery' as the 'power of the "master"',[57] the *Grundrisse* suggests that the same powers of science and technology can escape the master's control and crash capitalism.

To be sure, from the more mature perspective of *Capital*, the *Grundrisse*'s speculations are incoherent. Michael Heinrich has pointed out several important ways in which Marx's thinking in the *Grundrisse* clashes with his subsequent value theory in *Capital*. In the later work, Marx's concept of immediate labour finds its closest analogue in *concrete* labour, the specific, embodied expenditure of labour power that produces the particular use value of a commodity. Crucially, Heinrich notes that Marx makes a distinction in *Capital* that is missing in the *Grundrisse*, namely, that between the concrete labour embodied in a commodity and the *abstract*, or socially necessary, labour that the commodity crystallises.[58] *Capital* specifies that abstract labour, not concrete or immediate labour, is the substance of the commodity's value.

Given the conceptual importance of abstract labour in Marx's mature thought, it is unclear why the decline of immediate labour should have such a fatal effect on capitalism. The Marx of *Capital* would probably tell the Marx of the *Grundrisse* that even if concrete labour decreases, value does not collapse as a result. In this case, value represents a greater amount of the socially necessary labour embodied in machines, which pass on their value to their products. Indeed, in placing so much emphasis on immediate labour, and in claiming that this production-specific transformation is a point of systemic rupture, the Marx of the *Grundrisse* has not, in Heinrich's view, fully conceptualised the unity of production and exchange in the determination of value under capitalism. In *Capital*, Marx specifies that it is in commodity exchange that the concrete labour time expended in production is unintentionally homogenised and socially 'validated' as an instance of total social labour.[59] More broadly, we

---

56   Marx 1993, p. 705.
57   Marx 1990a, p. 549.
58   Heinrich 2013a, p. 204.
59   Heinrich 2004, pp. 50–1.

can say that the *Grundrisse*'s snapshot of collapse is inadequate to the more complex and interlocking totality of capitalism's contradictions, of which there is not just one, but *seventeen*, at least according to David Harvey.[60]

Heinrich's larger argument against the *Grundrisse* is that it reflects Marx's expectations of revolution prior to his full realisation of the significance of the Panic of 1857. Whereas Marx was initially concerned that capitalism would fall before he completed his book, he later saw the Panic as evidence that crises are normal elements of capitalism's reproduction. After capitalism survived the Panic, Marx 'no longer argued in terms of a theory of final economic collapse, and he no longer made out a direct connection between crisis and revolution'.[61] In Heinrich's view, the *Grundrisse* should not be considered a precursor to *Capital* but is instead a draft of a project that Marx *abandoned* in order to write *Capital*.

Yet if it is read as a dialectical companion of *Capital* and not as its substitute, the *Grundrisse* remains a valuable resource for thinking about the temporal dimensions of postcapitalism. In one of the most powerful readings of the *Grundrisse* to emerge from Frankfurt School critical theory, Moishe Postone argues that the *Grundrisse* underscores the potential victory in the unhanding of tools, namely, the replacement of the immediate unity of technical gesture and sensory-somatic experience with science, the 'accumulated collective knowledge of the species' (or what other traditions call the common).[62] 'One aspect of the development of large-scale industry', Postone writes, 'entails the historical constitution of socially general productive capacities and modes of scientific, technical, and organizational knowledge that are not a function of, and cannot be reduced to, workers' strength, knowledge, and experience'.[63] The contradiction that large-scale industry expresses is that while scientific and technological production reduces the social need for 'workers' strength, knowledge, and experience', capitalism nonetheless requires these workers as sources of surplus value. Expressed in terms of workers' labour time, capitalism's contradictory development of technology, 'presses to reduce labour time to a minimum, while it posits labour time, on the other side, as sole measure and source of wealth'.[64] While Postone stresses that there is no guarantee that

---

60 Harvey 2014.
61 Heinrich 2013b. Heinrich's views on Marx's crisis theory have sparked considerable controversy, especially with regard to his critique of the theory of the falling rate of profit. See the debate in *Monthly Review* at https://monthlyreview.org/features/exchange-with-heinrich-on-crisis-theory/.
62 Postone 1993, p. 298.
63 Postone 1993, p. 339.
64 Marx 1993, p. 706.

this contradiction will automatically destroy capitalism – and to this we should add the question of whether capitalism can ever dispense with its need for workers[65] – the *Grundrisse* moves beyond *Capital*'s negative critique and briefly envisions a communist future in which people have 'reappropriat[ed] [...] socially general capacities',[66] and in which the production of material wealth (and use values) is freed from the imperatives of value-producing labour time. In other words, by sketching an account of how automation, as the embodiment of common knowledge, could provide the basis for a restructuring of labour time, the *Grundrisse* complements Marx's famous claim in *Capital*, volume 3 that the 'true realm of freedom' is founded on '[t]he reduction of the working day'.[67]

4 The Orthodox Position

Considering these different strands in Marx's thinking about machinery raises the question how a less nuanced, more deterministic position came to be associated with Marxist orthodoxy for much of the twentieth century. According to Monika Reinfelder, 'it was Engels who first formulated the technicist version of the Marxian legacy' in his attempt to expand Marx's critique of political economy into a broader system.[68] Engels's *Dialectics of Nature* contains proclamations that 'in the most advanced industrial countries we have subdued the forces of nature and pressed them into the service of mankind'. From this mastery Engels derived the task of socialism as centralised, conscious planning of production, 'a task which is becoming "daily more indispensable", but also "with every day more possible"'.[69] Similar statements are found throughout *Anti-Dühring*, which was read even more widely by workers. While celebrat-

---

65   On this aspect, see Marx's comment in 'Wage Labour and Capital': 'If the whole class of wage workers were to be abolished owing to machinery, how dreadful that would be for capital, which, without wage labour, ceases to be capital!' (Marx 1977, p. 226). While Marx in the *Grundrisse* notes that '[f]or capital the worker is not a condition of production, only work is. If [capital] can make machines do it, or even water, so much the better' (Marx 1993, p. 498), value as tied to abstract human labour *is* a condition for capitalism, indeed its entire *raison d'etre*. The question of capitalism without workers is a key point for a Marxist evaluation of an imagined fully automated future that both Ferschli as well as Steinhoff, Kjøsen, and Dyer-Witheford address in this volume. See also Dyer-Witheford, Kjøsen and Steinhoff 2019.
66   Postone 1993, p. 357.
67   Marx 1991, p. 959.
68   Reinfelder 1980, p. 12.
69   Engels 1987b, quoted in Reinfelder 1980, p. 13.

ing the development of productive forces as a process of humanity's mastery over nature and proclaiming this process's completion only when the proletariat had seized the means of production and finally controlled its own destiny rather than being controlled by nature and history,[70] *Dialectics of Nature* also contains less triumphant realisations. In his chapter on 'The Part Played by Labour in the Transition from Ape to Man', Engels acknowledges the complexity and unpredictability of natural processes that continuously thwart humanity's plans of full control:

> Let us not, however, flatter ourselves overmuch on account of our human conquest over nature. For each such conquest takes its revenge on us. Each of them, it is true, has in the first place the consequences on which we counted, but in the second and third places it has quite different, unforeseen effects which only too often cancel out the first. The people who, in Mesopotamia, Greece, Asia Minor, and elsewhere, destroyed the forests to obtain cultivable land, never dreamed that they were laying the basis for the present devastated condition of these countries, by removing along with the forests the collecting centres and reservoirs of moisture.[71]

While Engels may have laid the groundwork for a teleological history of humanity's progressive mastery of nature through science and technology, this interpretation of the Marxian legacy throughout the first half of the twentieth century owes even more to Karl Kautsky, the Doyen of German Social Democracy, posthumous editor of some of Marx's unpublished notes, and a leading figure of the Second International. For Kautsky, the evolution of technology was not so much linked to social relations and economics as to our knowledge of nature. In dialling back the economic focus of Marx, Kautsky ends up with a materialism that focuses much more strongly on an understanding of nature (i.e. the development of the natural sciences) that seems to be independent of the economic basis. For Kautsky, the development of the 'material productive forces' is therefore in essence another name for the development of the knowledge of nature. The deepest foundation of the 'real basis' of the 'material base' of human ideology therefore appears as a mental process: the cognition of nature.[72] While different societies make different use of this knowledge and the technology it births, slaughtering, for instance, the American buffalo or whales in the arctic

---

70 Engels 1987a, p. 270.
71 Engels 1987b, pp. 460–1.
72 Kautsky 1927, p. 864; our translation.

'recklessly for a momentary benefit', technology when understood as a materialisation of the knowledge of objective natural laws seems largely value-neutral for Kautsky.[73]

Despite their opposition to Kautsky and the 'economism' and political 'reformism' of the Social Democrats,[74] the Bolsheviks' ideas on technology were likewise 'technicist'. Reinfelder attributes to the thinking of 'the Russian Kautsky', Georgi Plekhanov, a similar desire to refute the criticism of Marx's thought as 'economic determinism'. This led Plekhanov, like Kautsky, to 'concede [...] an absolute autonomy to (amongst other things) natural Science, where "a genius discovers laws the operation of which does not, of course, depend upon social relations"'.[75] For Plekhanov, the means of production directly define the social relations of production: '[O]n the basis of a particular state of the productive forces there come into existence certain relations of production'.[76]

For Plekhanov's pupil, Lenin, the development of the productive forces was likewise a means to advance the course of history towards the end of bourgeois rule. In a 1914 article for *Pravda* on Taylorism,[77] for instance, Lenin writes: 'The Taylor system – without its initiators knowing or wishing it – is preparing the time when the proletariat will take over all social production and appoint its own workers' committees for the purpose of properly distributing and rationalising all social labour'.[78] The idea behind this optimism was Lenin's conviction that the factory was not only 'a means of exploitation', but also 'a means of organisation' of the proletariat, which a socialist movement could build on.[79]

True to this conviction, and in light of the Soviet Union's lagging industrial development, Lenin and the other Bolshevik leaders remained dedicated to the development of (heavy) industry and technology. In *State and Revolution*, Lenin had outlined his belief that the proletariat could develop the productive forces more rapidly after seizing control of the means of production. This full development of the productive forces would erase the distinction between mental and physical labour and only then could society advance to the highest stage of

---

73   Kautsky 1927, p. 727.
74   Cf. Lenin 1960, Section III; Lenin 1964b, esp. Chapter VI.
75   Reinfelder 1980, p. 15. The 'production of nature' school based on Neil Smith's work, which proclaims that there is no external nature independent of human activity, shows how far the pendulum can swing in the other direction within a 'Marxist' tradition of thinking about the relations between society and nature (Smith 1984).
76   Plekhanov, quoted in Reinfelder 1980, p. 15.
77   See below for a short description of Taylorism through the work of Alfred Sohn-Rethel.
78   Lenin 1964a, p. 154.
79   Lenin quoted in Reinfelder 1980, p. 16.

communist society as sketched in Marx's 'Critique of the Gotha Programme'.[80] Lenin's most memorable quote showing his unceasing dedication to technological development may be his call to electrify the country in his 'Report on the Work of the Council of People's Commissars' (1920):

> *Communism is Soviet power plus the electrification of the whole country.* Otherwise the country will remain a small-peasant country [...] Only when the country has been electrified, and industry, agriculture and transport have been placed on the technical basis of modern large-scale industry, only then shall we be fully victorious.[81]

But Lenin was not unique in this respect. The Stalinist push for collectivization and the development of heavy industry is well known, as is its death toll. For what they are worth, Stalin's thoughts on the matter built on a crude understanding of Engels's ill-fated concept of 'dialectical materialism'. In *Dialectical and Historical Materialism*, Stalin's application of dialectical materialism to history leads to the realisation that 'the productive forces are not only the most mobile and revolutionary element in production, but are also the determining element in the development of production'. This inspires him to proclaim that the productive forces have outlived capitalist social relations and call workers to seize the means of production.[82]

Even Stalin's most eloquent adversary, Trotsky, expressed the belief that 'Marxism teaches that the development of the forces of production determines the social-historical process' and proclaimed Marxism as 'saturated with the optimism of progress',[83] including, apparently, an optimism in the development of machinery and modes of real subsumption. When Trotsky thus calls on workers not to end Fordism, but to 'separate Fordism from Ford and to socialize and purge it', he adopts the idea that technology does not bear the traces of the productive relations under which it was developed. It can simply be repurposed for a socialist way of production without, it seems, raising problems of working conditions or alienation (a question which Jeff Noonan addresses from

---

80 Lenin 1964b, ch. V.
81 Lenin 1966, p. 516; original emphasis.
82 Reinfelder 1980, pp. 17–8.
83 Trotsky, *Results and Prospects* and *The Revolution Betrayed* quoted in Malm 2018a, p. 172. Despite regarding him by and large as a productive-force determinist, Malm also notes Trotsky's 'moments of clarity', in which he expressed the opposite view: 'Technique and science develop not in a vacuum but in human society, which consists of classes. The ruling class, the possessing class, controls technique *and through it controls nature*'. Quoted in Malm 2018a, p. 177; original emphasis.

a different angle in this volume).[84] Andreas Malm wryly sums up the uncritical dedication to productive force development among the leading theoreticians from the 1880s to the Third and Fourth International: 'The first generation of Marxists could disagree on any number of things, but productive-force determinism was a sort of minimal unifying credo [...]; even Stalin and Trotsky could embrace each other on this point'.[85]

5 Beyond Orthodoxy

Productive-force determinism did not go unchallenged for long. In a 1925 review of Nikolai Bukharin's popular introduction *The Theory of Historical Materialism*, Georg Lukács criticised the reductive identification of the forces of production with what he called 'technique'. The latter was rather 'a *part*, a moment' of the forces of production. Lukács charged Bukharin with having fallen victim to a fetishistic vision of society. Marxists, the argument goes, need to remember that it is 'the social relations between men in the process of production' that serve as the principle which determines society in the final instance.[86]

Another critique of the orthodox view of technology can be found in Walter Benjamin's 'On the Concept of History', penned shortly before his suicide in 1940. In light of the rise of National Socialism, the defeat of the German workers' movement, and the beginning of World War II, Benjamin condemned the 'stubborn faith in progress' as well as the 'conformism' of German social democracy, which resulted in their embrace of the 'old Protestant work ethic':

> [The German working class] regarded technological development as the driving force of the stream with which it thought it was moving. From there it was but a step to the illusion that the factory work ostensibly furthering technological progress constituted a political achievement.[87]

---

84 Trotsky quoted in Reinfelder 1980, pp. 18–9.
85 Malm 2018a, p. 172. Monika Reinfelder (1980, pp. 24–5, 28) identifies Rosa Luxemburg as one of the few critics of this view. While retaining traces of 'technicism', she opposed Lenin's idea of the capitalist factory as a 'training ground' for a disciplined, revolutionary proletariat, and instead saw the foundation of a working class's new, voluntary discipline in breaking with the capitalist one.
86 Lukács 1966, p. 29; original emphasis. This critique was not published in English translation until 1966, however, and Lukács himself eventually ended up a defender of 'Soviet Marxist technicism', according to Reinfelder (1980, p. 28).
87 Benjamin 2006a, p. 393.

Instead of grasping technological development as a process which inevitably leads towards human emancipation, Benjamin challenges the very notion of 'progress' itself, suggesting that the 'progress in mastering nature' could come at the cost of a 'retrogression of society'.[88] In the 'Paralipomena' to the theses, he memorably calls for the activation of the 'emergency brake' precisely to stop the 'train' that is history, understood as a piling on of catastrophes, in its tracks.[89] It is hardly surprising that this 'openly *decelerationist* Benjamin'[90] has gained renewed popularity in recent years when the catastrophic consequences of the project of mastering nature in the service of capital accumulation can no longer be denied. Hence, critical theorists such as Andreas Malm and Benjamin Noys have appreciatively quoted Benjamin's demand to pull the emergency brake.[91]

After the Second World War, as Western European countries underwent processes of uneven (re-)industrialization, representatives of the workers' movement often embraced the introduction of labour-productivity-increasing technologies. The history of the Partito Comunista d'Italia (PCI), as told in Steve Wright's seminal account of post-war Italian Marxism, is a case in point.

> True children of the Comintern, for whom the organisation and form of production were essentially neutral in class terms, the PCI leadership saw no great problem in conceding – in the name of a 'unitary' economic reconstruction – the restoration of managerial prerogative within the factories.[92]

Thus, the PCI adhered to a position that can be called 'instrumentalist'. Technology for them was itself neutral and 'indifferent' to the ends for which it can be employed.[93] This vision did not remain unchallenged, however.

In the second issue of *Quaderni Rossi*, a journal published by dissident communists, Raniero Panzieri rejects what he calls 'objectivist' ideology, which treats technological progress, particularly automation,[94] as a politically neutral process that automatically results in the 'overthrow of existing relations'.[95] Instead, Panzieri argues that what Marx shows in *Capital* is that the develop-

---

88 Benjamin 2006a, p. 393. This line of thought is picked up by Theodor W. Adorno and Max Horkheimer in their *Dialectic of Enlightenment* (2002).
89 Benjamin 2006b, p. 402; Benjamin 2006a, p. 392.
90 Mueller 2021, p. 55; original emphasis.
91 Noys 2014, ch. 7; Malm 2016, p. 394.
92 Wright 2017, p. 8.
93 Feenberg 2002, p. 5.
94 Panzieri 1980, p. 48.
95 Panzieri 1980, p. 49.

ment of the technological division of labour is not a neutral, 'objective', and 'rational' development,[96] but 'a mode of existence of capital'.[97] The increasing scale of industrial production requires the 'capitalist's *authority*', which assumes the form of the 'capitalist's *plan*'.[98] Thus, Panzieri emphasises how capitalist production processes are inherently structured by relations of domination. The point for socialists is not to unearth the always-already existing 'rationality inherent in the modern productive process', but instead the creation of a 'new rationality' opposed to the one capitalism employs.[99] Panzieri does not, however, explain how this new rationality can emerge. Moreover, his essay later resorts to a more traditional position. Revolutionaries have to acknowledge the 'unity of the "technical" and "despotic" moments' that characterise productive processes: 'The relationship of revolutionary action to technological "rationality" is to "comprehend" it, but not in order to acknowledge and exalt it, rather in order to subject it to a new use: to the socialist use of machines'.[100]

In this passage, there is no longer talk of a 'new' rationality, or new machines, for that matter. Instead, it seems as if Panzieri is suggesting that existing technologies can be used for divergent – capitalist or socialist – ends. Wright, who praises Panzieri's 'truly pioneering' contribution,[101] remarks that he may have overlooked 'the material indivisibility of labour process and valorisation process'.[102] Regardless of these ambiguities, Panzieri's critique of 'objectivism' remains an important reminder that the capitalist production process qua valorisation process is not a neutral technical affair.

In the postwar United States, there was a similar discontent with trade union leadership among the rank and file, because the former had accepted the introduction of automation in return for tying wage increases to productivity hikes, which, however, meant giving up control over the production process.[103] In practice, productivity increases were often realised not through the introduction of new technologies, but through the intensification of work. Black workers, who had only recently begun to join the industrial workforce in increasing numbers, disproportionately bore the brunt of this. Consequently, it was often black militants who condemned automation and vowed to 'battle' it,

---

96   Panzieri 1980, p. 47.
97   Panzieri 1980, p. 46.
98   Panzieri 1980, p. 48; original emphases.
99   Panzieri 1980, p. 54.
100 Panzieri 1980, p. 57.
101 Wright 2017, p. 40.
102 Wright 2017, p. 39.
103 Mueller 2021, p. 66.

as Charles Denby, the editor of *News & Letters*, put it, with the ultimate goal of gaining '[c]ontrol over the conditions of labor'.[104] This was seconded by black militant autoworker James Boggs, who denounced union bureaucracies for reducing working-class struggles to the question of wage increases at the expense of challenging the 'conditions of work in the shop'.[105] The decline of the union movement was exacerbated by the introduction of automated machinery, which replaced workers, creating a vast army of 'outsiders' excluded from the wage relation, thus challenging the socialist belief that 'a mass of workers' would 'always remain [...] as the base of an industrialized society'.[106] Boggs' important contribution to the Marxist automation debate is further discussed in Jason E. Smith's contribution to this volume.

In the late 1950s, the humanist Marxist Raya Dunayevskaya struck up a correspondence with Herbert Marcuse. As Jason Dawsey recounts in this volume, Dunayevskaya sent Marcuse Denby's text on automation. Marcuse, however, made a distinction between the automation denounced by Denby – he called it 'pre-automation, semi-automation, non-automation'[107] – and the potential uses of automation in a non-capitalist mode of production. In a 1941 article Marcuse had already introduced a distinction between 'technology' and 'technics proper'. The latter, the 'technical apparatus of industry, transportation, communication', was only one moment of the notion of technology, which he conceptualised as a 'mode of production' and a 'mode of organizing and perpetuating (or changing) social relationships'.[108] Marcuse discerns a 'technological rationality' that has superseded the 'individualistic rationality' of the era of classical liberalism. Because the norms to which individual actions are subjected have become standardised and mechanical, forcing the individual to 'adjust and adapt' to technological rationality, resistance appears 'not only as hopeless but as utterly irrational'.[109]

And yet, a more hopeful perspective emerges in the text. The process of rationalisation of the technical division of labour, which has made the 'hierarchical distinction between executive and subordinate performances' increasingly arbitrary,[110] undermines the basis for social hierarchies, specifically the one between manual and intellectual labour. Under capitalist conditions, this

---

104   Denby 1960, p. 8.
105   Boggs 2009, p. 25.
106   Boggs 2009, p. 39.
107   Quoted in Dawsey, this volume, p. 25.
108   Marcuse 2004, p. 41.
109   Marcuse 2004, p. 48.
110   Marcuse 2004, p. 57.

'technical democratization of functions', however, clashes with their 'atomization', so that it appears as if only a 'bureaucracy' could guarantee their unification in a 'rational' manner.[111] Still, the article ends with an emphasis on the utopian potential of technological development, which would allow for a revolutionary break with capitalism:

> Technological progress would make it possible to decrease the time and energy spent in the production of the necessities of life, and a gradual reduction of scarcity and abolition of competitive pursuits could permit the self to develop from its natural roots.[112]

A proper break with capitalism would require more than a replacement of the anarchy of the market by a central plan while keeping the industrial mode of production intact. Instead, Marcuse hopes that automation will reduce the time humans need to spend at work, so that something like '[f]ree individuality, based on the universal development of individuals' becomes possible for the first time.[113]

Marcuse continued his critique of technological rationality in his influential study *One-Dimensional Man*. In the introduction, he emphatically rejects the thesis of technology's neutrality: 'Technology as such cannot be isolated from the use to which it is put; the technological society is a system of domination which operates already in the concept of construction of techniques'.[114] In advanced capitalist society technical rationality is 'embodied' in existing productive technology.[115] Again, however, Marcuse did not conceptualise (productive) technology as exclusively moulded in the image of domination. Recapitulating his earlier argument, he insists that '[c]omplete automation in the realm of necessity would open the dimension of free time as the one in which man's private *and* societal existence would constitute itself. This would be the historical transcendence toward a new civilization'.[116]

Yet, some ambiguities remain. On the one hand, Marcuse claims that a 'qualitative social change' would require a 'new technology', as the existing one

---

111 Marcuse 2004, p. 58.
112 Marcuse 2004, p. 64. As Ramírez (2012) has shown, Marcuse's argument for the postcapitalist affordances of automation was influenced more by Lewis Mumford than by Marx, whose *Grundrisse* he was to read only in the 1950s.
113 Marx 1993, p. 158.
114 Marcuse 1991, p. xlvi.
115 Marcuse 1991, p. 149.
116 Marcuse 1991, p. 40; original emphasis.

serves as an 'instrument of destructive politics'.[117] On the other hand, he writes that a 'break' with technological rationality would require 'the continued existence of the technical base itself', which merely needs to be redeveloped 'with a view of different ends'.[118] This ambivalence was noted by Jürgen Habermas, who pointed out that many passages of Marcuse's book demanded only that the 'institutional framework' and 'governing values' had to be transformed, leaving the 'structure of scientific-technical progress' intact.[119]

Alfred Sohn-Rethel, too, thought to have discerned a self-undermining tendency in the development of the productive forces under capitalism. While his pathbreaking *Intellectual and Manual Labour* is best known for his contention that 'in the innermost core of the commodity structure there was to be found the "transcendental subject"',[120] Sohn-Rethel also introduced an important distinction between 'societies of production' and 'societies of appropriation' as two distinct forms of 'social synthesis'. Societies of appropriation rest on the 'appropriation of products of labour by non-labourers'. Alternatively, in societies of production it is the production process that furnishes the 'form of synthesis'. Such a society 'is, or has the possibility of being, classless'.[121] It is the purpose of the book's third part to reveal how tendencies observable in existing capitalist society lead in this direction.[122]

Sohn-Rethel observes that since the 1880s, firms were increasingly compelled by competitive pressures to raise the rate of exploitation, using such methods as scientific management or time and motion studies. The reduction of workers' activities to minimal '"units" of motion' that could be timed 'with

---

117   Marcuse 1991, p. 232.
118   Marcuse 1991, p. 236.
119   Habermas 1971, pp. 88–9.
120   Sohn-Rethel 2021, p. xxi.
121   Sohn-Rethel 2021, p. 69.
122   Sohn-Rethel's book was first published in German in 1970. Two further editions, both containing revisions, were published in 1973 and 1989. An English translation came out in 1978. While the first two parts of the English version are roughly identical to the German original, the third part is substantially different. Part three in the German original is called 'Vergesellschaftete Arbeit und private Appropriation' (Socialised Labour and Private Appropriation); in the English translation it is 'The Dual Economics of Advanced Capitalism'. Moreover, the titles, order, and content of the individual chapters of part three differ in both versions. In 2018, the book was reissued in German in a two-volume set containing all writings relating to *Intellectual and Manual Labour* written between 1947–71, highlighting the differences between the various German editions. Unfortunately, the 2021 reissue in English follows the 1978 English translation without further commenting on the difference between the English and the German versions. In what follows, we mostly draw on the German 2018 edition. Where passages can be found in the English translation, we have used the latter. All other translations are our own.

the precision of fractions of a second'[123] allowed for the rigorous 'synchronisation of all partial operations belonging to a production process'.[124] Thus transformed along Taylorist lines, the production process can be automated. Interestingly, Sohn-Rethel believed that this transformation of the production process can elevate manual labour to the level of the 'scientific thought-form', thus abolishing the distinction between manual and intellectual labour[125] and affording workers a direct experience of the 'socialisation of labour'.[126] Consequently, they will realise that they can organise the production process themselves, and this would bring about the transition from a 'society of appropriation' to a 'society of production'.

Unlike the striking originality of his analysis of the exchange abstraction as a real abstraction, Sohn-Rethel's discussion of Taylorism and automation relies on a relatively traditional account of a 'dialectic of the productive forces and the relations of production', according to which the 'private appropriation' of the products of labour turns into a fetter that impedes the 'socialisation of labour'.[127] Following Marx, he points out that in the capitalist mode of production, private labours are only retroactively validated as part of total social labour in exchange, in an 'indirect', 'blind', and 'unconscious' manner. The 'Taylorist commensuration of labour', on the other hand, is 'direct' and 'conscious',[128] as it relies on the 'standards of uniform times measures' provided by time and motion studies.[129] In essence, Sohn-Rethel's suggestion that Taylorist principles could be applied at the level of society as a whole amounts to a sophisticated restatement of Lenin's belief in the possibility to use techniques of scientific management to rationalize the organization of labour.[130] Sohn-Rethel is adamant that this transition is not merely a 'technological' but a 'political' question and emphasises that workers need to take over production to prevent the establishment of a nominally socialist 'technocracy'.[131] Still, it is not exactly clear how workers could be convinced to follow the rigorous synchronisation of Taylorised labour in the absence of extra-economic constraints. Sohn-Rethel doubts that a 'material incentive to labour' would suffice,

---

123 Sohn-Rethel 2021, p. 140.
124 Sohn-Rethel 2018a, p. 383.
125 Sohn-Rethel 2018a, p. 401.
126 Sohn-Rethel 2018a, p. 397.
127 Sohn-Rethel 2018a, p. 391.
128 Sohn-Rethel 2018b, p. 482.
129 Sohn-Rethel 2021, p. 140.
130 Sohn-Rethel 2018b, p. 485.
131 Sohn-Rethel 2018b, pp. 491–2.

but has little to offer except the vague suggestion that the 'worker's consciousness' has to be developed to the 'standpoint of synthesized labour'.[132] Finally, Norbert Kapferer argues, Sohn-Rethel's model would leave in place the commensuration of labour by means of an 'abstract measure of time' and, hence, value relations, with the latter being *'consciously appl[ied]'*.[133]

Other Marxists writing in the early 1970s were less sanguine about Taylorism. In his path-breaking *Labor and Monopoly Capital* (1974), Harry Braverman argues that scientific management relies on the expropriation of knowledge about the labour process from workers and its 'centralization' in the hands of management. Consequently, the latter are able to use *'this monopoly over knowledge to control each step of the labor process and its mode of execution'*.[134] Braverman analyses the labour process as a relation of domination and argues that the introduction of new productive technologies results in the 'deskilling' of the working class, which results in a progressive loss of power over production. Yet, Braverman does not reject technology per se. What he condemns is the way 'science and technology [...] are used as weapons of domination' to reproduce class society. His own vision for an emancipated society looks both forward and backward – it is 'governed by nostalgia for an age that has not yet come into being', he quips, dreaming of a state in which technological progress benefits human beings while workers will remain able to derive pleasure from 'conscious and purposeful mastery of the labor process'.[135]

In addition to offering detailed analyses of the way the capitalist mode of production transformed labour processes, many passages in Braverman's book provide valuable methodological remarks on the critique of political economy that anticipate future developments. He insists, for instance, on distinguishing between machines regarded under their 'physical aspect' as 'instruments of production whereby humankind increases the effectiveness of its labor' and means of production which function as capital: 'The purely physical relationship assumes the social form given to it by capitalism and itself begins to be altered.'[136] This distinction between physical fact and social form informs socialists' perspective on a post-capitalist society. Many Marxists, Braverman claims, replaced the 'critique of the mode of production' with a 'critique of capitalism as a mode of distribution' as organisations representing the working

---

132 Sohn-Rethel 2018b, 491.
133 Kapferer 1980, pp. 94–5.
134 Braverman 1998, p. 82; original emphasis.
135 Braverman 1998, p. 5.
136 Braverman 1998, p. 157.

class struggled over wage increases rather than over control of the production process. Thus, he claims, they accepted the 'modern factory as an inevitable if perfectible form of the organization of the labor process'.[137]

Thus, Braverman anticipates Postone's critique of 'traditional Marxism', in which:

> ... a separation is made between class domination and private property as specific to capitalism, and industrial labor as independent of and non-specific to capitalism. *Once this basic framework is accepted, however, it follows that the industrial mode of production – that based on proletarian labor – is seen as historically final.* This leads to a notion of socialism as the *linear* continuation of the industrial mode of production to which capitalism gave rise; as a new mode of political administration and economic distribution of the *same* mode of production.[138]

This begs the question of the historical specificity of the capitalist mode of production and its relation to large-scale industry.

6 Form-Analytical Approaches

From the 1960s onwards, various new readings of Marx constituted an international phenomenon as radicals all over the world returned to Marx's writings to develop more robust conceptual understandings of capitalism and its political economy. When texts like the *Grundrisse* and 'Results of the Immediate Process of Production' became available in languages other than German, Marxists discovered a sophisticated critique of capitalism that was different from both the orthodox versions of historical materialism and Western Marxism's focus on Marx's early writings.[139] The ensuing debates focused on Marx's methodology, his value theory, and his *critique* of the discipline of political economy. While the contributions to these debates often arrived at markedly different conclusions, they are united to some degree by the focus on the question of social form. Failing to take into account the historical specificity of capitalist social forms such as value, money, and capital leads to a naturalisation of capitalist

---

137 Braverman 1998, p. 8.
138 Postone 1978, p. 741; original emphasis.
139 Ingo Elbe (2013) provides a useful classification of the different stages in the reception of Marx's work, focusing mostly on European Marxists. For the international discourse, see Hoff (2017).

social relations, as it is assumed that products of labour are always commodities or that labour always takes the form of abstract, value-producing labour.

A Marxist who highlighted the importance of a form-analytical reading earlier than most was Isaak Illich Rubin in his pioneering *Essays on Marx's Theory of Value* (1923). He explains:

> Political economy deals with human working activity, not from the standpoint of its technical methods and instruments of labor, but from the standpoint of its social form. It deals with *production relations* which are established among people in the process of production. But since in the commodity-capitalist society people are connected by production relations through the transfer of things, the production relations among people assume a material character. This 'materialization' takes place because the thing through which people enter definite relations with each other plays a particular *social role*, connecting people [...]. In addition to existing materially or technically as a concrete consumer good or means of production, the thing seems to acquire a *social* or *functional* existence, i.e., a particular social character through which the given production relation is expressed, and which gives things a particular *social* form.[140]

Rubin's reminder that the social function things assume in capitalist production relations gives them 'a particular social character, a determined social *form*',[141] is relevant for a critical theory of technology. It poses the question of whether particular means of production are form-determined by their function in the capitalist valorisation process – and how this affects the potential to direct production to other, non-capitalist ends.

According to Hans-Dieter Bahr, one of the weaknesses of Marxisms that had traded the critique of political economy for a 'Marxist Economics' was that they ignored the 'social form of use value qua means of labour'.[142] Bahr instead proposes to conceptualise the 'class structure of machinery'. Drawing on Marx's critical analysis of the fetish-like character of the commodity, Bahr argues that the 'genesis of technological development' is no longer visible in its 'result', giving rise to the 'illusion [...] that the individual tool, machine, apparatus, in fact the entire technology of the production process, is always a means, always an instrument, which in itself anyone can appropriate and use'.[143] In contrast, Bahr

---

140   Rubin 2008, p. 31; original emphases.
141   Rubin 2008, p. 37; original emphasis.
142   Bahr 1980, p. 126.
143   Bahr 1980, p. 139.

argues that the means of labour do not merely allow humans to appropriate and transform nature; they mediate between classes, which gives them a specific social form:

> If the means of labour, as means of production, come to mediate between the ruling and the subordinate class, they must acquire a dual social character in the course of their historical development: the means of labour are a means by which the ruling class can directly satisfy its wants, but they are also the 'purposive basis' for perpetuating the one-sided relation between worker and non-worker.[144]

The proletariat, separated from the means of production and, thus, from the conditions of possibility for autonomously realising its own purposes, confronts a process of labour that appears as 'rational' and 'subjectless'.[145] It takes the 'political understanding [*Verstand*] of the proletariat' to grasp the relationship between machinery as 'a "useful object" for the production of useful objects' and machinery as 'capitalist private property for the extraction of surplus-value'.[146]

In his *Die Krise der Revolutionstheorie* (*The Crisis of Revolutionary Theory*) (1977), Stefan Breuer offers a critical discussion of Marcuse's *One-Dimensional Man* that likewise turns to Marx's value theory. He distinguishes between an 'esoteric' and an 'exoteric' Marx.[147] The former rigorously analysed the abstract logic of domination based on the law of value, while the esoteric Marx stopped short of this insight. In contrast, this latter Marx assured himself that the 'development of the species through "labour"' was still active 'behind' the 'mere "appearance"' of the autonomization of a society mediated by value.[148] Yet, the lingering belief in a 'metaphysics of labour'[149] meant to guarantee the possibility of an eventual overcoming of capitalist relations of production offers false consolation. According to Breuer, Marx's analysis of the value-form is 'not a theory of revolution, but rather the notion of its impossibility'.[150]

---

144 Bahr 1980, p. 101.
145 Bahr 1980, p. 109.
146 Bahr 1980, p. 134.
147 Breuer 1977, p. 45. Marcel van der Linden even claims that Breuer may have introduced this distinction (1997, p. 448). In fact, Roman Rosdolsky already spoke of an esoteric and an exoteric Marx in 1957 (p. 348).
148 Breuer 1977, p. 45.
149 Breuer 1977, p. 11.
150 Breuer 1977, p. 15.

For present purposes, it is significant that Breuer arrives at this conclusion through a discussion of the real subsumption of labour under capital.[151] Marcuse, Breuer claims, rightly saw that the working class had become integrated into the capitalist mode of production by the 'disciplinary violence of the production-process'. Thus, labour as 'use-value production' no longer serves an antagonistic role vis-à-vis capital.[152] Unlike the formal subsumption of labour, which still required the capitalist's 'authority',[153] real subsumption transforms the 'techno-structure' of the process of labour in line with the requirements of value production.[154] This transformation renders the 'determined form of social labour' identical with its 'physical existence'.[155] Not only does living labour no longer control the process of production; the material transformation of the production process affects the capitalist class, too, as their authority is replaced by 'technical "objective constraints [*technische 'Sachzwänge'*]"', which are but 'capital's constraints'. When capital is materialised in the productive apparatus, Breuer contends, domination no longer needs a subject.[156] This state, in which no subject remains external to capital's overarching, abstract logic, is the true meaning of one-dimensionality, Breuer writes. His book therefore contains the ambitious attempt to conceptualise technology as being form-determined in its 'innermost structure'. It is also a resolute rejection of the instrumentalist view, which only sees the 'abuse' of technology for 'particular ends'.[157] However, as Ingo Elbe rightly remarks it is not exactly clear when capitalist society is supposed to have become one-dimensional.[158] The concepts of formal and real subsumption, as Patrick Murray points out, after all do not refer to distinct 'historical stages'.[159] While Breuer deserves credit for highlighting the role of real subsumption and for criticising the 'metaphysics of labour', ultimately his rejection of productive-force determinism comes at the cost of what could be called social-form determinism.

---

151   This concept is elucidated in the manuscript 'Results of the Immediate Production Process', which had originally been published in Moscow in 1933. It was reissued in 1969 by Verlag neue Kritik. Italian, French, and English translations soon followed. Thus, the 'Results' were first accessible to a larger readership (see Murray 2017b, p. 325, n. 1; and Hecker 2009).
152   Breuer 1977, p. 14.
153   Breuer 1977, p. 40.
154   Breuer 1977, p. 41.
155   Breuer 1977, p. 49.
156   Breuer 1977, p. 43.
157   Breuer 1977, p. 47.
158   Elbe 2010, p. 579.
159   Murray 2017a, p. 303.

Like Breuer, Postone criticises 'traditional Marxism' for relying on a 'transhistorical social ontology of labor', which posits labour as a force to be emancipated from capital.[160] For Postone, proletarian labour is, instead, deeply shaped by capital and its forms of 'abstract domination'. His *Time, Labor, and Social Domination* (1993) is an impressive attempt to reconstruct Marx's critique of political economy as a critical theory of society which takes traditional Marxism to task for its 'affirmative attitude toward industrial production'.[161] Most Marxists previously treated value as a 'category of the market' that merely regulated the distribution of the products of labour via commodity exchange, but Postone argues that value should be regarded as a 'critical category that reveals the historical specificity of the forms of wealth and production' in capitalism.[162] Production oriented towards creating wealth in the form of value creates a compulsion 'to produce in accordance with an abstract temporal norm', which affects the process of production when enterprises introduce technologies that increase labour-productivity.[163] Large-scale industry is not neutral; instead, it is the 'materialized form of capital'.[164] Unlike Breuer, Postone does not conclude that the material transformation of industrial production processes warrants a quasi-defeatist position. What Postone rejects is the attempt to transition from capitalism into socialism by transforming the mode of distribution while continuing to employ proletarian labour in large-scale industry, as happened in 'actually existing socialist' countries.[165]

The British value theorist Christopher Arthur is also interested in the way in which the '*materialisation of capital*' in the factory system can remain in place even in the absence of the capitalist mode of production.[166] This, he claims, characterised the Soviet Union. His argument relies on a distinction between form and content. As a social form, capital is 'self-valorising value', but it needs to be materially 'embodied'. The factory system is precisely the 'adequate' content.[167] However, without the compulsion exerted by the law of value, the Soviet factory system remained a 'mechanism' that was missing a 'spring', that is,

---

160 Postone 1993, p. 63. Somewhat characteristically, however, Postone does not mention Breuer by name, even though it is unlikely that he did not acknowledge *Die Krise der Revolutionstheorie*, which was published when Postone was writing his PhD at the University of Frankfurt.
161 Postone 1993, p. 9.
162 Postone 1993, p. 26.
163 Postone 1993, p. 301.
164 Postone 1993, p. 358.
165 Postone 1993, p. 40.
166 Arthur 2004, p. 208; original emphasis.
167 Arthur 2004, pp. 202–3.

an 'objective economic regulator'. Political directives could, accordingly, not guarantee the efficient development of the productive forces.[168] Since Arthur defines a mode of production as a 'stable, relatively harmonious combination of social form and a material content', he concludes that the Soviet Union was lacking precisely a proper mode of production: it was a 'self-aborting monstrosity'.[169]

In an earlier chapter of his *The New Dialectic and Marx's* Capital (2004), Arthur turns to the role played by the real subsumption of labour in the capitalist mode of production proper. Like Breuer, he entertains the possibility that capital is 'the real subject of production'.[170] This is because of the *'inversion* of subject and object' that occurs in the capitalist production process. Under conditions of real subsumption, workers become 'more like bees' whose labour supports the 'collectivity of production' while lacking 'any meaningful relationship to the enterprise as a whole', as the latter is managed by capital.[171] Still, Arthur emphasises that capital remains dependent on labour, whose "*agency*" is needed even by really subsumed production processes. That is to say, workers do not become actual objects. Their will must be '*ben*[*t*] [...] to alien purposes', which is to say their activity needs to be 'manipulat[ed]' without completely obliterating their 'subjectivity': 'They act for capital, indeed *as capital*, but in some sense *act*'.[172] Ultimately, the real subsumption of labour under capital might result in rendering the factory an 'adequate' content for the social form of capital, but it cannot abolish the possibility of resistance, because workers remain 'actually or potentially recalcitrant to capital's effort to compel their labour'.[173]

The Endnotes collective and its collaborators have devoted several articles in their eponymous journal to pondering the role of various technologies in capitalist society as well as the potential to repurpose capitalist technologies to non-capitalist ends. Their article 'Error' synthesises earlier discussions, drawing on Arthur's systematic dialectics to grasp the place of technology in the totality of capitalist social forms. Social relations, the authors claim, shape material forms and consequently 'lay down basic parameters – capacities and directionality – of activity', rendering some actions possible or probable and others

---

168 Arthur 2004, pp. 208–10.
169 Arthur 2004, pp. 210–11.
170 Arthur 2004, p. 50.
171 Arthur 2004, p. 47; original emphasis.
172 Arthur 2004, p. 52; original emphases.
173 Arthur 2004, p. 54.

unlikely or outright impossible.¹⁷⁴ Under capitalism, the existing 'practical-technical capacities and limits' are 'in large part defined by capital', whose logic has been 'crystallized' in infrastructures and architectures.¹⁷⁵ Endnotes speak of 'affordances' offered by the world. These provide a 'vast horizon of possibilities for action', but as they are ultimately shaped by capital, the latter seems to 'give [...] and foreclose [...] that horizon'.¹⁷⁶ At first glance, this may lead to an intractable problem for revolutionaries. If 'all the crud of the world' was, indeed, entirely form-determined by capital, a communist revolution could only be conceptualised as 'pure, total rupture'.¹⁷⁷ Given the myriad ways in which people's lives are integrated into and dependent on the social-technical infrastructure developed under capitalism, such a break would inevitably result in a 'gigantic global humanitarian disaster'.¹⁷⁸ Instead, Endnotes propose a more nuanced view which refuses to simply posit the non-neutrality of all technologies. Instead, not all terrains are 'intractable for struggle', and there exist technologies that can be used 'pragmatically' in the process of building communism – though others might not:¹⁷⁹

> [M]uch of the current technical structuring of the world is profoundly *anti-communist*, and struggles to come will have to work around such things until they can defeat or subsume them. Building that power will involve the establishment of new technical mediations and the repurposing of old, to the ends of a collective self-reproduction outside of class [...]. It will require as its first priority the establishment of collective control over the production and distribution of means of subsistence.¹⁸⁰

The approach offered by Endnotes, in short, decidedly rejects what Patrick Murray has termed 'use-value romanticism', which holds that socialists have to only expropriate the capitalist class in order to lay bare the innocent technologies that capitalists have merely used for their ends.¹⁸¹ At the same time, Endnotes avoids the dead-end of technological defeatism. Instead, what their approach calls for is a thorough assessment of individual technologies and the 'affordances' they offer for revolutionaries, as well as an acknowledgment that

---

174   Endnotes 2019, p. 115.
175   Endnotes 2019, pp. 116, 135.
176   Endnotes 2019, p. 117.
177   Endnotes 2019, pp. 115, 123.
178   Endnotes 2019, p. 117.
179   Endnotes 2019, p. 156.
180   Endnotes 2019, p. 159; original emphasis.
181   Murray 2017, p. 313.

the creation of new mediations and the repurposing of existing ones are practical problems that can only be solved in and through struggle.

Jasper Bernes has attempted to provide such an analysis of the affordances and challenges posed for class struggle by the global technological regime. Although Bernes does not focus on automation or technology explicitly, technology is central to his question of how the transformation of capitalism since the long downturn of the 1970s has changed the possibilities for a communist future. Bernes stands in opposition to both the optimism of accelerationism and the belief that the development of technology will make reaching communism inevitable. Rather, he asks: "[W]hat if these technologies actually make it harder?"[182]

Two elements stand out to Bernes. One is the restructuring of global capitalism following the logistics revolution of the 1970s, from a model dominated by production to one dominated by circulation.[183] This relies on a number of changes, including managerial reorganisation, technologies such as refrigeration, shipping, and the tracking technologies discussed by Steffen Reitz in this volume, and the reordering of capital through financialization, which passes the financial risks involved in production down to subsidised producers. This results in a shift of power from the factory and productive capital to the sphere of circulation: 'Under logistics, supermarkets become a new locus of power'.[184] An additional point of this global reordering and technologization are working conditions for those working in production, which Bernes calls '[p]rogram-dependent laborers', i.e. those who are 'dispossessed of usable knowledge by a technical system in which they appear only as incidental actors'.[185] The implication of both of these processes is that worker resistance has to find new ways, e.g. by blocking routes of transportation, whose susceptibility to disruption became visible during the Covid-19 pandemic.

'The Belly of the Revolution' introduces Bernes' second contribution to a Marxist technology debate: the idea of 'path-dependent' technologies, i.e., the realisation that existing technology has laid out certain paths for future innovation and change and blocked or made others difficult. As Bernes argues, we therefore 'need a new theory of technology, one that reckons with path-dependency'. Furthermore, Bernes draws attention to the material embeddedness of both Marx's materialism and technologies in nature when he calls for a 'return to an insight that has been lost but which was at the center

---

182 Bernes 2018, p. 333.
183 See Bernes 2013.
184 Bernes 2018, p. 352.
185 Bernes 2013, p. 190.

of Marx's thinking – technology is nature, an organization of natural elements and powers'. This realization includes the observation that 'the qualities and characteristics of natural forces themselves, along with social relations, determine the range of possible uses a technology affords'.[186] Against a simplistic way of thinking according to which it is possible to sort out 'good' (e.g. sustainable) technology from 'bad' (destructive) technology and only retain the former, once capitalist social relations have been overthrown, Bernes insists that technology is not 'a series of discrete tools' but rather 'an ensemble of interconnected systems' with fossil fuels as the source of energy written into most existing technologies, which would make it extremely difficult to replace fossil technologies with sustainable ones even if no political obstacles existed.[187]

7 The Twenty-First Century Resurgence of Automation Optimism and Its Critique

Bernes's arguments about the logistics revolution and path dependency stand in stark contrast to what has arguably been the most widely discussed leftist contribution to the automation debate in the early twenty-first century: a resurgence of optimistic speculation about automation that includes works like Paul Mason's *Postcapitalism*, Nick Srnicek and Alex Williams' *Inventing the Future*, Aaron Bastani's *Fully Automated Luxury Communism*, and Peter Frase's *Four Futures*, among others. While different in degree, these works are often structurally similar, reminding us of Joshua Clover's quip about books like David Harvey's *A Brief History of Neoliberalism*, which, after outlining the grim prospects of capitalism, fall into '[t]he temptation to end with prescription rather than empirical catastrophe'. Clover describes these books as 'Eleven Chapters of Marx, One Chapter of Keynes (or, alternately, Eleven Chapters of Communism, One Chapter of Social Democracy)'.[188] Analogously, the contemporary automation optimists' work could be described as 'Eleven Chapters of Neoliberalism, Unemployment, and Climate Change, One Chapter of Full Automation and Universal Basic Income'.

This is certainly the structure of Mason's *Postcapitalism*. After laying out aspects like the long waves of capitalism, its crisis tendencies, and the decline

---

186 Bernes 2018, pp. 335–6.
187 Bernes 2018, pp. 333–4; see also Pirani 2018.
188 Clover 2017, p. 540.

of the left over the twentieth century, Mason turns to climate change and over-aging as well as the market's inability, and the global elite's unwillingness, to deal with these issues as 'the rational case for panic'.[189] In the final chapter, Mason lays out suggestions for a transition to a postcapitalist society he calls 'Project Zero – because its aims are a zero-carbon energy system; the production of machines, products and services with zero marginal costs; and the reduction of necessary labor time as close as possible to zero'.[190] The way to achieve these conditions are highly dependent on technology ranging from computer modelling of the economy to the coordination of cooperative work by swarm intelligence and the internet, which Mason interprets through Marx's concept of the general intellect, and to which he ascribes a new power to act and resist. Mason ends by proposing universal basic income as a road to 'socialism', which seems a lot like Keynesian social democracy, and dreams about the potential of many of today's virtual commodities to become entirely free – never mind that 'immaterial' virtuality needs an ever-increasing amount of energy and material infrastructure to remain operational.[191]

At the beginning of *Four Futures*, Frase identifies 'two specters … haunting Earth in the twenty-first century: … ecological catastrophe and automation'.[192] He then sketches four types of future societies along the poles of scarcity / abundance and equality / hierarchy. Frase calls these societies, each of which addresses the challenges of technological unemployment and climate change differently, communism (equal and abundant), socialism (equal and scarce), rentism (hierarchical and abundant), and extremism (hierarchical and scarce). Once again, the 'capitalist road to communism'[193] goes through universal basic income and full automation, which, in line with the fettering thesis, is held back by wages being so low that it is cheaper to pay humans to do undesirable work than to automate it.

Possibly the most discussed contributions to the current automation debate on the left are those of Srnicek and Williams. In '#Accelerate: Manifesto for an Accelerationist Politics' (2013), they identify a host of contemporary problems beginning with climate change and extending to lingering financial crises, 'resource depletion, […] collapsing economic paradigms, and new hot and cold wars'. In this context:

---

189 Mason 2016, pp. 245–62.
190 Mason 2016, p. 266.
191 See Gratorp 2020; Huck 2020.
192 Frase 2016, p. 1.
193 Frase 2016, p. 55.

[i]ncreasing automation in production processes – including 'intellectual labour' – is evidence of the secular crisis of capitalism, soon to render it incapable of maintaining current standards of living for even the former middle class of the global north.[194]

Against the speed of neoliberal capitalism's creative destruction, 'a simple brain-dead onrush' constrained by the dictates of capital's need for self-valorisation, Srnicek and Williams call for 'an acceleration which is also navigational, an experimental process of discovery within a universal space of possibility'.[195] Once again, as Antonio Negri puts it in his review of '#Accelerate': '[T]he traditional refrain of Operaism returns': capitalism, which was once essential to the development of productive forces, is now fettering them.[196] In contrast to earlier productive force determinists, however, Srnicek and Williams, acknowledge that technological development is not enough and that social change needs to go hand in hand with it; but technology remains central to their project. It is a means and a necessary precondition for overcoming capitalism from within, both by making possible advanced economic planning in a post-capitalist society and by allowing a new 'intellectual infrastructure' among the left that can develop and spread new ideas about economic and social models for a future society.[197] To the extent that technology incorporates capitalist class rule, this union seems to Srnicek and Williams temporary and dissolvable. Technology can and must be repurposed for the accelerationist, post-capitalist project. In *Inventing the Future*,[198] Srnicek and Williams further elaborate their demands for more automation and the eventual end of wage labour. These include a redistribution of remaining work, a diminishment of the work ethic, and, once again, universal basic income. Like Frase, they argue that raising the cost of labour will incentivise further automation and call on the left to repurpose the structures of capitalism for a post-work politics.

In the wake of Williams' and Srnicek's project but addressing a relative neglect of domestic and other care work in accelerationist writings, the feminist collective Laboria Cuboniks has given automation optimism its own twist by connecting it to Marxist feminist thought, particularly that of Shulamith Firestone, and redirecting the largely productivist automation debate to the sphere of social reproduction and gendered identities in their concept 'xenofemin-

---

194 Williams and Srnicek 2014, p. 349.
195 Williams and Srnicek 2014, p. 352.
196 Negri 2014, p. 367.
197 Williams and Srnicek 2014, p. 359.
198 Srnicek and Williams 2015.

ism' (or XF). Although not explicitly Marxist in methodology, the 'Xenofeminist Manifesto' shares an anti-capitalist stance and makes use of the concept of alienation, which the authors redefine 'as an impetus to generate new worlds'.[199] While noting technology's stinted (or might we say 'fettered') use under capital, the collective sees technology as a way to confront not only economic inequalities, but also those in gender and body politics. Embracing intersectional feminism as well as trans and queer politics, Laboria Cuboniks define XF as 'gender- [and race-] abolitionist', and call for 'costruct[ing] a society where traits currently assembled under the rubric of gender [or race], no longer furnish a grid for the asymmetric operation of power'. Technology is one (sometimes it seems the main) way along this road to 'combat' issues such as 'unequal access to reproductive and pharmacological tools, environmental cataclysm, economic instability, as well as dangerous forms of unpaid/underpaid labour'. Consequently, they end their manifesto for (rather than against) alienation with the call: 'If nature is unjust, change nature!'[200]

The final contribution to the recent line of manifestos imagining automation as a road to post-capitalist freedom is Bastani's *Fully Automated Luxury-Communism*. Bastani's central idea is that 'post-scarcity' is made possible by revolutions in production and information technologies.[201] In a world freed from the stranglehold of capitalism's artificially created scarcity, Bastani offers the 'solution' of hyperbole: the sun and wind's energy are clean and abundant, and even 'the limits of the earth won't matter anymore – we'll mine the sky instead'.[202] Much of what Bastani writes is somewhere between hopeful conjecture and blatant disregard of the evidence that it will not be so easy to wean production off fossil fuel-based technologies. Bastani, however, seems uninterested in or ignorant of decades of eco-socialist scholarship and Marxist analyses that explore how capitalist relations are manifested in technology and how machinery is often an instrument of class struggle. Despite Bastani's casual dismissal of present-day Leninists,[203] his vision is a promethean utopia of technology unfettered from the limits of capital, which will solve all of today's geo-political and ethical problems through techno fixes like asteroid mining, robot workforces, genetic engineering or synthetic meat.[204]

---

199 Laboria Cuboniks 2018, n.p.
200 Laboria Cuboniks 2018, n.p.
201 Bastani 2019, pp. 53–181 and *passim*.
202 Bastani 2019, p. 119.
203 Bastani 2019, p. 196.
204 Indeed, Bastani even evokes Prometheanism positively in one place (as do Williams and Srnicek 2014, p. 360) when he claims: 'Our ambitions must be Promethean because our

There is another, usually more historically informed, line of writing within automation theory of the last decade. The economic historian Aaron Benanav, for instance, discusses Bastani and other automation theorists as 'late-capitalist utopians',[205] while remaining sceptical of the adequacy of what he terms the 'automation discourse' by both leftists such as Bastani, Frase, Mason, Srnicek, and Williams as well as liberal economists and pundits, who likewise assume that a largely automated society is imminent. This, the automation theorists claim, is bound to have catastrophic results such as mass technological unemployment, unless the connection between work and income is severed through the provision of universal basic income. In his book *Automation and the Future of Work* (2020), Benanav rejects this narrative and argues that, contrary to these theorists' belief, labour productivity is not to blame for an impending 'job apocalypse'.[206] Indeed, labour productivity only appears to rise quickly because of global economic stagnation: 'Since 1973, both output and productivity growth rates have declined, but output growth fell much more sharply than productivity growth rates', which is why employment growth has slowed down,[207] leading to 'chronic *under*employment'.[208] Drawing on Robert Brenner's influential account of global industrial overcapacity, Benanav offers a powerful explanation why the 'industrial-economic growth engine'[209] has run out of steam: not because of the development of the technological forces of production, but because of the crisis-prone nature of global capitalism itself. The cause of underemployment and precarious living standards is not due to job destruction through technological change, but a tottering global economy in which too few decent jobs are created. As a result, technological development cannot be the solution, either, as automation theorists assume.

Like Benanav, Jason E. Smith treats contemporary automation in the context of capitalist stagnation. Smith, too, argues that the global economy has long been affected by a protracted 'crisis of profitability' which stems from an expansion of 'unproductive labor'.[210] This work, much of which is located

---

technology is already making us gods' (Bastani 2019, p. 189) – thus conjuring the very spectre eco-Marxists in particular have long tried to overcome, see e.g. Burkett 1999, pp. vii, 5–14, 147–73; Hornborg 2019b, p. 148; 162–74. For a different perspective see Grundmann 1991a.

205 Benanav 2020, p. 11.
206 Benanav 2020, p. 46.
207 Benanav 2020, p. 19.
208 Benanav 2020, p. 46; original emphasis.
209 Benanav 2020, p. 57.
210 Smith 2020, p. 11.

in the so-called service sector, often requires forms of 'intuitive knowledge and decision-making' not easily replaced by machines.[211] Smith's book *Smart Machines and Service Work* (2020), like his contribution to this volume, further discusses the implications of these sectoral shifts for the working class and its capacity to struggle for change.

The latter is also the focus of Gavin Mueller's recent contribution to the automation debate, the 'neo-Luddite' *Breaking Things at Work* (2021). For Mueller, the automation of the American automobile industry after World War II needs to be seen as a part of a longstanding struggle over the control of the production process, which aligns the original Luddites with militant auto workers of the 1950s, who likewise 'battled' automation, as Charles Denby put it. Mueller insists on the distinction between 'perfect automation' and 'actually existing automation' and remarks that the latter usually reorganises rather than replaces labour, recomposing the workforce in the process often with the aim to 'wrest control away from the machinists'.[212] Against accelerationist dreams, Mueller advocates for working-class autonomy that does not set its hopes in technological development but rather sees technology as a crucial site of class struggle:

> Instead of imagining a world without work that will never come to pass, we should examine the ways historical struggles posited an alternative relationship to work and liberation, where control over the labor process leads to greater control over other social processes, and where the ends of work are human enrichment rather than abstract productivity.[213]

The historian Jason Resnikoff likewise examines automation as an ideology. Resnikoff argues that the concept of automation was used in the post-World War US as a discursive weapon against workers. As Resnikoff shows, references to necessary and inevitable technological 'progress' and automation hid and continue to hide among other things the intensification of work as well as the loss of power by workers.[214]

Despite the differences in approach and emphasis, these critical theorists of automation all share a rejection of productive-force determinism and, instead,

---

211   Smith 2020, p. 122.
212   Mueller 2021, p. 64.
213   Mueller 2021, p. 29.
214   Resnikoff 2021.

note the necessity of understanding technology as a structuring part of social relations, whether it is the way commodity production and exchanged are organised globally or locally through the hierarchies that structure a given workplace. Several also note the ideological dimension of 'automation' as not merely a label for actually occurring technological changes, but a larger narrative informing technological innovation and class dynamics.

8 Ecosocialist and Energy Humanities Contributions to the Automation Debate

The last group of thinkers we want to discuss, ecosocialists, by and large share this view of the importance of social relations behind production but focus on a dimension often still side-lined in Marxist technology and automation debates: the materiality of production and the environmental impact of modern fossil-fuel capitalism. Gaining momentum with the beginnings of the green movement of the 1970s, early ecosocialist thinkers, such as James O'Connor or Ted Benton, tried to bridge a disconnect they saw between ecologists and Marxists by criticising Marx for his oversight of ecology and suggested concepts such as the inclusion of a 'second contradiction' between capitalism and its material basis in the earth's resources or 'natural limits'.[215] A second generation, around John Bellamy Foster and Paul Burkett, set out to defend Marx against charges of being ecologically blind or having a weak understanding of nature-society relations. Both movements tried to reconcile the red and green camp by pointing to a common cause and the benefits of combining the analysis of class and ecology rather than giving precedence to one over the other.[216] While many in the ecosocialist camp are not interested primarily in technology, their perspective can still serve as an important intervention into automation optimist discourse, which – while often beginning with an admonition of the danger of climate change as one reason for the need to move beyond capitalism – frequently forgets about climate change, pollution, resource depletion, etc. in favor of dreams of riches for all resulting from an unfettering of the forces of production.[217] Ecosocialist thought can also serve

---

215 As an introduction to O'Connor's central arguments see O'Connor 1988; for a critique of Marxism as ignoring 'natural limits' see Benton 1987 and 1992. On the disconnect between 'green' and 'red' thinkers see e.g., Benton 1987, Grundmann 1991a, Burkett 1999, Haug 2003, p. 27.
216 E.g., Burkett 1999.
217 There are also more extreme cases. Phillips (2015) tries to unmask 'the counter-Enlighten-

as a corrective lens for Marxism's traditional focus on social relations as if they existed not in a metabolism with nature, but in a vacuum created entirely by humans.

Against this tuning out of nature, ecosocialists warn that *technology* 'is the vehicle in and through which ecologically damaging behaviour is embodied and effected', and that only blaming capitalist social relations for environmental problems *directly* is too easy.[218] This is not to say that ecosocialist writers agree that technology is the means, not the root cause for environmental damage or that changes in technology, as promised by green growth enthusiasts, are likely to solve ecological problems without social transformation. Instead, John Bellamy Foster points out that 'environmental degradation is not the result of increased population, or increased accumulation, or the introduction of less environmentally benign technology. It is a product of all three'. Since the impact of technology can never reach zero, any increase in population and affluence will have an effect on the environment.[219] According to Foster there are two common mistakes in mainstream environmental discussions. Firstly, there is the Malthusian argument that puts an emphasis on population growth, which is far outweighed by the factors of affluence and technology, particularly on a global level. Secondly, there is a romantic imagination that all problems began with industrial technology and that pre-industrial societies created no ecological problems.[220] Foster therefore stresses the need to criticise the 'tendency to blame technology rather than the social systems underlying the technology' as well as ideas of 'return[ing] to a mythical state of preindustrial ecological harmony'.[221] The only way out of our social and ecological predicament is a revolutionary transformation of the social relations that can order production and distribution more justly *and more* sustainably.[222]

In a recent piece co-authored with Brett Clark, Foster addresses more concrete questions of ecology and technology, namely technological fixes such

---

ment origins of anti-modernist green thought' in the hopes of reviving a 'pro-industrial, pro-growth left [for] those who are frustrated with the predominance of hair-shirted, anti-development greenery' with an idea of growth based on a sustainability (which seems to be mainly about the use of nuclear power) based on an overcoming capitalism that cannot even be called vague, since it is essentially non-existent in his book. For a critique of this line of 'techno-optimist' thinking, including Phillips's book, see Foster and Clark 2020, pp. 269–87.

218 Grundmann 1991a, p. 106.
219 Foster 1999, p. 30.
220 Foster 1999, pp. 31, 36, 114–18.
221 Foster 1999, pp. 35–6.
222 Foster 1999, p. 148.

as geoengineering through solar radiation management held up by techno-utopian socialists like Frase, Bastani, or Leigh Phillips. For Foster and Clark, such fixes would only 'infinitely expan[d] commodity production and capital accumulation', while bringing with them 'immense, unforeseen repercussions'.[223] Foster and Clark are not impervious to uses of technology, however: 'There is no doubt that the current planetary ecological crisis requires technological change and innovation', including 'alternatives to fossil fuels [...]. It is not true, however, that all the technologies needed to address the planetary emergency are new, or that technological development alone is the answer. The wonders of smart machines notwithstanding, there is no solution to the global ecological crisis as a whole compatible with capitalist social relations'.[224]

While ecosocialists mostly agree with Foster's scepticism of capitalism's ability to transform its technologies into ones that are non-destructive of the non-human environment,[225] many see an end of capitalist production as a necessary precondition to solving social and ecological problems, not as the solution itself. Contrary to the optimistic voices that equate the end of capitalism with an end of ecological problems, we need to see technologies as 'a feature of modern societies, which they must live and cope with'.[226] Since technology is the main way in which humanity interacts with external nature,[227] technology will remain a major part of our being in the world and this will continue to involve a transformation of the human and non-human environment through technologically mediated labour, which may be more or less destructive even under an egalitarian organisation of social relations among humans. This is the conundrum of technology's environmental impact: since technological advances expand both the scope and complexity of our transformation of the environment and since 'social institutions and technology permeate each other', the 'cause' of a problem is not always easy to identify.[228] As Grundmann points out, the aspects of technology's ecological impact include

---

223 Foster and Clark 2020, p. 284. See also Soper 2020, p. 40.
224 Foster and Clark 2020, p. 284; 285.
225 Foster 1999, pp. 32, 125–42. It is worth noting, however, that some critics like Andreas Malm regard the mitigation of the most damaging environmental effects of current production, distribution, and consumption patterns as so pressing that it cannot be deferred to a post-capitalist future (Helle Panke 2021).
226 Grundmann 1991a, p. 106. See also Grundmann 1991b, pp. 41–3.
227 Grundmann 1991b, pp. 109–13.
228 Grundmann 1991b, p. 29. On this point see also Bernes's speculations about the challenges past choices in technological development pose for a communist future discussed above.

not only the consequences of economic growth with ever greater production, but unintended consequences of new technologies and industrial accidents as well. All of these aspects are exacerbated socially by capital's tendency to externalise costs, such as environmental damage and degradation, into 'social costs'.[229]

The introduction of energy into a materialist analysis certainly constitutes another major theoretical intervention often ignored by Marxist automation discourse. While energy is usually taken to be an underlying enabler of technology, the so-called energy humanities point to energy as 'one of the "main blind spots in Marxist thought"'[230] and insist that: 'Cheap and abundant energy has probably done more to shape the human-environment relationship than anything else in the last 150 years'.[231] Matt Huber makes a compelling argument to examine capitalism through its main source of energy. Citing E.A. Wrigley, he describes the Industrial Revolution as a shift from 'an "organic" economy wherein the bulk of energy, food and fiber was derived from land-intensive resources (e.g., cotton, wheat, and livestock) toward a "mineral-based energy economy ... freed from dependence on the land for raw materials."' Alongside this shift, first coal, then electricity and oil provided more and more energy feeding ever growing machines that replaced human and animal muscle power as 'the core *productive force of production*'. Rather than make labour obsolete, however, this energy shift led to a transformation of the labour process itself. It also hastened the generalisation of wage labour relations, not in the sense of a technological determinism, but as based on 'industrial capital's peculiarly mammoth levels of productivity', which necessitated the greater generalisation of labourers' skills away from providing muscle power while cementing class relations though the control of energy sources and machinery by the capitalist class.[232]

Among those working in a Marxian tradition, Andreas Malm has done perhaps more than any other thinker in tracing the intertwined social and ecological dimension behind the energy shifts occuring since the Industrial Revolution. Working historically, Malm began by exploring the process and reasons behind a change from water power to steam and fossil fuels, which at first seems uneconomical until one considers the greater mobility and constant availability and reliability of fossil fuel technologies that made it more attractive to

---

229 Grundmann 1991b, pp. 30–41.
230 Huber 2008, p. 105, citing Debeir and Hémery.
231 McNeill 2019, p. 493.
232 Huber 2008, pp. 107, 108, 110; original emphasis.

factory owners.²³³ Malm traces the transition in Marx's own thinking from an early productive force determinism that would, as outlined above, shape the direction subsequent generations of Marxist thinkers took, to a realisation that the historical sources Marx studied indicated that it is was not technology that determined society but social relations that led to the breakthrough of different technologies and sources of energy such as steam power that had been known and used for centuries. Malm traces this recognition in Marx's notebooks and parts of *Capital*, while also noting Marx's unwillingness to fully abandon his earlier programmatic theory in light of his empirical findings. There is then, according to Malm, a contradiction in Marx's accounts between his and Engels's model from the 1840s and his realisation that it was not a revolution of technology but of property relations that brought about the transition from feudalism to capitalism and that relations of production shaped the productive forces according to their needs rather than the other way around.²³⁴

Turning his attention to the present and the *longue durée* of fossil fuel-dependent capitalist production, Malm highlights that the development of productive forces with their steadily increasing productive power also increases the environmental impact of production. This leads Malm to speculate that given capital's continued reliance on fossil fuels analogously to Marx's 'law' of the rising organic composition of capital there is 'a law of a rising *fossil composition of capital*', which over time 'translates into a law of a *rising concentration of CO_2 in the atmosphere*'.²³⁵ While, as Malm notes, capital is not the only factor contributing to the rising concentration of CO_2, 'it constitutes the *main propulsive force* of the fossil economy'.²³⁶ Any lasting contribution to climate change will therefore have to confront fossil capital.

Malm's colleague and former mentor, Alf Hornborg, adds yet another dimension to the technology debate. Hornborg has long examined how global ecologically and economically unequal exchange is dependent on and perpetuated by an unequal distribution of technology that is concentrated in the economic centres. This centralization of dead labour (materialised in technology and machines) leads to a highly uneven exchange of money, energy, and entropy, in which energy is funnelled to the centre, while entropy is 'exported' to the least developed corners of the earth.²³⁷ Modern technologies are therefore 'products

---

233   Malm 2016.
234   Malm 2018a.
235   Malm 2016, p. 354; original emphases.
236   Malm 2016, p. 355; original emphasis.
237   E.g. Hornborg 2001; Hornborg 2019b; Dorninger, et al. 2021. For a discussion of entropy in Marxist and mainstream economics, see Burkett 2006.

not just of engineering – viewed as innocent research into the physical regularities of the material world – but finally also of unequal social relations of exchange' that are 'made invisible by the ostensibly neutral operation of market mechanisms'.[238] As Kate Soper paraphrases, '[t]echnological "progress" has to be seen not simply as an index of ingenuity but as a social strategy of appropriation' that continues a neo-colonial imbalance wreaking havoc on people and the environment, particularly in the global peripheries.[239] As such, Hornborg opposes a view he attributes to Marx that sees 'capitalist machines as intrinsically nonsocial products of engineering, that is, as "productive forces" detachable from their social context', a view he calls 'machine fetishism', i.e. the inability to see that technology depends on unequal exchange.[240] What is more, Hornborg understands '[g]lobalized technologies' as 'a zero-sum game', a 'wa[y] of saving time and space for some at the expense of time and space lost to others' that depends on and perpetuates 'an unequal or asymmetric societal exchange of resources such as embodied labor, energy, land, or materials', or 'an arrangement for redistributing resources in global society'.[241] This technologically mediated, (ecologically) unequal exchange is obscured by the illusion of equal exchange that is built into the market through the form and idea of general purpose money. As Hornborg insists, 'money cannot neutralize ecological damage in a physical sense', which means, in thermodynamic terms, 'money cannot compensate for entropy'.[242] Concepts such as 'underpayment' therefore miss the point by comparing apples (bio-physical energy) with oranges (value as a social, economic concept) and thus fall prey to the illusion that 'everything has a correct price' while failing to acknowledge the difference between 'asymmetrical flows of values' and 'asymmetrical flows of biophysical resources'.[243] It is Hornborg's conviction that technology and money form a 'complex' that needs to be broken conceptually and materially: 'To curb asymmetric global resource flows, and to avoid the most disastrous scenarios of the Anthropocene, our only chance is to critically rethink and redesign the artifacts – money

---

238   Hornborg 2019b, p. 28.
239   Soper 2020, p. 17.
240   Hornborg 2019b, pp. 146–7.
241   Hornborg 2019b, pp. 147–8.
242   Hornborg 2019b, pp. 59, 144. See also Hornborg 2019b, pp. 152–76 for Hornborg's critique of Marx's and later (ecological) Marxists' conception of use and exchange values and his alternative conception of unequal exchange and the necessity for a fundamental critique of general-purpose money. Hornborg 2016 provides a short account of technology and unequal exchange over history.
243   Hornborg 2019b, pp. 138, 143.

and technology – that currently rule our thoughts and lives'.[244] In a sense, Hornborg is thus one of the most radical and original critics of technology, even if he has over time abandoned many Marxist concepts and more orthodox Marxists will certainly object to his developments or critique of some Marxian concepts.

There are, finally, two problems that ecosocialist thinkers remind us of, but which are too often forgotten in techno-enthusiast writing. Firstly,

> 'the end of capitalism' does not necessarily equate with the coming of a revolution that will bring about a new, morally, and ecologically superior form of production that is not based on capitalist accumulation. After all, the end of capitalism could just as easily lead to 'the common ruin of the contending classes' as Marx and Engels noted long ago.[245]

Secondly:

> Our customary idea of the transition to socialism is the abolition of the capitalist order within the basic conditions European civilisation has created in the field of techniques and technology [...]. Marxists have so far rarely considered that humanity has not only to transform its relations of production, but must also fundamentally transform the entire character of its mode of production, i.e. the productive forces, the so-called technostructure.[246]

Some have begun this thought process, but it will be a long road. As Soper reminds us in her call for a post-growth hedonism, '[t]he definition of progress in terms of capitalist-driven technology and industrialisation can no longer be left unchallenged; [...] A less techno-driven and growth-oriented organisation of nature has now to be viewed as offering more advanced norms of welfare and modes of providing it'.[247] Bernes's work on the 'path dependency' of technology or Malm's on a shift from fossil fuels give us an idea of how hard it would be to get out of the choke hold of the fossil economy even if capitalism were to end tomorrow and everybody were to agree on the importance of achieving sustainability as fast as possible.[248] What we may need from a radical ecological as well as a social perspective, ecosocialist thinkers remind us, is what Kohei

---

244 Hornborg 2019b, p. 10.
245 Caffentzis, 2018, p. 96.
246 Bahro 1982, p. 27, quoted in Soper 2020, p. 31.
247 Soper 2020, p. 32.
248 Bernes, 2018, pp. 333–5 and *passim*.

Saito has recently called a 'de-growth communism',[249] which will – with Soper – require a radical rethinking of our acquired notions of progress and wealth, where the real wealth of society lies not in the value of material goods it creates, even if the obfuscations of capitalist social relations seem to suggest so.

9   Contributions

The contributions of this collection are divided into three thematic sections. The first section, 'Histories of the (De)Automation Debate', examines specific moments in the long history of conceptualising automation to show that the anxiety over technological unemployment has not arisen in recent years, but has long concerned radical thinkers. It begins with a chapter in which Jason Dawsey compares attempts made by Herbert Marcuse and Günther Anders to develop a framework to gauge the impact of automation and the Left's appropriate response to it. Dawsey identifies in their often-oppositional approach the central tension in the evaluation of automation between what Moishe Postone has termed industrial technology's 'manifest form' and its 'latent potential'. Marcuse saw an emancipatory potential in automation, claiming that the problem was not automation *per se*, but automation in its current capitalist form. Indeed, Marcuse called on the Left to fight for a thoroughly automated society. Anders, in contrast, viewed automation with far greater hostility. For him, automated production was even worse than the Fordist-Taylorist labour regime, bearing only joblessness and passivity for those who remained employed. Anders thus encouraged the Left to see automation as an enemy.

The second chapter is an updated reprint of Jason Smith's two-part essay 'Automation: Then and Now: Nowhere to Go', originally published in *The Brooklyn Rail* in 2017. Beginning with James Boggs's account of how automation in the automobile industry all but destroyed workers' ability to resist the demands of capital, Smith turns his attention to fears about automation and mass unemployment, which he sees as recurring cyclically since the 1930s. Smith then examines the latest version of this fear around the rise of IT in the 1990s and puts it in the context of global capital and its long falling profit rates. Against those who extrapolate from specific processes of automation in certain branches, Smith draws attention to the specificity of different sectors, particularly the service sector, calling into question whether scenarios taken from automation and job losses in e.g., the oil extracting industries – a relatively

---

249   Saito 2022.

small sector of employment – can be extrapolated in the largely unautomated care and service sectors. Indeed, as Smith shows, many truisms of the automation debate do not stand up to scrutiny. The automation of a lot of services, for instance, merely changes the tasks of those employed. On top of this, as productivity rises and services become cheaper, demand for these services increases. At the same time, however, there is a tendency of service jobs to shift from skilled to low-waged, unskilled labour resistant to processes of real subsumption. This leads Smith to return to the question of worker resistance under this new labour regime in which the strikes in production of former times have been made all but impossible in many sectors.

The second section, 'Key Concepts in the Automation Debate', collects contributions that address important theoretical issues that arise when automation is approached from a Marxist perspective. These include the implications of automation for the labour theory of value, the dependence of automation on preparatory manual labour, the concept of deskilling, and, finally, alienated labour versus the importance of retaining non-alienated labour under socialism. The contributions in this section discuss the usefulness or limitations of key concepts in the Marxian critique of political economy when it comes to making sense of recent technological developments.

Christina Gratorp's chapter opens this section by questioning automation's supposed overall reduction of labour time. Automation is a highly selective process, argues Gratorp, in which labour time is reduced only for some processes and in some locations, but at the same time additional, often hidden, un- or underpaid labour fills the gaps created by current computer-driven automation, such as the labour necessary to train AI. Two ideological narratives mask this labour: the narrative of labour reduction through automation and the lesser valuation of the kinds of labour that digital technologies at once create and make ever harder to detect. In this way, automation cements inequalities along class, gender, and racial lines. Building particularly on Alf Hornborg's work, Gratorp highlights uneven flows of energy and matter from the global peripheries to the industrial core. Automation is a global spatiotemporal rearrangement that relocates labour from the highly industrialised centres to locations where it is much cheaper. This importation of labour compresses time at the centres. Gratorp then turns to the popular narrative that technology is a 'solution' to social problems. A look at reproductive work shows that 'time-saving' household appliances have not reduced labour, but merely redistributed it. In the final section, Gratorp confronts the idea underlying fantasies of full automation as a way to achieve a 'leisure revolution': the idea that automation itself can be automated. Here, too, Gratorp is sceptical. While 'full' automation might be a dream for some, her examination suggests that every technology creates new forms of labour necessary to uphold it. Automation is thus more likely to perpetuate existing inequalities rather than alleviate them.

Benjamin Ferschli's contribution goes back to Marx's labour theory of value and takes it as a basis to examine the viability of recurring fantasies of a 'capitalism without workers'. Building on Marxian notions of the source of value as connected to human labour and the problem of a rising organic composition of capital explored in *Capital* volume 3, which he contrasts with bourgeois political economy as well as marginalist theories, Ferschli traces the contradictory tendencies in capitalism to, on the one hand, automate to get rid of troublesome workers and, on the other, cut into surplus value and profit through automation.

Amy E. Wendling's 'Deskilling: Automation and Alienation' proposes the concept of skill as central to Marx's concept of labour and the role of automation and alienation in the restructuring of work processes. Beginning from the ambivalence of automation as both a blessing and a threat, Wendling notes that labour is not simply replaced, but restructured from complex to simple labour through automation. In this process, the definition of skill becomes central. As Wendling shows, linking skill exclusively to intellectual capabilities owes more to capital's interest than to an inherent characteristic of the concept or a feature of automation. It is, furthermore, arbitrary and contradictory in its application. As Marx recognizes, skill can be related to both mental and bodily capabilities. Taking the Royal Institution's attempt to keep knowledge from simple labourers, Wendling discusses the struggle over the general intellect, a force that capital at once depends on as a 'free gift' and wishes to constrain and control to force it into its logic of maximising profits. Building on Dyer-Witheford, Kjøsen, and Steinhoff's *Inhuman Power*, Wendling finally turns to the future of automation and deskilling, using a seemingly simple tool, the calendar in everybody's phone, to propose an alternative to the posthuman, fully automated future Dyer-Witheford, Kjøsen, and Steinhoff imagine.

In contrast to the examinations of aspects of automation under capitalism, the final essay in this section, Jeff Noonan's 'De-alienated Labour, Technology, and the Social Heart of Socialism', poses the question about the future of labour and automation *after* the end of capitalism. Contra a long history of automation enthusiasts, beginning with Paul Lafargue, Noonan insists on the importance of the young Marx's conviction that labour in a non-alienated form will remain central to our species being. Distinguishing between a necessity to do wage labour – that can be reduced through automation and a transformation of society into a socialist one – and a freedom to develop one's talents and interests, Noonan argues against a conception of socialism that builds only on individual self-fulfilment and *for* forms of labour that are meaningful because they do justice to our existence as social beings. Breaking from the individualization of capitalist society is necessary, writes Noonan, to ultimately overcome alienation: we can only realise our full potential by contributing to a larger community.

The third and final section, 'Automation, Labour, and Resistance: Production, Distribution, Representation', provides two case studies of the role of automation in the retail sector and along global supply chains, one that examines automation in higher education, and a final essay on automation in art. It starts with Larry Liu's examination of automation at Amazon's warehouses. Building on Braverman's concept of monopoly capitalism, Liu uses field notes and online commentary by warehouse workers to examine how Amazon employs technology to maximise profits and control workers. Robotic and computer technology, like shelf-moving robots and hand-held scanners, is used to deskill tasks and keep workers and their productivity under constant surveillance, allowing Amazon's management a tight control over the work process while maintaining precarious employment conditions that make workers easily replaceable (if they are not hired seasonally to begin with) and their resistance risky, a demonstration of how technology and class struggle go hand in hand.

Like Liu, Steffen Reitz focuses not on a production compant, but a retail giant relying on subcontracting: the "fast fashion" corporation Inditex, parent company of Zara and other brands. Reitz's chapter examines the impact of the introduction of Radio Frequency Identification Technology (RFID), which makes individual goods trackable. As Reitz shows, RFIDs influence production, distribution, working conditions and possibilities for worker struggle in a globalised economy increasingly organised around supply chains. RFID thus enables the switch from a push model, in which commodities are produced then sold, to a pull model, in which demand controls production – outsourced to multiple smaller companies – made possible in the wake of the logistics revolution. As Reitz argues, the distributed network of an RFID-controlled pull model transforms the power between employer and workers not least by making many traditional models of resistance by workers no longer viable. In line with Jasper Bernes' sober description of the logistics revolution,[250] Reitz also highlights the necessity for workers to understand the new technology of RFIDs and the new model of just in time pull production as well as its breaking points. Due to the company's ability to flexibly change its production to circumnavig-

---

250   Bernes 2013.

ate strikes by production workers and its simultaneous reliance on fast shipping, places of disruption have shifted from the sphere of production to that of circulation.

Leaving behind the retail sector, Robert Ovetz shifts to a different locale not usually considered in debates of automation: higher education. Ovetz critically examines the implementation of learning management systems (LMSs) by universities as an attempt to deskill the formerly highly individualised task of teaching at universities and introduce something like real subsumption by dividing course design, teaching, mentoring, and grading into segments. At the same time, LMSs like Canvas as well as the teleconferencing app Zoom make monitoring and control of both students' and professors' activities easier while outsourcing acquisitions to teachers and making particularly adjunct faculty into 'self-disciplined precarious platform workers.' Ovetz ends by suggesting a need to resist this development – and not for reasons of quality control, but as a way to resist the rationalisation of academic labour in accordance with the demands of global capitalism as well as the university's own restructuring along profit principles. As long as the teacherless classroom remains a fantasy by university administrators, however, teaching and assessment remain 'critical choke points for disrupting the reorganisation of higher education.' As Ovetz asserts, only collective action to slow the shifting of teaching online (significantly sped up by the Covid-19 pandemic) can help higher education teachers bundle their forces to collectively resist the transformation of universities in capital's favour.

Yet another highly individualised and therefore unusual area for an examination of automation is art production. Jens Schröter examines art created by machine learning algorithms. Schröter begins by revisiting attempts made in the 1930s and 1960s to formalise what makes an artwork aesthetically pleasing and thus establish the foundations for automating art production. Schröter goes on to show why art production nevertheless does not seem particularly threatened by automation. The reason lies in the specificities of what Pierre Bourdieu would call 'the field of cultural production',[251] in which art is created in part through the recognition of its creator. Writes Schröter: 'Artistic work is tied to a specific body, socially and historically located – and this location means that the body is an artist because he or she is enmeshed in a network of institutions and other bodies that collectively produce his or her authorness'. Curiously, this connection to a creator can sometimes survive the separation between conception and execution, which Braverman saw as central to auto-

---

251 Bourdieu 1993.

mation, through the historical situatedness of an artist's body of work – even if an artist uses others' labour to create the work. In closing, Schröter objects to the idea that art production is the non-capitalist, non-alienated activity *par excellence* that some have made it out to be. While art in its 'radical historical singularity disrupts capitalist rationalisation', it does not follow that art is an anti-capitalist utopia that can serve as a model for the organisation of post-capitalist labour.

The volume concludes with an afterword by James Steinhoff, Atle Mikkola Kjøsen, and Nick Dyer-Witheford, who engage the contributions through their own work on automation. In particular, they stress the role of AI in automation particularly in the sphere of circulation. Against the line of thought that full automation would undermine the law of value production, which is dependent on human labour, Steinhoff, Kjøsen, and Dyer-Witheford expand the argument made in *Inhuman Power*.[252] They ask which conditions would have to be fulfilled for advanced AI to be able to labour, i.e. become variable capital, and thus exploitable and capable of value generation, identifying two prerequisites: machines would have to possess a 'general intelligence' that is transferrable between domains, and they would have to become 'proletarianized', doubly free in the Marxian sense: separated from the means of production and legally free to sell their labour in the market. A proletarianised machine obviously would also have to 'be made dependent on purchased commodities for its continued existence'. In the final section of their afterword, the authors turn to algorithmic management in corporations such as Uber, Amazon, and Walmart where algorithmic devices control workers in various ways, and to the potential of blockchain technology to drive further the tendencies of automation. The authors develop a vision of AI's potential to disrupt capitalism as we know it, not in the sense of a crisis that leads to the end of capitalism, as is common among automation enthusiasts, but one that might lead to the end of capital's reliance on human labour.

---

252 Dyer-Witheford, Kjøsen, and Steinhoff 2019.

PART 1

Histories of the (De)Automation Debate

CHAPTER 1

Production and De-humanization: Herbert Marcuse, Günther Anders, and the Marxian Response to Automation

Jason Dawsey

In the summer and fall of 1978, Günther Anders, living in Vienna, and Herbert Marcuse, then in La Jolla, California, resumed their correspondence following nine years without any contact.[1] After reconnecting, the two philosophers filled their letters with humour, mutual admiration, and genuine elation that advanced age had not yet prevented them from continuing their philosophical endeavours. Marcuse expressed excitement that Anders was back at his desk writing a second volume of his great work, *Die Antiquiertheit des Menschen* (The Obsolescence of Human Beings), which he took to be Anders's 'best book'.[2] In particular, Marcuse looked forward to an essay in Volume 2 titled 'Die Antiquiertheit der Arbeit (The Obsolescence of Labor)'. 'That is almost the key to the whole! Also for me!' he declared.[3]

Marcuse's sudden death from a stroke in Starnberg, West Germany, in July 1979, abruptly closed any further exchange between them. This is quite regrettable. Both former students of Martin Heidegger and left-wing German-Jewish refugees from Nazism who fled Germany in 1933–34 and eventually settled in the United States, Marcuse (1898–1979) and Anders (1902–1992) shared a friendship that lasted roughly four decades. The links between them were thus long and rich, and yet remain extremely understudied.[4]

Barry Kātz's contention that the 'practical imperatives' of Marcuse's critical social theory were 'imposed by the events of the twentieth century' easily applies also to Anders.[5] While Marcuse held academic positions in the US and

---

1 My thanks to the editors, to the anonymous reviewer, and to Christopher John Müller for their feedback.
2 Herbert Marcuse to Günther Anders, 27 August 1978, in Anders 2022, p. 131.
3 Marcuse to Anders, 7 November 1978, in Anders 2022, p. 134. The entire correspondence between Marcuse and Anders is also available on Harold Marcuse's webpage about his grandfather Herbert: http://marcuse.faculty.history.ucsb.edu/projects/anders/AndersMarcuseCorrespondence1947_1978.pdf.
4 One exception is Fuchs 2002.
5 Kātz 1982, p. 12.

Anders built a life in Austria as an independent writer, they shared a commitment to write outside of established leftist parties and organizations and trade unions. Both sought to expose the 'affluent society' of Western Europe and North America in the three decades after World War II as totalitarian and stultifying.[6] As long-time antifascists, they joined the worldwide struggle against the counterrevolutionary American war in Southeast Asia. As is well known, the two theorists also exerted broad influence over young people in social movements (student, environmental, peace) distinct from the socialist and communist workers' movements and concerned about the ultimate survival of humanity.

Most importantly for my purposes here, Anders and Marcuse shouldered the challenge of thinking through what the twentieth century's technological revolutions entailed for the cause of international socialism. Although their approaches suffered conflation with the reactionary mystifications of the later Heidegger, each defended the necessity of the critique of technology within frameworks grounded in or at least heavily indebted to Marxism.[7] For both, the emergence of automated technologies after the Second World War posed an especially sharp challenge to revolutionary working-class politics. If they never directly commented on each other's writings about machinery and the proletariat, Marcuse and Anders quickly realized the enormously disruptive power automation entailed for the classic workers' movements they proudly supported. These thinkers did so as they witnessed a steep decline in proletarian militancy and the domestication of socialist and communist parties. Thus, they called on the international left, traditionally more focused on questions of property relations and class struggle, to confront the ramifications of this latest onslaught of mechanization.

∴

This is an opportune moment to return to Marcuse and Anders's analyses. Since the 1950s, widespread apprehension over the impact of automation has recurred in the advanced capitalist countries. We are again living through such a moment, a moment documented extensively in this volume's introduction.

---

6   The reference is obviously to John Kenneth Galbraith 1958. Marcuse and Anders both used the title of Galbraith's book for their distinctive but not dissimilar indictments of this period of capitalist prosperity.
7   For the discussions about the ostensible left-wing Heideggerianism of Marcuse, see Richard Wolin 2001, ch. 6; Abromeit 2010. For similar debates surrounding Anders, see Hildebrandt 1990; Woessner 2011, pp. 66–78; Dawsey 2017.

My aim here is not to directly intervene in current debates on the left about the possibilities or perils of automation, but to expedite their historicization. While some commentators on the subject, such as Paul Mason, Nick Srnicek, Alex Williams, Peter Frase, and Aaron Bastani, have striven to reclaim futurity for a Left mired in deep pessimism, much of the existing literature on automation is, conversely, often myopic historically, with the Great Recession the major point of reference.[8] Given the recrudescence of job-liquidating mechanisms during the Age of Capital, a longer historical perspective is exactly what is needed so that old debates are not blindly refought.

With an eye to the contemporary, trans-Atlantic discussions on the left, this piece raises the question – and it can only be posed here – of an adequate Marxian framework to the impact of automation by juxtaposing the positions of Anders and Marcuse on the trajectory of automated production. It does so through a comparative analysis of a selection of writings they authored between 1955 and 1980, the first quarter of a century after automation emerged as a general concern. While his theory of the Atomic Age has overshadowed his analyses of mechanized labour processes, the approach to automation Anders laid out disputed many of the most essential aspects of Marxist theory on the social effects of automatic technologies, hence my characterization of it as a post-Marxist approach preceding and distinct from that of Ernesto Laclau and Chantal Mouffe.[9] Having experienced factory work first-hand from his time of American exile, he viewed automation with great hostility. In his two-volume major work *The Obsolescence of Human Beings* (1956, 1980) and supplementary writings, he contended that automated production was even worse than the Fordist-Taylorist labour regime.[10] Automation spelled only joblessness for the many and a reduction to absolute passivity for the few who clung to employment. For Anders, the left should therefore deem automation an enemy. He believed that its expansion represented a nail in the coffin for the hopes of proletarian socialism. Only a fundamental critique of technology could respond to the spread of automation.

In contrast, Marcuse discerned real emancipatory potential in automation for the abolition of unnecessary toil, an aim of Marxism since the time of Marx and Engels. The problem was not automation *per se*, but automation in its cur-

---

8   Mason 2016; Srnicek and Williams 2015; Frase 2016; Bastani 2019.
9   For an approach to Anders's critique of modern technology as a post-Marxist response to the perceived inadequacies of traditional Marxism in an era of technological revolution, one quite different from the focus on the Left's essentialism set forth by Laclau and Mouffe in their 1985 *Hegemony and Socialist Strategy*, see Dawsey 2019.
10 Anders 1956; Anders 1980.

rent stunted, capitalist form. Against much of the anxiety of the 1960s about the spread of automated processes in industry, Marcuse criticized any kind of blanket rejectionism. The Marxist Left should fight for a thoroughly automated society, a society where the maximum time away from heteronomous labour would be achieved. Such a perspective, as I will show, fed directly into Marcuse's famed 1964 *One-Dimensional Man*.[11]

This piece is an attempt to examine their vastly differing perspectives through a lens both knew well: Karl Marx's still unsurpassed dialectical critique of the infusion of scientific knowledge and machinery into social production during the Industrial Revolution. In the first volume of *Capital*, Marx carefully separated the 'spontaneously developed, brutal, capitalist form' of industrial production where the machine system 'becomes a pestiferous source of corruption and slavery' and the 'worker exists for the process of production, and not the process of production for the worker' from its potential form where, 'under the appropriate conditions', industrial technology turns 'into a source of humane development'.[12] Recovering, rethinking, and deploying this framework for a reconstituted Marxist politics and theory is of immense importance for twenty-first century socialism.

Moishe Postone, in his brilliant interpretation of the later Marx, borrowed the lexicon of Freudian psychoanalysis and characterized Marx's approach as distinguishing between the '*manifest form*' and the '*immanent potential*' of industrial technology.[13] The latter, a 'nonidentical moment', 'comes increasingly in real contradiction to its capital form-determination', yet there is 'no smooth linear progression to a new form'.[14] Only a radical transformation of society could accomplish that. Postone's harnessing of Marx's seminal insights obviates, on the one hand, a descent into total dystopianism, while demanding critical assessment of new technologies based on whether and how they engender social emancipation. Such a dialectical form-potential distinction could further the reconstitution today of a socialist imagination at a time when mass yet often inchoate discontent, susceptible to extreme right-wing pleas, arises in country after country.

As I hope to show, Anders's post-Marxist position stressed the 'manifest form', while Marcuse's critical Marxist theory tended to accent the 'immanent potential'. Both philosophical outlooks became one-sided, and Anders ulti-

---

11   Marcuse 1964.
12   Marx 1990a, p. 621.
13   Postone 1978, p. 777; original emphasis. For the extremely rich elaboration of this argument, see Postone 1993, pp. 360–1.
14   Postone 1978, p. 778.

mately fell into a fetishism of technology. Through a comparison of Marcuse and Anders's vastly differing perspectives, I hope to illuminate how crucial is the distinction Marx and Postone adduced between form and potential when evaluating technologies tied to capital's apparatus of production, distribution, and consumption. This is particularly poignant for a phenomenon like automation where its effects on livelihood in commodity-producing society are so explosive.

∴

When one looks at his long career as a philosophical writer and political activist, it is no surprise that Günther Anders chose to examine the implications of automated labour operations. Since at least the late years of the Weimar Republic, he had been an ardent philosophical adherent to the ideal of *Homo faber* (Man the Maker); the classic definition of humans in terms of their capacities for the conceptualization and fabrication of worlds, external and internal. Originally, Anders embraced this concept as he worked out, after extensive study of the writings of Max Scheler and Helmuth Plessner, a philosophical anthropology of the 'human being's estrangement from the world'.[15] With no fixed, instinctual structure to orient it to its environment as animals possessed, *Homo sapiens* was free to create (and destroy) its own worlds. For Anders, this radical ontological freedom could and did encompass labour as an essential activity.

Before he fled Germany in 1933, shortly after the Nazis took power, a crucial shift in Anders's anthropological thinking occurred via a burgeoning interest in Marx and Marxist thought. In his critical writings on modernist literature and the visual arts, he analysed characters who were '"people without a world" (*Menschen ohne Welt*)'.[16] By this term, he meant those who were forced to dwell in a world in which they were not at home, a world they largely constructed but which confronted them as something alien and cast them out. Initially, this shift to 'worldlessness' crystallized through an examination of the victims of the devastating mass unemployment brought on by the Great Depression. Anders intended this new line of argumentation as an 'expansion of the basic Marxian thesis that *the proletariat* does not possess *the means of production*, by means of which it produces and keeps operating the world of the ruling class'.[17]

---

15   For his early philosophical anthropology, see the impressive collection of texts in Anders 2018.
16   See the introduction to Anders 1984, p. xi.
17   Anders 1984, p. xii; original emphasis.

Marxist social critique – and this was a major theme of his philosophy – was too narrowly defined and had to be broadened to encompass what might be called the existential deprivation inflicted on the working class by the capitalist mode of production. Ultimately, Anders's attempts to overcome the problems he discerned in Marxism (while still utilizing many insights from Marx, Georg Lukács, Bertolt Brecht, and the Frankfurt School of Critical Theory) led him to examine instances of domination tied primarily to mechanization.

It was in his superb *The Obsolescence of Human Beings* where Anders elaborated a 'philosophical anthropology in the age of technocracy', a post-Marxist conception of modern social life.[18] In the first volume he warned of the threat posed to human beings by a 'technocratic totalitarianism' and, framing contemporary history for his audience, announced that 'in no different sense than Napoleon had asserted it 150 years ago of politics, and Marx 100 years ago of the economy, technology is now our fate'.[19] Finally completing the second volume of *Obsolescence* after years of participation in the politics of nuclear abolitionism and the resistance to the Vietnam War, Anders issued even bolder formulations about the agentive force machines had acquired. '*Technology has now become the subject of history* with which we are only still "co-historical,"' he wrote.[20] As the capabilities of machines improved and expanded, humans appeared stunted, outdated, and useless. Capitalism and 'economic' concerns, which he frequently assessed quite narrowly, thus receded for Anders behind 'the question of the transformation or liquidation of human beings by their own products'.[21]

While the essay concluding the first volume of *Obsolescence*, 'Über die Bombe und die Wurzeln unserer Apokalypse-Blindheit' (On the Bomb and the Roots of Our Blindness to Apocalypse), featured a new and powerful conception of the Atomic Age, Anders certainly did not confine his critique of modern technology to this most extreme case of 'liquidation'.[22] Frequently overlooked is his brilliant set of reflections on the terrible moral and physical consequences of Taylorism, the assembly line, and automated technologies for workers.[23]

In several fiercely critical, public and private commentaries, Anders examined how *Homo faber*, 'the human being who is "essentially" built for laboring',

---

18 Anders 1980, p. 9; original emphasis.
19 Anders 1956, pp. 82, 7.
20 Anders 1980, p. 9; original emphasis.
21 Anders 1956, p. 7.
22 Anders 1956, pp. 233–324.
23 For some significant exceptions, see Delabar 1992; Hoffmann 1992; von Greif 1992; van Dijk 2001, pp. 66–76, 128–33.

experienced the twentieth century as a time of relentless and catastrophic degradation.[24] Thinking about the genesis of the age of human antiquatedness, Anders pointed to the assault the working class underwent with Fordism (though he rarely, if ever, used the term) and Taylorism. Under the Fordist-Taylorist organization of labour, the worker 'allows himself to be incorporated by the running of the machine', and learns to 'function loyally like a "wheel."'[25] The physical pressure exerted on workers on an assembly line, which was captured, Anders emphasized, in the widely used and deceptively innocuous phrases like '"*Anpassung* (adjustment),"' could, in the beginning, all but overwhelm.[26] 'Whoever has once been confronted at one time with a new assembly line task', he described, 'knows what exertion it costs to turn this first confrontation into coordination (*Gleichschaltung*) with the motion of the machine, thus to keep step with the running machine; and they know the fear of not being able to keep step'.[27] Work on an assembly line was, thus, 'something worse', 'much more accursed than any earlier labor had been'.[28] One could date, then, the historically formative moments for Anders' theory of human obsolescence to the revolutionizing of capitalist production that transpired with Fordism and Taylorism between 1910 and 1930.

Holding in view the longer history of the debasement of work in the twentieth century, Anders' critique of machine-dominated labour culminated with a detailed and damning commentary on automation and rationalization. Throughout his examination of automated industry, Anders never veered from an outlook of total opposition. If Taylorist labour had been 'alienated' and 'inhuman', automation proved to be even, if that were possible, worse.[29] Already in 1956, he had noted, in a terse remark, how the onset of automation contributed to the disintegrating sense of responsibility among workers for what they produced.[30]

In 1959, Anders revisited the topic in a short text for a journal, revealingly titled *Homo ludens* (Human Being at Play), after the 1938 book by the Dutch historian Johan Huizinga.[31] In these 'journal entries', Anders reported on an automated operation he had recently witnessed. Completely passive, the

---

24   Anders 1980, p. 103; original emphasis.
25   Anders 1956, pp. 90–1.
26   Anders 1956, p. 90.
27   Anders 1956, pp. 89–90. 'Gleichschaltung' was the phrase the Nazis used to describe the 'coordination' or 'synchronization' of Germany's institutions under Hitler's rule.
28   Anders 1980, p. 93.
29   Anders 1980, p. 94.
30   Anders 1956, p. 350, n. 290.
31   Anders 1959, pp. 8–9.

worker Anders watched took his cues from the machines' signals. 'His behavior resembled more that of the leisurely moviegoer than any kind of activity which we could designate as "labor."'[32] Furthermore, Anders observed, one could not truly speak of 'collaboration (*Zusammenarbeit*)' among the workers. The sole form of *Zusammenarbeit* the labourer experienced was 'collaboration' with the production process itself that required very little human intervention and proceeded, he said quoting Marx, 'behind his back'.[33] This type of labour did not coordinate the men but, on the contrary, separated them, a situation that they did not even realize. For Anders, the deplorable combination of 'actionless' assignments for workers and isolation in their tasks in this automated plant heralded 'a general "trend"' in the industrialized world.[34]

These first investigations of automation did not emphasize the loss of jobs. Rather, the erosion of solidarity and the forced passivity among workers grabbed Anders's attention. One can see a transition to profound concern, though, about looming unemployment four years later, when Anders corresponded with Danilo Dolci, the Italian peace activist and tireless advocate for the poor in Sicily, on the effects of automation.

> Today already, especially since the introduction of automation, a new type of unemployment, at least partial unemployment, has begun in the highly industrialized countries such as the USA. Through a most peculiar dialectical development, there are there today thousands upon thousands, no millions, who do not know what they should go about doing with their free time, thousands of non-doers, who are subjected to (*ausgesetzt*) to total demoralization, who plunge themselves, 'to kill their time', into the most senseless pseudo-activities ('do it yourself') – in short: millions of brothers of those millions, who loiter dully in Sicily or in other underdeveloped countries or pursue pseudo-activities or crime or vice. This new poverty is, of course, a poverty of luxury. While the impoverished in Palermo or Calcutta who go to ruin spiritually and physically because the opportunity to work has not been given to them, because they do not yet enjoy the right to work and because they remain robbed of the products indispensable for their lives, indeed of the will to a better life – their comfortable brothers go to ruin because production is so enormous, that they need to participate, no are permitted to participate, in produ-

---

32   Anders 1959, p. 8.
33   Anders 1959, p. 8.
34   Anders 1959, p. 9.

cing only in ever lesser proportion, in short: these people (one can no longer name them proletarians and they also no longer name themselves so) are no longer permitted to work.[35]

In these perceptive remarks, Anders distinguished the older impoverishment in Sicily and India from unemployment wrought by automation. Tellingly, he did not specify what part modern technology should have in alleviating the poverty caused by underdevelopment. He understood the latter, a 'poverty of luxury', as symptomatic of the prosperity that had recently swept the First World. Integral to this new poverty, he related to Dolci, was a 'new type of non-doing', where 'thousands upon thousands who may not still even be proletarians, do not yet enjoy the bygone right to the rightlessness of the proletarian'.[36] These criticisms, while not yet fully worked out in the letter, were the basis for his later theory of automation.

Subsequently, when Anders recommenced work on *The Obsolescence of Human Beings* in the mid-1970s, mass unemployment had again become a reality in the capitalist world. The undoing of the 'affluent society' amidst the economic downturn of that decade (e.g. the energy crisis, stagflation) led him to identify an underlying tension within the Fordist-Keynesian welfare state, a central '*contradiction between rationalization and full employment*' characterizing industrial society.[37] This was not a contradiction in the traditional Marxist sense in terms of countervailing tendencies that might prove emancipatory. Both tendencies Anders construed as heteronomous. A new turn to automation resembled for him the drive toward 'rationalization', which he took to be the elimination of jobs in the name of greater 'efficiency'. However, sustained attention to the dynamics of capitalism precipitating such a slashing of employment was conspicuously absent from his commentary.

During the resumption of writing on *Obsolescence*, Anders broadened his assessment of the introduction of automated production techniques and accorded them a central place in his critique of labour. He designated automation as one of the 'internal revolutions' unfolding within a Third Industrial Revolution (the first was the classic Industrial Revolution; the second

---

35   Günther Anders to Danilo Dolci, 12 October 1963, Nachlass Günther Anders, Literaturarchiv der Österreichischen Bibliothek, Vienna, Signatur: 237/B84. The phrases 'to kill their time' and 'do it yourself' were inserted into the letter in English.
36   Günther Anders to Danilo Dolci, 12 October 1963.
37   Anders 1980, pp. 99–100. The term 'rationalization' pointed back to debates in Weimar Germany about the importation of American business models, such as Fordism and Taylorism. For these debates, see Nolan 1994.

dealt with massive changes in advertising).[38] In these passages, he warned of the impact of automation on the industrialized world in terms of a growing '"trend"' toward the superfluity of workers. So alarmed was Anders by reports in a 1978 issue of the high-profile West German newsmagazine, *Der Spiegel*, of the Japanese company Kawasaki's experiments with a totally automated factory, that he feared, once firms elsewhere replicated such experiments, an *'existence without labor'* awaited the industrial working class.[39]

Much of this ultra-critical disputation Anders put forward in 'The Obsolescence of Labor', the essay from the second volume Marcuse so wished to see. There he asserted that automation, when implemented, inflicted a twofold damage. It generated a new form of unemployment, a 'second unemployment' (the Great Depression was the first), and a new type of inhuman work. With an eye to the former, Anders predicted the worst scenario, that by the year 2000 all those fortunate to have a job would be employed in automated pursuits.[40] Refusing to accept the promises of politicians that retraining programs would ameliorate the loss of jobs, he denounced those leaders as *'salami tacticians'*, who either out of ignorance or duplicity, neither realized that *'rationalization reduces absolutely the number of work spots'* nor fathomed the 'iron rule' that *'with the increasing amount of automation the number of necessary workers sinks'*.[41] The notion of a '*"workers quotient"* (WQ)', he urged, had to be interjected into discussions of automation's impact.[42] This WQ would measure the number of workers regarded as 'indispensable' to the maintenance of a firm, whether blue-collar or white-collar. According to Anders, in the advanced industrial nations, this quotient dropped steadily toward zero.[43]

Recognition of this tendency led Anders to formulate another ostensibly 'iron rule' – *'the postulate of full employment thus will be all the less fulfillable,*

---

38   Anders 1980, pp. 26–31.
39   Anders 1980, p. 27; original emphasis. See also Anders 1980, p. 97. For the special issue, see 'Die Computer Revolution – Fortschritt macht arbeitslos', *Der Spiegel*, 17 April 1978. The front cover of the magazine depicted a robot carrying off a worker. It was the article 'Uns steht eine Katastrophe bevor', with its description of the Kawasaki plant, that elicited so much attention from Anders.
40   Anders 1980, p. 94.
41   Anders 1980, pp. 100–1, 94; original emphasis.
42   Anders 1980, pp. 27, 100–1; original emphasis. Anders admitted that such a WQ could not be applied to the Indian economy in the same way that it might to the American. He did recognize the immense (in the 1970s) differences in levels of development, infrastructure, and growth.
43   Anders 1980, p. 27.

the higher the technological status of a society'.⁴⁴ Communist regimes, which named themselves workers' states and glorified proletarian labour, were not immune to this revolution, though Anders, for reasons he did not elucidate, indicated they would succumb to it more slowly.⁴⁵ With respect to the advent of the 'second unemployment' in the capitalist countries, he conceded that the driving impulses for recourse to automated techniques of production were sheer profitability and the weakening of workers' organizations.⁴⁶

The men and women who secured or retained work in automated plants constituted essentially a new and most peculiar aristocracy of labour, to borrow the classic phrase from Lenin. This 'type of the automation worker is entirely new', Anders argued.⁴⁷ The removal of rigor, of exertion, from the labour process, in motion since the start of the twentieth century, reached its most extreme in automated systems.⁴⁸ Under automated conditions, the worker worked in isolation, like a cobbler, but could not touch the product. Instead, the machines signalled to the worker what the next task was to be. Such an employee could not be classified even as an 'intellectual worker', but 'rather merely a machine policeman (*Polizisten der Maschine*)', a policeman, incidentally, who hoped never to have to do anything.⁴⁹ The sole activity the production process permitted was a '*waiting*', a waiting for the signals from the machinery and a '*solitary watching*'.⁵⁰ Anders decided, after proposing these sundry appellations, that '"*object shepherds*,"' with its play on Heidegger's 'shepherd of Being', fit the automation worker best.⁵¹

For those who supported (as he himself had) a radical politics of labour, Anders bore especially grim news. Unlike assembly-line laborers, he contended, automation workers had little contact with their colleagues during working hours. Reiterating claims from his 1959 essay, he stated, the 'forced *asociality*' of this sort of 'waiting' obstructed the formation of shop floor identity and, hence, mass action against employers.⁵² Anders's aversion to automated rationalization was most evident when he compared assembly-line workers with their counterparts in automated factories. In comparison – and only in comparison – the former were subjected to decent work, still worthy of a human

---

44   Anders 1980, p. 99; original emphasis.
45   Anders 1980, p. 94.
46   Anders 1980, p. 100.
47   Anders 1980, p. 436, n. 5.
48   Anders 1980, p. 103.
49   Anders 1980, p. 437, n. 5.
50   Anders 1980, pp. 95, 96; original emphasis.
51   Anders 1980, p. 95; original emphasis.
52   Anders 1980, p. 96.

being. By contrast, the increasing prevalence of automation, for Anders, spelled the definitive liquidation of *Homo faber*.[53]

'Aversion' may yet understate Anders's contempt for rationalization. His remarks on the role of automation in the obsolescence of labour flowed into two quite radical commentaries on an automated future. By the late 1970s, after several years of economic dislocation, the situation faced by workers had gotten so desperate that it could only be compared to the English weavers and stockingers devastated by the invention of the power loom in the early nineteenth century. Clearly referring to the Luddites, Anders recalled their miseries, how critics deliberately distorted the nature of their struggle, and, with pride, named them 'our ancestors'.[54] It was illusory, he pronounced, to think that such open antagonism between people and machinery had forever passed. Expecting another wave of rationalization, Anders wrote, 'after one-and-half centuries of latency, the crisis has now become virulent again. The machine has again become a competitor and enemy'.[55] Rage against the disappearance of jobs and, he added, for those with jobs, the elimination of a fulfilling work experience, would likely play out in a reprise of the violence and machine-smashing of the Luddite revolt.[56]

In the second instance, Anders asked a very different question about the social implications of automation. He speculated about how political elites would respond to the prospect of mass joblessness. What would governments do with the surplus of people? First, he reminded his audience that an unemployment crisis precipitated the electoral successes of National Socialism in Germany and ridiculed the German people for being so 'gullible' for accepting Hitler and Goebbels's promises to solve the crisis.[57] Then, in one of his more harrowing, if not downright paranoid moments, he warned it was 'not at all impossible' 'that the (at the time economically absurd) Auschwitz ovens' could supply the 'models for the "coming to terms" (*"Bewältigung"*) with the fact that there are, compared with the opportunities for work, *"too many people."*'[58] Like Hannah Arendt, who had made similar statements in *Eichmann in Jerusalem*, Anders linked the phenomenon of superfluity with the preconditions for gen-

---

[53] Anders 1980, p. 27.
[54] Anders 1980, p. 29.
[55] Anders 1980, p. 29.
[56] Anders 1980, p. 102.
[57] Anders 1980, pp. 91–8.
[58] Anders 1980, pp. 98–9; original emphasis. On the other hand, this does not seem so paranoid when one envisions where the militarization of police, mass incarceration, and drone technology might lead. For another scenario of 'exterminism' as a possible future, see Frase 2016, pp. 128–43.

ocide.[59] The elimination of people from work through rationalization during a period of crisis might possibly prefigure, his argument suggested, the repetition of physical elimination of people jettisoned by advancements in workplace technology.

That is why the 1978 *Spiegel* issue, which appeared shortly before he informed Marcuse about his progress on the second volume of *Obsolescence*, distressed Anders so much. The findings reported therein about Kawasaki's '"unmanned"' factory in Japan led him to think the 'trend' toward rationalization, if pressed forward, would make people '*superfluous*' in the production process.[60] Such developments, long underway, overturned, he believed, the longstanding Biblical depiction of work as a universal curse. Having a job could very well become a privilege cherished by an elite few.[61]

This horrid, possible future entailed for Anders a new form of scarcity, a machine-induced scarcity of livelihoods that would constantly jeopardize the full employment attained and once celebrated by the developed countries. Inexorably, the First World gravitated towards its 'secret ideal', the '"unmanned factory,"' he later said.[62] Even if the welfare state could manage the dislocation caused by extreme rationalization, the transition to it could only eventuate in an existentially impoverished condition for people. What would men and women do without the possibility for real work? Curiously, Anders wondered how they would fill their supposedly 'free' time, a seemingly much less worrisome development, given his apprehension about the annihilation of the jobless?[63]

On all of these points, Anders opposed the thrust of Marxian theory with respect to the effects of machinery on production. Notably, he targeted Marx's statements that the shortening of the working day served as the precondition for an emancipated, communist society. In his discussion of the 'Trinity Formula' in Volume 3 of *Capital*, Marx had argued that the development of socially productive forces could facilitate the reduction, though never the complete elimination, of labour. Therefore, the 'true realm of freedom, the development of human powers as an end in itself', would be able to expand in hitherto unimagined ways because of the dramatic decrease in socially necessary labour

---

59 For Arendt's similar statement, see Arendt 1977, p. 273.
60 Anders 1980, pp. 26–7, 97; original emphasis.
61 Anders 1980, 27. This expectation of Anders has found a dreadful reality in the thousands who have perished fleeing to other, more prosperous countries in the hope of securing decent employment. See Srnicek and Williams, 2015, pp. 101–2.
62 For this, see Anders 1982, p. 272.
63 Anders 1980, pp. 27–8.

facilitated by the importation of science and technology into the production process.⁶⁴ Anders's post-Marxist reaction to this line of thinking is startling. Such statements, he believed, sounded like something 'from a different age'.⁶⁵

Anders largely dropped the subject of technology and mass unemployment and underemployment when he became, understandably, preoccupied once more with the nuclear arms race after 1980. Soon thereafter he became embroiled in a major controversy within the Central European peace movement over the use of violence.⁶⁶ Waning health and productivity intervened after 1987. Nevertheless, the writing Anders left behind remains one of the most radical indictments of automation we have, one borne from a fetishization of technology that prefigured much of the current disquiet.

∴

Herbert Marcuse's dialectical approach to the issue of automated processes provides an invaluable theoretical and political contrast to that of his friend Günther Anders. It must be remembered that Marcuse had contributed vitally to Marxist discussions of technology and labour reaching back to the late 1920s. Between 1928 and 1932, before his break with Heidegger and affiliation with the Frankfurt School of Critical Theory, he had written several essays where he examined the concept of alienated labour, including 'New Foundations for Historical Materialism', one of the first treatments of Marx's *Economic and Philosophic Manuscripts*. While these texts still show too much of the influence of Heidegger and are quite philosophically abstract in how they deal with the question of labour under capitalism, they indicate his path toward a fuller Marxist encounter with industrialism.⁶⁷

After several years of exile in the United States, Marcuse in 1941 published *Reason and Revolution* and 'Some Social Implications of Modern Technology', two works not only central to his own philosophical development, but tremendously fertile for any future Marxist theorization of mechanization.⁶⁸ In

---

64   Marx 1991, p. 959. It should be pointed out that what Marx called the 'realm of necessity', the requirement of men and women to work and transform nature for the satisfaction of human needs, would continue to exist under communism.
65   Anders 1980, p. 438, n. 2.
66   For this controversy, see Anders 1987.
67   These texts can be found with very good contextual material in Marcuse 2005.
68   J. Jesse Ramírez has also argued that Marcuse's engagement with fully mechanized production predated the 1960s. Ramírez contends that Marcuse, long before the coining of the term 'automation', was first fascinated by its prospects via his study of figures associated with 'Left Technocracy', such as Lewis Mumford. See Ramírez 2012.

the former, he explored the links between Hegel's idealism and Marx's historical materialism. One of the most salient points Marcuse advanced there concerned social freedom: how the organization of society (and not only political institutions and the law) might 'realize' human freedom. 'The study of the labor process', he argued, 'is, in the last analysis, absolutely necessary in order to discover the conditions for realizing reason and freedom in the real sense. A critical analysis of the process thus yields the final theme of philosophy'.[69]

The essay 'Some Social Implications of Modern Technology' truly broke new ground, expanding the purview of contemporary Marxist thought beyond the imposing forays into the impact of 'technological reproducibility' on art and mass culture (radio, film, the gramophone) recently initiated by Walter Benjamin and Theodor W. Adorno. In doing so, Marcuse refused to cede the fundamental systemic critique of technology to right-wing critics. Unequivocally repudiating romantic, reactionary approaches to industrialism, he contended:

> All programs of an anti-technological character, all propaganda for an anti-industrial revolution serve only those who regard human needs as a by-product of the utilization of technics. The enemies of technics readily join forces with a terroristic technocracy.[70]

Crucially, 'Some Social Implications of Modern Technology' cemented a Marxian, double-sided understanding of the effects of technological development. If technology in its present configuration grounded a new and redoubtable form of social domination, there were still emancipatory possibilities in modern technology that allowed human beings to 'decrease the time and energy spent in the production of the necessities of life, and a gradual reduction of scarcity and abolition of competitive pursuits', in favour of 'free human realization'.[71] A different technology, Marcuse indicated, might enable rather than restrict the full development of human beings in a future post-capitalist epoch.

It is quite illuminating to trace the path of Marcuse's thinking from the early 1940s to his later, bold assessment of automation.[72] That path led from his reading of the recently published 1857–58 manuscript by Marx, the *Grundrisse*, through his interest in post-war French thought, primarily the writings of

---

69   Marcuse 1960, p. 273.
70   Marcuse 1978, p. 160. Taking the term 'technics' from Lewis Mumford, Marcuse in this essay characterized the Third Reich as a form of technocracy.
71   Marcuse 1978, p. 160.
72   For more on Marcuse's treatment of automation, see Granter 2009, pp. 70–82.

Gaston Bachelard, Maurice Blanchot, Gilbert Simondon, and Serge Mallet, as well as the later Jean-Paul Sartre.[73] As he worked on *Eros and Civilization*, his synthesis of Hegel, Marx, and Freud, Marcuse was deeply affected by Marx's analysis of the contradictory character of industrial production in the *Grundrisse*.[74] Like many other socialist thinkers, he found incredibly powerful the passages in the manuscript where Marx asserted that, with the infusion of science and technology into industry the 'creation of real wealth comes to depend less on labor time and on the amount of labor employed ... but depends rather on the general state of science and on the progress of technology or the application of this science to production', making possible, then, a new form of production 'created by large-scale industry itself', with a massive increase of '*disposable time*' for human beings.[75]

The *Grundrisse* also featured the description of how 'the human being comes to relate more as watchman and regulator to the production process itself', a statement which proved immensely significant for Marcuse as the debate around the introduction of automated technologies swirled. Based on his study of the *Grundrisse* and related arguments in *Capital*, Marcuse began to reflect on what automation could accomplish, if freed from the grip of capital.

Marcuse's contacts with Raya Dunayevskaya (1910–1987) certainly furthered and deepened his study of automation. After reading *Reason and Revolution*, Dunayevskaya, the former Trotsky supporter and founder of Marxist Humanism, asked Marcuse, then a professor at Brandeis University, to pen a preface to her 1958 major work of Marxist thought, *Marxism and Freedom*.[76] Marcuse agreed. The book included startling and inspiring accounts of workers' opposition to the spread of autonomous machines, such as the 1955 Westinghouse Strike.[77] Consequently, Marcuse communicated with Dunayevskaya at length about the likely impact of automated processes on the trade-unions.

Marcuse did not come to these exchanges as a novice but with a theoretical perspective already well-developed. In August 1960, he cited Serge Mallet's studies on the sociology of labour in the French left-wing journal *Arguments*

---

73 For the version of the *Grundrisse* Marcuse used, see Marx 1953. The first German edition had appeared in 1939–41. Marcuse's many connections to French thought require more study.

74 Marcuse 1955. For more on the impact of the *Grundrisse* on Marcuse, see Rockwell 2013.

75 Marx 1993, pp. 704–5. It would be interesting to compare Marcuse's appropriation of the *Grundrisse* with that of Mason 2016.

76 Dunayevskaya 1958.

77 For these accounts, see Dunayevskaya 1958, ch. 16, 'Automation and the New Humanism'.

(later the basis for Mallet's *La nouvelle classe ouvriére* (The New Working Class) three years later). Marcuse requested good analyses of American conditions for his own new book project – what would become *One-Dimensional Man*. 'One of my problems', he related to Dunayevskaya, 'will be the transformation of the laboring class under the impact of rationalization, automation and particularly, the higher standard of living'.[78] Dunayevskaya eagerly complied with his request, sending him a huge list of materials which, she believed, provided a corrective to the overly academic literature Marcuse was examining.[79] One of them was 'Workers Battle Automation', a special issue of the Marxist-Humanist periodical *News and Letters* she had founded in 1955.

'Workers Battle Automation' is an extraordinary text.[80] It reads almost like a preview of Anders's commentary on automation, though there is no evidence Anders knew of *News and Letters* or Marxist-Humanism. The editor of *News and Letters*, Charles Denby, an African-American production worker, bluntly stated that the subject of the issue not 'what Automation could be if we lived under a different system but what Automation is right here and now'.[81] Throughout the issue, the voices of workers, facing the new mechanized system in the mining, auto, steel, tire, electrical, and meatpacking industries, put forward their stories. They spoke of the loss of control of the pace of work, the isolation of this type of labouring activity, worry, particularly for black workers, about the future of their jobs, and of fear of injury or death because of the machines. Anger at labour leaders like John L. Lewis of the United Mine Workers or Walter Reuther of the United Auto Workers for accepting capital's position on automation filled the pages. For Denby, automation ultimately meant the 'elimination of the laborer' and, for those workers who retained their positions, being 'subjected to the inhuman speed of the machine'.[82] Denby called for workers' control of production, a much shorter working week (forty hours' worth of pay for thirty hours' worth of work), and health and retirement benefits.[83] Studying the special issue greatly provoked Marcuse.

Marcuse immediately related to Dunayevskaya his response to 'Workers Battle Automation'. Taking the workers and Denby at their word, he did not challenge any of the accounts in the special issues. Instead, he sharply distinguished his dialectical position, grounded in Marx's *Grundrisse*, from Denby's

---

78   Marcuse to Dunayevskaya, 8 August 1960, in Anderson and Rockwell 2012, p. 59.
79   Dunayevskaya to Marcuse, 16 August 1960, in Anderson and Rockwell 2012, pp. 60–5.
80   Special issue of *News and Letters* 1960.
81   *News and Letters* 1960, p. 1.
82   *News and Letters* 1960, p. 1.
83   *News and Letters* 1960, p. 7.

interpretation. The distinction between automation 'under a different system' and 'what Automation is right here and now' was thus absolutely crucial for him. 'What is attacked' in 'Workers Battle Automation', Marcuse insisted, 'is *not* automation, but pre-automation, semi-automation, non-automation'.[84] The stunted character of automated production under capitalism engendered the horrors presented in *News and Letters*. While currently constrained, automation nonetheless represented 'the explosive achievement of advanced industrial society' and augured 'the practically complete *elimination* of precisely that mode of labor which is depicted in these articles'.[85]

What exactly prevented the implementation of what Marcuse called 'genuine automation' or 'complete automation' in his missive to Dunayevskaya? This is most important for understanding his assessment. Both the capitalists and the working class were complicit, though for very different and hardly equal reasons. The issue of profitability of course dampened the former's enthusiasm, as did government regulation of the use of machines. For workers, opposition stemmed, understandably, from the spectre of technological unemployment.[86] While he did not ignore the alarm of workers, like those quoted by Denby, Marcuse consistently stressed the potential in these technologies over the likelihood they, in their present form, would precipitate joblessness. 'It follows that arrested, restricted automation saves the capitalist system, while consummated automation would inevitably explode it', he told Dunayevskaya. The hopes for a better, more fulfilling life he glimpsed in the pages of the journal could only be realized by what he memorably defined as an automation-facilitated 'total *de*-humanization' of 'material production'.[87] What, for Anders, was the inevitable and horrific outgrowth of the expansion of automation Marcuse embraced as the goal a socialist labour movement should fight for with everything it had. His analysis implied that only a proletarian revolution could bring about such a transformation.

Clearly, Marcuse feared that a conflation of automation in general with its current, truncated deployment by the ruling class would be an enormous mistake for Marxists. Subsequently, he told Dunayevskaya that the goal of a '*technological Aufhebung* of the reified technological apparatus', i.e., an overcoming of alienated labour by realizing its latent potential, should be the goal of the Left.[88]

---

84   Marcuse to Dunayevskaya, 24 August 1960, in Anderson and Rockwell 2012, p. 66.
85   Marcuse to Dunayevskaya, 24 August 1960, p. 66; original emphasis.
86   Marcuse to Dunayevskaya, 24 August 1960, p. 66.
87   Marcuse to Dunayevskaya, 24 August 1960, p. 66; original emphasis.
88   Marcuse to Dunayevskaya, 22 December 1960, in Anderson and Rockwell 2012, p. 75.

Marcuse continued to clarify his position. In 1961, addressing 'historical alternatives' immanent to the social-democratic welfare-state, he envisioned the 'progressive automation of material and routine production to the point where the traditional ratio of (necessary) working time to free time is reversed'; such a momentous outcome would mean 'free time becoming a "full-time occupation" *at the disposal of the individual*'.[89] This 'trend' was generated by the internal dynamics of what he dubbed 'advanced industrial society' or 'technological society', categories which he tended to already use, in a very problematic way, interchangeably with 'capitalism'. The revolutionary possibilities of automation resulted from the degree to which the society was '*forced* to consummate technical progress: forced by the need for continuously raising productivity – a need which is in turn enforced by the necessity of internal growth and security, and by the external contest between capitalism and communism'.[90] Strikingly, Marcuse said little about the capitalist and working-classes' opposition to the spread of such technologies. Rather, he accented structural pressures and constraints endemic to the societal formation itself which, in contradictory fashion, deployed technology to curb and contain the possibilities for vast social change to which it gave rise.

It was *One-Dimensional Man*, his epochal 1964 book, where Marcuse presented his most fully-elaborated outlook on the promise and peril of automation.[91] As he had in 1941, Marcuse advocated in the book for a 'critique of technology' which 'aims neither at a romantic regression nor at a spiritual restoration of "values"'.[92] A Marxian critical social theory, which countered romantic criticisms of mechanization, 'must be a historical position', he declared, 'in the sense that it must be grounded on the capabilities of the given society'.[93] The point of departure for this incredibly influential work was the assertion that 'our society distinguishes itself by conquering the centrifugal social forces with Technology rather than Terror, on the dual basis of an overwhelming efficiency and an increased standard of living'.[94]

Although he referred to *News and Letters* and cited Denby and 'Workers Battle Automation', Marcuse did not mention Dunayevskaya at all in *One-Dimensional Man*. Contact between them had ceased in 1961 after she dir-

---

89 See 'The Problem of Social Change in the Technological Society', in Marcuse 2001, pp. 42–3; original emphasis.
90 Marcuse 2001, p. 43.
91 For some recent analyses of the book, see Reitz 2014; Aronson 2014; Maley 2017; Jay 2020.
92 Marcuse 1964, p. 57.
93 Marcuse 1964, p. xvi.
94 Marcuse 1964, p. x.

ected some hyper-critical remarks at the formidable Marxist biographer of Trotsky and Sovietologist, Isaac Deutscher, remarks which angered Marcuse.[95] Still, Marcuse owed a great debt to 'Workers Battle Automation' and his 1960 exchanges with Dunayevskaya.

One can see this readily in Chapter 2 of *One-Dimensional Man*, where Marcuse advanced a series of formulations about 'possibilities', 'tendencies', and 'alternatives' intrinsic to advanced industrial society that 'overshoot' the present. 'The oppressive features of technological society are *not* due to excessive materialism and technicism', he wrote.

> On the contrary, it seems that the causes of the trouble are rather in the *arrest* of materialism and technological rationality, that is to say, in the restraints imposed on the *materialization* of values.[96]

Automation, Marcuse contended, comprised one of the key 'centrifugal tendencies' 'inherent to technical progress itself'.[97] Referring back to the special issue of *News and Letters*, Marcuse acknowledged that the spread and entrenchment of automated or quasi-automated processes led to a particularly onerous form of 'drudgery' –

> ... exhausting, stupefying, inhuman slavery – even more exhausting because of increased speed-up, control of the machine operators (rather than of the product), and isolation of the workers from each other.[98]

Unlike in 'The Problem of Change in Technological Society', he did not neglect the voices of workers collected by Denby. Under the existing conditions, automation undoubtedly threatened their livelihood and degraded their experience of labour. However, he refused to let Denby and Dunayevskaya have the final word on this subject.

Marcuse mustered an extensive scholarly literature and his study of the later Marx, especially the *Grundrisse*, to respond, including Charles Walker's 1957 *Towards the Automatic Factory* and the AFL-CIO's report from 1958 on *Automation and Major Technological Change*. For Marcuse, the problem with auto-

---

95 See the March 1961 exchanges between them in Anderson and Rockwell 2012, pp. 82–84. Their correspondence did resume in the summer of 1964, shortly before the appearance of *One-Dimensional Man*.
96 Marcuse 1964, p. 57; original emphasis.
97 Marcuse 1964, p. 35.
98 Marcuse 1964, p. 25.

mation under capitalism (or, for that matter, Soviet-style socialism) remained, however, was that it was '*arrested, partial* automation', the 'coexistence of automated, semi-automated, or non-automated sections within the same plant'.[99] He enthusiastically quoted Marx's statements in the *Grundrisse* about the worker transformed into 'supervisor and regulator' and the trend toward the 'development of the societal individual'.[100] The new technologies should not be condemned by the left. Rather, they opened new vistas for a society without domination and unnecessary repression.

This conviction justified Marcuse's claim that automation was the 'great catalyst of advanced industrial society' and 'the technical instrument of the turn from quantity to quality'.[101] In the book, he imagined a future socialism where automation would be established as '*the* process of material production'.[102] According to Marcuse, 'complete automation in the realm of necessity would open the dimension of free time as the one in which man's private *and* societal existence would constitute itself. This would be the historical transcendence toward a new civilization'.[103] 'Disposable time', then, would become a universal boon for human beings.

To Marcuse, automated labour systems, unshackled from the imperatives of capital accumulation, could free people from unnecessary toil. Yet 'complete automation' would not happen automatically, even after the end of capitalism. Revolutionary forces could not simply take over the existing apparatus. Marcuse criticized the salience certain sectors of the left placed on nationalizing or socializing the productive forces. 'Neither nationalization nor socialization alter *by themselves*' completely mechanized social production, but were a 'precondition' for socialism, he wrote.[104] Marcuse then went on to invoke Marx's vision of the 'immediate producers' transforming 'production toward the satisfaction of freely developing individual needs'.[105] Seizing and preserving the 'technical base' of capitalist society, he elaborated later in the text, was not sufficient. Building socialism entailed the 'reconstruction of this base – that is, in its development with a view of different ends'.[106]

---

99   Marcuse 1964, p. 25; original emphasis.
100 Marcuse 1964, p. 36. I have retained Marcuse's own translations of these passages from the *Grundrisse*.
101 Marcuse 1964, p. 36.
102 Marcuse 1964, p. 36; original emphasis.
103 Marcuse 1964, p. 37.
104 Marcuse 1964, pp. 22–3; original emphasis.
105 Marcuse 1964, p. 41.
106 Marcuse 1964, p. 231.

Thus, Marcuse seemingly tempered his own considerable pessimism about the trajectory of advanced industrial society with admonitions about the truly liberatory possibilities, though currently suppressed under capitalism, of experiments in automation. A careful re-reading of *One-Dimensional Man* reveals, though, some pressing questions for Marcuse's critical theory. While he did recognize the solitary character of labouring in automated plants, it did not really trouble him. Yet for Marx and Engels the massing of workers in factories, mills, docks, shipyards, and mines established a central precondition for class consciousness.[107] If automation in its current form could dramatically reduce the number of workers in any given operation to skeleton crews, what could compensate for this 'de-humanization' of the workplace?

Moreover, Marcuse's own broader analysis demonstrated little faith in the parties, organizations, and trade unions of the traditional Left. As was and still is often brought up, he turned to the 'substratum of the outcasts and outsiders, the exploited and persecuted of other races and other colors, the unemployed and the unemployable' to resist the 'comfortable, smooth, reasonable, democratic unfreedom' of the prevailing social order.[108] The task of executing the 'Great Refusal', surely one of the most indelible phrases in the long history of leftist politics, Marcuse transferred in this text from the proletariat to this 'substratum'. Though he did not totally relinquish hope in a rejuvenated socialist labour movement, the fight to realize the promise of automation rested on extremely shaky ground in *One-Dimensional Man*. Marcuse so much as admitted that when he stated, 'the critical theory of society possesses no concepts which could bridge the gap between the present and its future; holding no promise – and showing no success, it remains negative'.[109]

Nonetheless, Marcuse differed tremendously from Anders in regarding automation as the contemporary instance of a tendency immanent to capitalism preparing the conditions for the system's demise. This assuredness in the revolutionary implications of automated production carried over into 'The Containment of Social Change in Industrial Society', a talk Marcuse gave at Stanford University the year after the release of *One-Dimensional Man*. There, he described how automation 'undermines the basis of scarcity, toil and repression, and threatens to do away with the need for earning a living in full-time occupation. *This is the final threat to domination and the existence of the estab-*

---

107 For examples, see Marx and Engels 1978, pp. 479–80; Engels, 2015, pp. 57–8.
108 Marcuse 1964, pp. 256, 1.
109 Marcuse 1964, p. 257. See also Marcuse 1967.

lished civilization is at stake'.[110] To be sure, Marcuse thought the social order countered this threat with every political, economic, and psychological means available to it. Yet 'for the first time in history society has the material and intellectual resources to create a life without fear, a life in peace', a time when 'the very dynamic of the society has reached the stage where its own productivity undermines its own basis', he insisted.[111] Such a radical contention about the potential of the new technologies of labour looks all the more striking when one recalls Marcuse's deepening conviction at the same time that post-1945 capitalism had fully integrated the proletariat and denuded it of its militancy. Thus, his optimism about the emancipatory aspects of automation was subsumed within a larger framework of deep theoretical pessimism.

In 1968–69, Marcuse altered his position. By that point the coalescence of radicalized students and youth in Europe, in both Cold War blocs, in the United States and Mexico, and in Japan demanding systemic change and constantly quoting his writings, forced him to briefly question his pessimism. Anders joked in a letter to Marcuse that only with 'fear and trembling' did he dare approach one in the company of 'Marx, Mao, the Maccabees, [Douglas] MacArthur, and Marat'.[112] He referred, with healthy sarcasm, to the iconic status Marcuse had won among young people around the world. Marcuse reciprocated, seeing the New Left and Counterculture, largely, though not exclusively, comprised of bourgeois youth, as the only legitimate opposition in the capitalist and communist blocs.[113] As he supported, critically, the 'new sensibility' he discovered in these young men and women, he resumed his arguments about automation and emancipation.

The 1968 essay, 'Aggressiveness in Advanced Industrial Society', showcased how Marcuse extended his critique of much automation discourse. He objected to the panic engendered by the 'threat of the "bogey of automation" '.[114] This panic-mongering fostered both the 'perpetuation and reproduction of technically obsolete and unnecessary jobs and occupations' as well as legitimizing 'the education and training of the managers and organization men of leisure time, that is to say, it serves to prolong and enlarge control and manipulation'.[115] On

---

110 'The Containment of Social Change in Industrial Society', in Marcuse 2001, p. 88; emphasis added.
111 Marcuse 2001, pp. 87–8.
112 Anders to Marcuse, 29 December 1967, in Anders 2022, p. 124.
113 See Marcuse 1969. The notion of a 'new sensibility' among young people is central to the book.
114 'Aggressiveness in Advanced Industrial Society', in Marcuse 1968, p. 256.
115 Marcuse 1968, pp. 255–6.

the eve of the world-historical events of 1968, of 'demanding the impossible', Marcuse continued to affirm the progressive dimension of automation.

As the 1970s began, Marcuse moved away, though, from the theorization of automation. While he never completely abandoned the subject, he discerned a transition to a new socio-historical era in several urgent, interrelated phenomena. The rightward turn in American politics, most shockingly the re-election of Richard Nixon in 1972, worried him that 'on the road from laissez-faire to monopoly and state capitalism, bourgeois democracy in its present form marks the stage where only two alternatives seem possible: neo-fascism on a global scale or transition to socialism'.[116] Over the next decade, similar shifts soon transpired in the United Kingdom, Israel, and West Germany, all countries he visited in this period. In some of his most thoughtful moments of the last decade of his life, Marcuse devoted enormous energy to construing how Marxists could engage productively with the New Social Movements (feminism, ecology) and to rethinking the role of subversive art in social liberation.[117] While he witnessed the unravelling of the post-war prosperity and the retrenchment of joblessness, he did not systematically consider how this gigantic moment of socio-economic transformation might affect his earlier thinking on automation and the abolition of superfluous work. The 'affluent society', the basis for so much of his (and Anders's) philosophical output, collapsed in the time running up to the renewal of Marcuse's correspondence with Anders and his excitement about the latter's new thinking about the 'obsolescence of labour'.

∴

Several decades have passed since Herbert Marcuse and Günther Anders proffered their analyses of automation and de-humanization. Virtually all of their expectations and predictions did not come to pass. This fact, though, is far less important than their radically diverging assessments of automated production for the contemporary left. A simplistic bifurcation of 'optimism' versus 'pessimism' does not comprehend the disagreements over automation the two men never aired.

Both philosophers reached an impasse about the disheartening prospects for proletarian revolution, nowhere more evident than in their essays, writ-

---

116 'The Historical Fate of Bourgeois Democracy', in Marcuse 2001, p. 165. The success of George Wallace's campaign with working-class voters also startled Marcuse. See also Marcuse 2001, pp. 24–9.
117 Marcuse 2001, chs. 2–3; Marcuse 1974; Marcuse 1978; Marcuse 2007.

ten in the late 1970s, on 'The Obsolescence of the Proletariat' (Anders) and 'The Reification of the Proletariat' (Marcuse).[118] Each essay made the case that the 'proletariat' had expanded dramatically beyond Marx's original conception – Anders stretched the category to the point of breaking when he placed all of humanity within it and considered the old Marxist focus on exploitation too narrowly framed to grasp the many unfreedoms endured by 'proletarians'. If Marcuse held to the working class as 'potentially revolutionary force', he, like Anders, no longer evinced much faith it might lead a revolutionary opposition.[119] These two texts hint at how the 1978 exchange about labour's obsolescence they discussed might have gone. The two intellectuals took contrasting positions about automation within, for all the varying inflections, similar outlooks of thoroughgoing theoretical and political pessimism, outlooks confirmed by the emphatically rightward turn in capitalist politics (Thatcher, Reagan, and Kohl), the onset of a new arms race between the superpowers, and the ascent of neoliberalism of the late 1970s and early 1980s.

Revisiting the framework devised by Moishe Postone, one can note how Anders consistently accentuated the 'manifest form' of automated production. He found absolutely no progressive potential in these experiments. Anders's critique of production – and this is one of its greatest weaknesses – almost always posed the machine against the human being as a competitor, a thief, a usurper, an enemy. Given his fetishisation of technology, for Anders, one could not expect automation to ever function very differently than under its current form.[120] Automated labour processes, for him, necessarily brought mass unemployment and literal de-humanization of workplaces; these trends, for Anders, did not point to a classless society.

With Marcuse, we encounter a representative theorist of the 'immanent potential' of industrial technology and a forerunner of much 'post-work' theory. He almost always emphasized the discrepancy between mechanization under the dominance of capital and what it might become under socialism. It would not be fair to say that Marcuse ignored how automation exacerbated un-and underemployment, but he certainly did not foreground these problems either. They remained peripheral. Instead, Marcuse again and again stressed

---

118   Anders 1992; Marcuse 1979.
119   Marcuse 1979, p. 20; Anders 1992, pp. 10–1. Anders actually criticized what had become of Marcuse's 'Great Refusal' by the late 1970s, with young activists immersed in lifestyle leftism like vegetarianism. See Anders 1992, p. 8.
120   My own thinking about the fetishism of industrial technology owes much to Leo Kofler and Ernest Mandel, without endorsing all the specifics of how they applied their arguments. See Kofler 1971 and Mandel 1978, ch. 16.

the possibility within automatic production for freedom from labour rendered historically superfluous.

Using Postone's interpretation, one can understand both of these perspectives as one-sided. A critical, Marxian perspective on automation would have to encompass the phenomenon as contradictory, as a 'pestiferous source of corruption and slavery' *and* a 'source of humane development'. Mechanization under capitalism is perpetually marked and rent by these countervailing tendencies. Marx, as Aaron Bastani maintains, contended 'changes in technology, production and social life would come to form the basis of an entirely new society'.[121] Yet such changes would never automatically lead to socialism. Revolutionary praxis was necessary.

Marx's confidence in the capacity of the proletarian masses for self-organization simply cannot be imported into the present, however. It will not be easy to mobilize value-producing labour around such a double-sided perspective. After all, the question of bridging the gap between the onerous present and the possible future demands a radical politics, where the working class, which bears the brunt of technological innovations of this sort, has to play the decisive role in its own abolition as a class. To complicate the situation even more, it is conceivable that the COVID-19 pandemic will mask the erosion of jobs through automation for some time.[122] Current ideologies such as the search for 'work-life balance' and 'self-care', tenacious remnants of neoliberalism, will hinder comprehension and mobilization around revolutionary demands for more and more mechanization without penury. But at the very least Marx's dialectical position can honestly pose the dilemma technologies such as automation have thrust upon us for decades. To redeem the best aspects of Anders and Marcuse's work on industrialism, one will have to envisage a possible liberated future without losing sight of a horrendously unfree present.

---

121 Bastani 2019, p. 35.
122 For some recent works on the pandemic, see Davis 2020; Brenner 2020; Malm 2020.

CHAPTER 2

Nowhere to Go: Automation, Then and Now

Jason E. Smith

> It is in this serious light that we have to look at the question of the growing army of the unemployed. We have to stop looking for solutions in pump-priming, featherbedding, public works, war contracts, and all the other gimmicks that are always being proposed by labor leaders and well-meaning liberals.
> JAMES BOGGS, *The American Revolution*[1]

⁂

---

[1] Boggs 2009, p. 17 [Note: This essay was originally published in two parts in *The Brooklyn Rail* in the Spring of 2017. Its primary objective was to criticize the way that 'automation' had been discussed in the popular and financial press, as well as in a few well-publicized books, since around 2011 or so. My method was to place the claims made for and about automation – that the advanced economies of North America and Europe were witnessing a rapid extension of automated labor processes, especially in the service sector – in context: by comparing these claims to similar ones made on a cyclical basis since the Second World War and, more importantly, by measuring their rhetoric against prevailing trends in labor markets, wage growth, rates of business investment and labor productivity growth. Readers of newspapers like *The Financial Times* or *The New York Times* in the 2010s were likely to be struck by the apparent contradiction between refrains warning of the 'rise of the robots' and those lamenting a long-standing stagnation in labor productivity growth. Indeed, the 2010s were notable for the resurgence of what Robert Solow referred to in the 1980s as the 'productivity paradox': all the advances in computing power since the invention of the silicon chip, he noted, had produced little evidence of productivity growth in advanced economies in which the vast majority of employment was concentrated in the service sector. Because my point in this essay was primarily polemical, I was only able to address the thornier conceptual problems raised by the figure of 'automation' in haphazard and sometimes inadequate ways. As a result, I decided to write a short book – my *Smart Machines and Service Work* [2020] – that tackled these problems head-on, and in a more systematic fashion. While I spend some time in that book criticizing the concept of the 'service sector', or 'service work', the core theoretical problem addressed in it is the concept of productivity itself: the difficulties involved in measuring it and the contradictions that arise in its use in mainstream economics. Above all, I argue that mainstream economics neglects to distinguish between productive and unproductive labor, a conceptual pair that was essential in classical political economy, and particularly important in Marx's value theory. In *Smart Machines*, I use that theory to address some of the riddles of mainstream economic theory, not least the productivity paradox, while arguing that the

In 1963, James Boggs, a black autoworker employed for over two decades at a Chrysler plant in Detroit, published a short book focused on the nefarious effects of automation on class struggle in the United States. The story told in *The American Revolution: Pages from a Negro Worker's Notebook* begins with the early 1930s, the decomposition of the old craft unions, and a global economy in the throes of an unprecedented near-collapse; it arrives at a high point with the late 1930s, with a now-forgotten wave of sit-down strikes that tore through the tire and auto industries between 1933 and 1937, most famously at the Flint General Motors plant in early 1937. This was, in Boggs's estimation, the 'greatest period of industrial strife and workers' struggle for control of production that the United States has ever known'.[2] But this period also gave rise, under the reformist efforts of the New Deal and in a climate of mass unemployment, to the Wagner Act and the institutionalization of class struggle. The United Auto Workers, which just a few years earlier organized the sit-down strikes in the auto industry, had by 1939 banned the tactic in the plants. In the cast shadow of imminent war, the union's no-strike pledge, along with the inevitable encrustation of a bureaucratic stratum more at home in the offices of management than on the workbenches, left workers to wildcat their way through the war. The Second World War witnessed thousands of work stoppages: an astonishing 8,708 strikes implicating over four million workers took place, according to Boggs, over one two-year period while war production was in full swing. Union pledges of discipline notwithstanding, order did not therefore always prevail. Workers, many of them from the rural South, and new to the world of the factory, consistently bucked against the dictates imposed by management and enforced by their own representatives. The wildcat strikes were not, however, always defections from the dictates of union bureaucrats and the boss. In 1943, a UAW-organized Packard plant was the site of a 'hate strike' organized by white workers to push back against the influx of black workers into the factories, and the integration of assembly lines. Soon after, a tumultuous "race riot" broke out in the city, as white workers attacked black workers who now competed with them for housing. Dozens were killed, hundreds wounded; mostly black, and primarily at the hands of police and the National Guard. The city would be occupied by federal troops for a full half year after. Such was, for better and for worse, the American workers movement at its most militant.[3]

---

distinction between productive and unproductive labor helps us understand better the tendency of the rate of profit to fall, both in Marx's theoretical sketch of this tendency, and in the actual 'performance' of the capitalist mode of production over the past half-century.]
2 Boggs 2009, p. 17.
3 In his *Riot.Strike.Riot*, Joshua Clover – who also discusses Boggs at some length – emphasizes

The onset of the post-war economic boom – with its soaring growth, surging wages, and near-full employment – did little to dampen the combativeness of workers on the line. The wildcat waves continued well into the 1950s, with the movement cresting, in Boggs's reckoning, in the middle of the decade. The movement and its off-and-on open conflict with union brass ('porkchoppers', to rank-and-file) was chronicled in a series of broadsides (*Punch-Out, Union Committeemen and Wildcat Strikes*) by the irrepressible Martin Glaberman, Boggs's long-time comrade in the Detroit-based Correspondence Publishing Committee. At stake in these struggles was what *The American Revolution* specifies as 'control over production', the ability of workers on the shop floor to dictate the pace and intensity of work through collective action and novel tactics. Chrysler's management responded to this volatile situation with a weapon hitherto mostly under wraps: 'A new force ... entered the picture,' Boggs writes, as management, with union blessing, 'began introducing automation at a rapid rate'.[4] Where prior efforts to speed up work rhythms met with fierce opposition from thousands of workers concentrated in massive production sites, this capacity for interruption depended upon worker control over the machinery set in motion during production. The stunning productivity gains made possible by the introduction of large-scale machinery and the moving assembly line still depended in large part on worker oversight of the production process. The lure of automation, from the perspective of Chrysler management, was obvious: many tasks performed, and decisions made currently by workers could be replaced by programmable computers and cybernetic control systems. The promise of rising productivity in the workplace also entailed compromised worker control over the pace of production, threatening an outright swapping out of labour for capital on the other, with computer-assisted machines replacing potentially tens of thousands of workers almost overnight. It was precisely this threat of substitution that, Boggs concludes, was decisive in the quashing of the strike movement in the middle of the 1950s: 'since the advent of automation there has not been any serious sentiment for striking'.[5]

Warnings about the perils of automation are as old as the capitalist mode of production. The first revolts of workers' movement produced the myths of General Ludd and Captain Swing, and the insurrectionary forays of the canuts of

---

that 'the history of race riots in the United States begins with whites disciplining insubordinate other populations' (Clover 2016, p. 112). Oddly, Boggs makes no mention of these white supremacist race riots, even as his account focuses in large part on the fate of the 'Negro' worker and anticipates the 'coming explosions' of the mid-to-late 1960s.

4 Boggs 2009, p. 23.
5 Boggs 2009, p. 33.

Lyon. In their wake were left wrecked shearing frames and looms; barns, buildings, and goods were targeted by proletarian arsonists. Yet the development of the productive forces, and the implementation of large-scale machinery in capitalist factories, never quite made workers purely and simply redundant. To the contrary, over the course of more than a century, the demand for labour had grown exponentially, even as millions of peasants poured into cities, and entered into the wages system and the urban cash nexus. But this time, Boggs warned, was different: 'Automation replaces men. This is of course nothing new. What is new is that now, unlike most earlier periods, the displaced men *have nowhere to go*'.[6] These men and women, many of whom, like Boggs, had left the deep South for the industrial North and its factories and great cities, were loath to return to the countryside, to Jim Crow, rural isolation, and hardscrabble miseries. And the countryside wouldn't have them: advances in mechanized farming across the South dramatically augmented agricultural productivity during the 1920s and after, in a matter of a few decades eliminating what jobs were left in the field. There was no turning back, in any case; these workers would not dare leave the cities, unless it was to 'get away from the Bomb'.[7]

∴

Over the past hundred years or so, laments over an impending purge of workers by technological innovation have come in cyclical pulses, once every third decade. James Boggs's variant of this complaint remains something of an exception: such concerns have largely been voiced, in the twentieth century, by the emissaries of the dominant class charged with implementing automation, rather than by those at risk of replacement. Keynes notably wrote, in 1930, of the 'new disease' of 'technological unemployment' visited upon a society otherwise enjoying the productivity gains reaped from a cluster of breakthroughs: the wholesale electrification of industry (Lenin's definition of communism as 'soviets with electrification' was no idle quip), the widespread use of internal combustion engines and newly paved road networks, the marvels of indoor plumbing, and the availability of cheap, plentiful steel. Another round of hand-wringing commenced in the mid-fifties – Boggs was far from alone – as technological leaps broached in the 1930s began to come online, the prospect of atomic power loomed, and primitive computers were coupled with large-

---

6 Boggs 2009, p. 36; emphasis added.
7 Boggs 2009, p. 40.

scale machine production. Essays, studies, and books devoted to the marvels of 'cybernation' abounded. The fascination with technological forces, typical of the capitalist class, was spoiled only when distracted by the fate of those potentially expelled from production. A booming industry in popular sociology speculated with optimism on the just-out-of-reach society of leisure delivered by these technological advances. Many worries, however, centred on a future of mass unemployment, with an attendant widespread immiseration, and even an uptick in class antagonism. Above all, those viewing this situation through the lenses of the capitalist class feared a crisis of underconsumption, as workers, deprived of the wage, would not be able to buy up all of the cheap commodities produced by such wonderful machines.

Within a decade, however, by the late 1960s, many of these same commentators would herald a coming post-industrial society and its rapidly expanding service sector, which would quickly soak up the vast majority of those Boggs claimed would have 'nowhere to go' (he spoke of 'surplus' people). In the 1970s, tens of millions of women began pouring into labour markets in the U.S. alone, often finding work in clerical and business services. One effect of this wholesale entry of women into workplaces was to accelerate the commodification of personal services as well, previously carried out in the form of unwaged, domestic or socially reproductive labour. Then, again, in the mid-1990s, just as the 'New Economy' was said to be taking off – the dot.com bubble began to swell – another wave of worry washed over the chattering classes, with impeccable timing. Typical was this 1994 article from *The Wall Street Journal*, which breathlessly recycled the old tune: 'technological advances are now so rapid that companies can shed far more workers than they need to hire to implement the technology or support expanding sales'.[8] Jeremy Rifkin's 1995 book, *The End of Work*, which counted on a 'nearly automated' service sector (in 1995, almost 70 percent of unemployment by the economists' calculations) by the mid-twenty-first century, was as ubiquitous in the discourse of intellectuals as in the business papers. Since its publication, millions more workers have entered the service sector in high-income countries, as manufacturing has contracted still further. In the meantime, a hundred million Chinese peasants have made their own Great Migration, moving into mushrooming cities across that vast country, exchanging their labour power day in and out for *yuan*.

Since the global economic meltdown of 2008, and especially over the last five years, there has once again been – in perfect sync with the cyclical pat-

---

8   Cited in Rifkin 1995, p. 141.

tern – an outpouring of articles and books detailing the wonders and pitfalls of an imminent rise of the robots. It is held we are living through a 'second machine age',[9] dawning half a century after what was already, in the 1960s, called the 'third industrial revolution.' But where *that* promised technological leap, to be unleashed by the conjoining of automation and atomic power, was proffered in the midst of a veritable explosion of economic growth, here the hyperbole comes on the heels of a near-fatal financial crisis, and at the end of a decade that registered 'the slowest growth in productivity of *any decade in American history*'.[10] Recent trends suggest this torpor has not been shaken. Indeed, since 1999, the height of the dot.com bubble, private investment in software and computer equipment has fallen precipitously, by a full quarter: it is, today, as low as it was in 1995. This state of affairs is not lost on many commentators, who struggle to reconcile the marvels and menace of machine-learning algorithms (able, it is said, to 'write their own programs') with the prevailing conditions on the ground. Unemployment rates have only begun to ease in the US as millions simply drop out of the labour market.[11] Abroad, especially in southern Europe, they remain historically high. But these job losses are due less to the revenge of the robots than to a plethora of capital idling on the sidelines.

The presumption held by most contemporary discussions of automation is that new digital technologies constitute a revolutionary innovation on a par with electricity, whose cheap, networked availability by the 1920s spurred a half-century round of economic expansion. A handful of sceptics (such as Robert Gordon) contend that whatever IT-induced productivity gains are to be had were already reaped during a short period in the 1990s, tailing off by the end of the decade. The stakes of such a claim are sizable, since the implementation of any 'truly general-purpose technology' across the economy – not only in manufacturing, but in the massive service sector as well – should, through the productivity gains they promise, bring the global economy out of its doldrums. If this new explosion in productivity were to follow the pattern set in the middle of the twentieth century, we should expect not a crisis of employment but rising demand for the cheap commodities (goods or services) pumped out by newly

---

9   Cf. Brynjolfsson and McAfee 2014.
10   Gordon 2016, p. 529; emphasis added.
11   'The unemployment rate is not wrong, but it does not tell us much about the festering crisis of worklessness in America. For that, you need to look at the rising share of people in their prime years (between 25 and 54) who are neither working nor looking for work: a figure that now stands – as it happens – at about 20 per cent'. 'America's "jobs for the boys" is just half the employment story' (O'Connor 2017).

automated production processes, with corresponding bumps in both demand for labour and wage levels: such was the 'Golden Age' of the post-war boom. In *The Rise of the Robots*, Martin Ford claims that older automation technologies 'tended to be relatively specialized and to disrupt one employment sector at a time, with workers then switching to a new emerging industry';[12] today, we are warned, information technology is spreading across all sectors simultaneously, including a huge swath of service sector occupations in health, education, and retail, leaving displaced workers – as Boggs put it fifty years before – *nowhere to go.*

Despite these imagined threats of a new round of technological employment hovering on the horizon, Ford, like most commentators on the subject, is at pains to explain the lag in the implementation of this 'truly general-purpose technology', even as he bemoans its potential fallout. And yet at one point late in *Rise of the Robots*, he puts his finger on a peculiar inversion characteristic of the ongoing global recession – an inversion that could provide a key to understanding the puzzle of the present moment. While in most economic slumps' productivity tends to drop off rapidly, with output falling faster than jobs can be shed, in the opening round of the recent crisis something else happened entirely. Firms on average registered modest *gains* in productivity, despite the hostile climate. Yet they did so despite rapid *drop-offs* in output: total output was shrinking, but payrolls were being slashed even faster. The uptick in productivity, in this case, was likely due not to technical innovations, but to longer, more stressful, days on the job for those who kept them. Ford: 'during the Great Recession ... productivity actually increased. Output fell substantially, but hours worked fell even more ... The workers who kept their jobs (who certainly feared more cuts in the future) probably worked harder and reduced any time they spent on activities not directly related to their work; the result was an increase in productivity'.[13] Lest we imagine these patterns to be those of a cyclical if atypical downturn, a 2014 study by a group of researchers at MIT – like the authors of *The Second Machine Age*, though in this case, tellingly, not from the department of management, but from economics – detected a similar, longer-standing pattern in IT-intensive industries. Backdating this trend to the pre-crisis period, they find 'little evidence of faster productivity growth in IT-intensive industries after the late 1990s'; when this evidence does appear, it is traced not to rapid productivity gains through the implementation of automation, but is instead said to be 'driven by declining relative output accompanied

---

12   Ford 2015, p. xvi.
13   Ford 2015, pp. 207–8.

by even more rapid declines in employment'.[14] Lacklustre performance like this is surely one reason investment in IT has fallen off so precipitously since the late 1990s; it rhymes, moreover, with Gordon's claim that the period between 2004 and 2014 exhibited the slowest productivity growth over a decade in U.S. history. It also gives us a hint as to why, even with central banks holding the choke open on the global economic engine, flooding it with free money, surplus capital has been shunted into short-term, speculative fixes – real estate, finance – rather into new lines of production.

This is the forbidding environment in which firms today operate. Predictions of rapid replacement of millions of jobs by machines must contend with these longer-term tendencies. Under such conditions, it is hard to imagine a sudden surge of growth in the manufacturing sector itself, even if certain lines find ways to undercut their competitors with temporary technological fixes. The 2014 MIT study just cited – the authors' express purpose was to refute Brynjolfsson and McAfee's 2011 book, *Race Against the Machine* – bears the pointed title 'Return of the Solow Paradox?', invoking the notorious comment offhandedly made by economist Robert Solow in a 1987 *New York Times Book Review* article: 'what everyone feels to have been a technological revolution … had been accompanied everywhere by slowing-down of productivity growth, not by a step up. You can see the computer age everywhere but in the productivity statistics'. Thirty years later, the needle hasn't moved. If it is true that the staggering productivity gains of the 1920s and after can be attributed in part to the widespread use of electricity and the internal combustion engine, the real revolution was in the networking of these technologies, through the expansion of power grids and paved roads. Yet in a world in which seven in ten Haitians has a cell phone, the unimaginable density of global communication networks – even the planet's poorest inhabitants are now 'networked individuals' – has yet to put a dent into what many mainstream economists are calling a long-term, even 'secular', capitalist stagnation. Seen in this light, the anxious exhilaration surrounding contemporary machine-learning algorithms can feel hyperbolic. Measured against the potentially terrifying forces tapped by nuclear energy in the mid-twentieth century, Google Glass might seem a modest venture. Google's parent company Alphabet speaks in exalted tones of technological moonshots, but ninety percent of its revenue and almost all of its profits still come from advertising, most of it via search engines. It is buying up smaller robotics and AI firms, but not necessarily to ramp up investment: it is to establish monopoly conditions that will guarantee super-profits and higher market share within

---

14   Acemoglu et al. 2014, pp. 394–9.

these stagnant conditions. Today, high profits are assured for firms able to disrupt market dynamics and its price signals. Such firms are often 'more adept at siphoning wealth off than creating it afresh'; they thrive less through innovation than through exorbitant market shares, and streams of technological rent.[15]

A cursory look at the global economy over the past four decades indicates that, after the deep recession of the early 1970s, promised returns to levels of growth typical of Boggs's time never materialized. Growth rates not only in the U.S. but in most OECD countries have on average remained listless for over forty years, expanding at less than half the rate of the so-called 'Golden Years'. What accounts for this sluggishness? Many analysts point to declining profit rates for capitalist firms throughout this period. As profit rates fell, beginning as early as the mid-1960s, less capital was available for investment, both in existing and new lines of production; this blockage led in turn to job losses and high rates of unemployment.

Explanations vary on why the initial decline in profit rates occurred. Some accounts suggest a high level of worker militancy in production explain the initial downturn, as full employment and high wages 'squeezed' profit margins from below, leading to dwindling returns and a subsequent shakeout. Under these conditions, private firms set about restoring their profit rates through a variety of fixes, but above all by slashing wages, which have remained on average stagnant for this entire period, buoyed temporarily by a dizzying rise in consumer debt over this same period. And yet the 'profit squeeze' theory cannot account for a crucial detail. If wages were slashed beginning in the 1970s, and have flatlined since, why hasn't the aggregate profit rate been restored to pre-1970s levels, relaunching in turn productivity gains (as rising profits are reinvested in production) and expanding employment? Robert Brenner and Fred Moseley, among others, have attempted to respond to this question in different ways (global overcapacity in manufacturing, a rising ratio of unproductive-to-productive labour, and so on). In the current climate, and in certain sectors, monopoly-like conditions for specific firms can engender abnormally high returns for firms and their shareholders. In other sectors, select companies can invest in technologies to win competitive advantage long enough to capture a larger market share, even as that market, and total output in a given sector, remains static, or even declines. This is not soil in which new shoots will grow.

∵

---

15   *The Economist* 2016.

Most alarms rung over the course of the twentieth century regarding the perils of automation have been torn between a fascination with technological development, which promises a tendential spread of worklessness to the whole of society, and a shopkeeper's anguish over just who might consume the mountains of cheap commodities disgorged by the machines. Historically, many approaches to automation on the Left have emphasized the way the deployment of technological breakthroughs, and the substitution of constant capital for labour, constitute strategic moves in a raging war at the point of production. Boggs's observation that the capitalist use of automation allowed plant owners to recapture control over production, putting paid to a long wave of strikes, is just one example in a rich vein of analysis that emphasizes the specifically capitalist nature of the complex machinery and organizational refinements characteristic of contemporary production processes. Writers as varied as Raniero Panzieri, Harry Braverman, David Noble, and Moishe Postone have all made important contributions to this strain of thought, emphasizing the way patterns of technological development increasingly reflect capitalist value-relations, making any future 'socialist use' of much of this machinery onerous at best.[16]

Some recent examinations on the Left of the structural drive toward the replacement of labour by machines have taken a different tack. Writers like Antonio Negri have seen changes in the composition of capital in an altogether positive light, reading the rising organic composition of capital through the lens of Marx's 1858 'Fragment on Machines': the 'monstrous disproportion' between the productive capacity of large-scale, computer-controlled machinery and the diminishing quantities of labour time required to set this system in motion.[17] A common version of this position imagines an automation-induced 'abolition of work' that would, as this worklessness initially takes the form of mass unemployment, be offset by the implementation of a state-administered 'guaranteed basic income'; such payments supposedly would, as they stimulate effective demand and keep capitalist production ticking over, gradually sever the sacred tie between income and the time of work.

Nick Srnicek and Alex Williams's recent *Inventing the Future*, heralding a post-capitalist 'world without work', takes up this legacy in its way, putting forth as its core programmatic demand the total automation of the 'economy' (a term they leave unexamined). They, like the mainstream accounts they are reprodu-

---

16 Recently, Jasper Bernes (2013) has made an important contribution to this important if relatively marginal tradition, with specific reference to the contemporary logistics 'revolution'.
17 Marx 1993, p. 705.

cing, are compelled to grapple with the 'return of Solow's paradox': for all of the hype about big data, the internet of things, and workerless factories, aggregate growth rates remain as we saw lacklustre at best, especially compared with their mid-century peak. It is for this reason that the substitution of machines for human labour *should* be, they write, 'enthusiastically *accelerated* ... as a political project of the left' (my italics). Here we hit the rub in their vision of the future. They make a half-hearted stab at accounting for automation's 'diffusion lag', but twist themselves into knots in doing so.[18] 'It is highly likely', they write, that 'low wages are repressing investment in productivity-enhancing technologies'.[19] This is undoubtedly one factor that must be considered: why would business owners invest in fixed capital that depreciates over years, when loose labour markets allow cheap labour to be picked up and dropped at a moment's notice? Following this line of reasoning, Srnicek and Williams argue that 'in the effort to bring about full automation, fighting for higher global wages is a crucial complementary task'.[20] Leaving aside the Herculean task of organizing a struggle across the planet for higher 'global wages' – narrowing wage differentials on a global level might be a more plausible objective, but this would require *lowering* wages of U.S. workers, as those in East Asia rise to meet them – this proposition is a puzzling one. Rather than considering why low-wage jobs and the people compelled to work them are so plentiful in the first place, or why these workers are incapable of organizing these low-wage sectors in order to demand higher wages, the authors suggest that higher wage levels must be implemented by political fiat, or bureaucratic decree. But this would, according to this logic, be an intermediate step: since compelling employers to raise wages will require them to deploy automation, imposing higher wages on employers will have as their 'desired' effect mass unemployment: putting those who've just won bigger paychecks out of work. Such is the strategic vision offered by social democratic accelerationism.

---

18   An important historical account comparing the 'diffusion lags' entailed in full-scale implementation of electricity and that with computers is Paul A. David's 'The Dynamo and the Computer: An Historical Perspective on the Modern Productivity Paradox' (David 1990). This essay is very rich in historical information and is a valiant attempt to solve the puzzle of Solow's paradox. While it makes quite convincing arguments about the delay in the diffusion of electricity (the technology for which had been available for decades), it is worth noting that David's paper was written in 1990: 25 years later, we are still waiting on the productivity explosion. Gordon argues, to the contrary, that the modest gains registered in the 1990s is all we will get.
19   Srnicek and Alex Williams 2015, p. 112. Other important 'left' discussions of automation are found in Peter Frase's *Four Futures* (2016) and Paul Mason's *Postcapitalism: A Guide to Our Future* (2016).
20   Srnicek and Williams 2016, p. 188.

We must stand this problem back on its feet. The lag in implementing wholesale automation across all sectors of the economy, with the corresponding and long-standing lag in productivity gains, must be considered from the perspective of the dynamics of global capitalism as a whole. Current speculations on both the promise and threat of automation are confronted with an ongoing crisis of accumulation. In this climate, a fragmentary implementation of automation is unlikely either to liberate large fractions of humanity from work, or produce mass unemployment of the sort envisioned over and again by commentators for the past century. The conviction held by many on the Left, here following tech enthusiasts like Martin Ford, is that the technical capacity to automate most if not all occupations is virtually present. Any lag in implementation is, by this reckoning, due to 'failures of government policy': with the right cocktail of social democratic adjustments (shorter work weeks, higher minimum wage, basic income, etc.), with the correct 'political choices', a world of 'tight labour markets' and a decent standard of living for all could be won.[21]

My own investigation starts from a different place. I want to ask why, for all of the froth churned up around the productive potential lurking in test labs, the pattern exhibited over the past fifteen years has been one of declining investment in information technology, and falling output for IT-assisted manufacturing? Why has almost all growth in employment – *ninety-six percent* – since 1990 'come from sectors known to have low productivity ... and sectors where low productivity is merely suspected in the absence of competition and proper measurement techniques'? Why has some *ninety-four* percent of new employment in the U.S. since 2000 been in education, healthcare, social assistance, bars, restaurants, and retail, that is, in the vast, motley, and above all technologically stagnant service sector?[22]

∴

On February 19, 2017, *The New York Times* ran a feature story on recent changes in the U.S. oil industry.[23] The focus was on the recent 'embrace' of technological innovation in the industry after the 2014 plunge in the global oil market. This was just one of a rash of such pieces in the popular press, relying, as is typical of such writing, on a smattering of skewed, decontextualized data, a healthy serving of the anecdotal, and a host of the worst tech journalism clichés

---

21   Frase 1016, p. 17.
22   Klein 2016.
23   Krauss 2017.

('a few icons on a computer screen', 'a click of the mouse', video game marathons as job training, a compulsory reference to drones). Zeroing in on the effects of these changes on workers in west Texas, the article's upshot is unobjectionable enough: as oil prices recover, and output rises, many workers who lost jobs in the downturn will be replaced by machines as production becomes more capital-intensive. These workers, often Latino, are sure to be forced out of these semi-skilled, relatively well-paid jobs into other sectors of the labour market, where their skills and experience will serve little purpose. At first blush, the situation seems dire. We are told that some thirty percent of jobs in the industry were lost after the oil market crash of mid-2014, when employment in the industry was at its peak. But dating these losses from 2014, at the height of a boom in the industry, crops the picture too dramatically. Employment in the oil and gas sector exploded between the turn of the century and the oil market collapse – an historic one, by all measures – a few years ago: by some estimates, employment in the industry shot up by 150 percent during this period.[24] What is more, while the paper of record warns of 'jobs left behind' as prices rise and output picks up again, other reports anticipate another surge in employment in the field, and even a dearth of qualified workers ('oil and gas industry could hire 100,000 workers – if it can find them', warns one headline from late 2016).[25]

My intention is not to adjudicate these matters, but only to make the following points: First, the effects of automation on employment are never straightforward, but depend on the relationship between output and job replacement: if output rises more quickly than jobs are replaced, the rising ratio of capital-to-labour will nevertheless result in a growing demand for jobs, rather than their scarcity. More important, for my purposes, is the atypical character of the oil sector, with respect to the relation between automation and the larger labour market. Whereas the *Times* insinuates there is something exemplary about the situation it describes ('As in other industries ....'), other commentators argue that the domestic oil industry is in many ways singular. In an October 24, 2016 post on the *Financial Times*'s excellent Alphaville blog with the title 'The robot revolution may be exaggerated, globalisation edition', Izabella Kaminska insists that '*aside from the oil industry* ... there is little evidence' that we face the 'loss of millions of middle-class jobs [in the near future] as algos and robots displace not just blue-collar workers but middle management and intellectual jobs as well'.[26] It need not be emphasized that the oil industry is unique in still other,

---

24 Bureau of Labor Statistics data indicate a 65 percent increase from 2004 to 2014; with support activities, the increase is closer to 150 percent.
25 DiChristopher 2016.
26 Kaminska 2016.

more self-evident, ways: It is subject to monopoly-like conditions, as oil cartels artificially raise or lower global crude prices by releasing or holding back reserves. It is, moreover, highly politicized, as even the most casual observer might conclude, and as the devastating attack on Iraq in the interest of commanding this sector – just the latest in a long litany of twentieth-century wars with this objective – attests. The political pressure to ramp up domestic production undoubtedly had effects on the changing composition of oil production, as directional drilling and hydrofracturing techniques were refined by U.S. producers. The oil industry is undeniably a key strategic node in the global economy, with the rising and falling prices of energy inputs affecting almost all other economic sectors. But, perhaps for this very reason, patterns of technological change and employment characteristic of this subsector are hard to generalize.

Even more pertinent for our purposes is this: according to the occupational employment statistics compiled and published by the Bureau of Labor Statistics, the Oil and Gas Extraction subsector currently employs 178,000 workers. Of these, only about 90,000 employees are characterized as holding 'production and nonsupervisory' positions: a crowd small enough to fit easily in the Pasadena Rose Bowl. Put more pointedly, this same agency pegs the total number of workers making up the current U.S. labour force at 152 million (not accounting for the many millions who have dropped out of the workforce entirely): all employees in this subsector, including supervisors, constitute .001 percent of the current U.S. workforce. By contrast, the number of U.S. workers employed in 'leisure and hospitality' jobs, per 2014 statistics, was 15 million; yet another 15 million were working in the so-called retail sector. The arithmetic is as simple as it is grim: the size of the American workforce serving as cooks, waiters, bartenders, and cashiers is roughly 170 times that of the total number of workers employed in the oil industry; if we consider just those workers in oil production, the blue-collar semi-skilled laborers featured in the *Times* article, the ratio of restaurant and retail jobs to those in oil and gas extraction rises to 333:1.

There is nothing surprising in this disparity. Currently some 80 percent of the labor force in the U.S. is classified as working in what are commonly called services. While employment in manufacturing declined between 2004 and 2014 at a rate of 1.6 percent annually, 96 percent of net employment gains in the U.S. since the turn of the century have come in the broadly defined service sector, a hotchpotch of occupations that lumps together retail and restaurants with other, vast sectors like education and healthcare. The trend will remain relentless. BLS projections predict that 19 of 20 jobs added between 2014 and 2024 will be in services. The preponderant majority of these jobs will be the least attractive ones, requiring few if any skills, and paying poverty-level wages. Under the heading 'Occupations with the most job growth' for the period 2014–

2024, U.S. statisticians list fifteen job profiles, eleven of them requiring less than a bachelor's degree, and the majority – including *four of the top five*–requiring 'no formal education credential' at all. Not surprisingly, the average median income for these positions' peaks at $31,000 per year, with the lowest dipping to $18,000, and most in the low-$20,000 range. Since 1990, the U.S. economy has registered a net gain of 34 million jobs. But these jobs are almost universally poorly paid and precarious, pooling at the very bottom of the labour market.[27]

Thus, perennial fears of mass unemployment have once again been refuted by the facts, as wealthy, complex economies such as the U.S. and the U.K. continue to add jobs to the payroll. The United Kingdom, in particular, has shown incredible facility in job creation since the Great Recession of 2007, when the bottom last fell out of the job market. But real wages have lagged far behind 2007 levels, and 'many of these additional workers are doing little to boost real living standards [...] their continued employment is effectively the product of subsidies extracted to provide make-work, rather than the result of competitive market conditions'.[28] These 'subsidies' take the form of government payouts and what remains of the social wage, as food stamp programs and meagre tax credits allow large firms to hire unskilled workers at cut-rate compensation. Many of the fasting-growing occupations in the U.K. as in the U.S. are not in what the *Financial Times* writer calls 'make-work' – state-subsidized employment in retail, restaurants, sales, and security – but in the least desirable forms of care work, often involving cleaning, washing, and disposing of waste, or simply standing watch over the elderly, the idle, or small children. A recent report by Deloitte, a consultancy group surveying labour market shifts in deindustrialized Britain, noted a nine-fold increase in nurses' assistants in the past ten years, and a six-fold jump in teachers' assistants. On the US side of the Atlantic, the two occupations expected to have the highest rate of growth over the next decade are 'home health aide' (projected to grow at a 38% clip) and what is euphemistically called a *personal care aide*, according to the U.S. Bureau of Labor Statistics. This is the capitalist labour market of the early twenty-first century: a world of domestic servants, resembling, in this way, the archaic world of the mid-nineteenth century. For many, often women, and more often still women of colour, the employment on offer can truly be called *abject*.

∴

---

27 See the summary table published by the BLS here: https://www.bls.gov/news.release/ecopro.to6.htm.
28 Klein 2015.

Productivity gains are still to be had in agriculture and manufacturing, but – as in the example of the oil sector – these sectors make up a tiny fraction of employment in high-income countries, and a declining fraction even in otherwise industrializing nations. As a result, the impact on jobs of still further automation in these shrinking sectors will be minimal by comparison. So it is not surprising that the core claim made by all recent commentators on the coming wave of automation is that those jobs that stand to be replaced *en masse* are almost entirely located in the service sector. This sector can in fact almost be defined as that mass of occupations and labour processes that, whatever the disparity in wages and skill level among them, have as their common trait that they are *technologically stagnant*, to use William Baumol's useful term: they exhibit, unlike capital-intensive manufacturing and agriculture, consistently anaemic productivity growth.[29] As a result, what all the recent writing warning of the imminent automation of massive chunks of the economy presume is that the pattern witnessed with the manufacturing sector in the twentieth century will return to ravage services in their turn: in other words, that a sector resistant to technological innovation and perennially registering minimal growth in labour productivity, will be transformed into dynamic, technologically progressive lines of production.

Jeremy Rifkin's late twentieth-century prediction of an almost fully automated service sector by mid-century is received as near gospel not only by most popular writing on this theme, but also by the many on the Left who have picked up on the theme:

> The service sector, while slower to automate, will probably approach a nearly automated state by the mid-decades of the next century … . Hundreds of millions of workers will be permanently idled by the twin forces of globalization and automation.[30]

Peter Frase's 2016 book *Four Futures*, a social-democratic exercise imagining the social effects of near-full automation, throws up its hands and simply:

> … takes for granted the premise of the automation optimists, that within as little as a few decades we could live in a Star Trek-like world where … a large amount of the labor currently done by humans is in the process of being automated away.[31]

---

29 On the distinction between technologically progressive and stagnant sectors, see Baumol 2012.
30 Rifkin 1995, p. 291.
31 Frase 2016, p. 9.

And yet Frase, like Srnicek and Williams, devotes little energy to examining the specificity of the service sector itself. Along with most commentators on the subject, these writers rely on a single 2013 study put out by Oxford University's Martin School predicting that some forty-seven percent of U.S. jobs are 'at risk' of automation.[32] Other studies pile on with even more dramatic prognostications, raising the bar closer to eighty percent in the not-too-distant future.[33] These accounts, which are shared by almost all commentators whatever their political orientation or ambition, all rely on a single unexamined assumption: that the sector in which nearly all new job creation over the past quarter century has taken place will soon be decimated by a gathering legion of 'intelligent' machines.

The leap in productivity and dynamism required for such a conversion of stagnant labour processes into technologically progressive ones would, moreover, take place in the midst of an epoch of historically low growth in labour productivity. A January 2017 report issued by the Bureau of Labor Statistics notes that growth in labour productivity in the U.S. for the current business cycle – dated back to the fourth quarter of 2007 – is the lowest in the post-World War II era. To give a sense of how deep the damage really is, the author of the report indicates that since 2011 labour productivity has grown at a historically low level of 0.7 percent. Put in perspective, this would mean that in order for productivity levels to return to the historical pre-crisis trend rate over 60 years – a rate that already incorporates a steady decline in productivity over the past forty years – it would have to register an astronomical surge of 7.7 percent over the next two years: *eleven times the growth rate seen between 2011 and 2017*.[34] A full half of the overall productivity gains seen in this cycle occurred in a single year, between late 2008 and late 2009. But this modest bump in productivity did not result from an uptick in investment, or the coming online of long-promised innovations. As I have already indicated, the short spurt in productivity seen during this slim interval actually resulted from *rapidly falling output* (the greatest since the 1930s) combined with an even more precipitous collapse in hours worked (a full 10 percent drop). In the U.K. the pattern has been slightly different, but with the same results. Employment has recovered to pre-2007 levels, but wages have remained stagnant, and productivity gains are non-existent.

Much hay has been made of this among British politicians and economists, who speak of a 'productivity puzzle' that as yet has found no solution. The *Fin-*

---

32 Frey and Osborne 2013.
33 Elliott 2014.
34 Sprague 2017.

ancial Times has nevertheless emphasized that since almost all of these new jobs are in low-productivity, low-wage occupations – in short, the 'service sector' – the aggregate productivity rates for the economy as a whole are dragged down by the addition of so many unproductive hours to the denominator of the ratio defining labour productivity (output in money terms divided by labour hours). Many of these jobs, as mentioned, have been created with the help of strategically distributed state subsidies, allowing companies to quickly hire – but also release at the drop of a hat, or a market downturn – workers with few appropriate skills and for poverty-level wages. But the alternative is daunting in its turn. For the same article notes that the U.S. did indeed register a modicum of growth for a spell during the recovery, but only by means of '*savage* cuts in employment'; Spain has continued to raise output per hour since 2008, but this remarkable upturn comes at the price of 'its *horrific* drop in employment', to the tune of over 50 percent unemployment among young people.[35] In the U.S., flatlining labour productivity is undoubtedly due in large part to the staggering number of workers parked in low-productivity service work, exchanged against sub-subsistence wages. And even this picture is a distorted one. To truly grasp 'the festering crisis of worklessness in America', one would also have to account for the full fifth of those proletarians in their prime working years who are out of work, or not looking anymore.[36]

∴

Any honest reckoning with the so-called service sector must begin by underlining that the category itself barely stands up to scrutiny. Said to comprise four-fifths of employment in the U.S. and similar high-income countries, it lumps together an enormous number of economic activities that differ in wage and skill level, location, size of enterprise, and capital-to-labour ratios. Its definition is largely negative, including anything deemed neither agriculture (farming, but also forestry and fishing) nor industry (manufacturing, but also construction and mining). A deep rift runs through the vast range of service sector occupations, between so-called business or professional services, on the one hand, and consumer or personal services on the other. The former are 'intermediate inputs' provided directly to businesses, often manufacturing firms; the latter are sold to consumers able to afford them. Among the first we find an array of activities, be they design, accounting, custodial, or clerical activities,

---

35   Klein 2015; emphasis added.
36   O'Connor 2015.

not to mention transportation services. Historically, these tasks were organized "in-house" by large manufacturing firms, rather than contracted out to autonomous firms specializing in them. Over the past forty years or so, capitalist enterprises have tended to refine the detail division of labour in ways that allow them to externalize these tasks. Thus Apple, to take a major example, owns no factories; only a tiny fraction of the retail cost of its products is derived from their outsourced assembly in China from parts produced elsewhere. This leads statisticians who collect these data to assimilate these activities to services, though many of them – research and design are prime examples, but so are trucking and shipping – are part and parcel of an extended manufacturing process.[37] If a significant fraction of what are counted as services by economists are in fact externalized segments of manufacturing, the statistical shift toward employment in services must be seen as at least in part an effect of an ever more ramified global division of labour.[38]

On the other hand, many nominally manufacturing firms – like Apple – today outsource production while focusing on the provision of consumer services related to these products, since the rate of return on such services is often higher than that associated with the manufactured good itself. As such examples suggest, the statistical segregation of manufacturing from services prevents us from understanding how they form an articulated whole. The history of automation suggests a complex dynamic between these two sectors, whose cyclical or spiral-type pattern requires that we think of them in strict correlation with one another. This can be seen also if we look closely at the other side of the divide within 'services', those aimed at individual consumers. If we remember that in the nineteenth century personal service occupations were more numerous than those in manufacturing, we can observe that the diminution of this sphere of personal service over the course of the twentieth century can be attributed in large part to the fact that many of those services were 'automated' in a very peculiar sense: they were transformed into discrete manufactured goods.[39] But this substitution of manufactured goods for services – dishwashers for domestic servants, individual automobiles for collective train service (in addition to the cart or carriage), mass-produced hamburgers for home-cooked meals – gave rise to a profusion of closely-related

---

37 This trend, sometimes awkwardly called 'servitisation', is a key contemporary trend that would require a separate treatment to explore its genesis and its implications.
38 Walker 1985. I thank Paul Mattick for this reference, and for his salutary criticisms of earlier drafts of this essay.
39 Gershuny (1978) theorizes this replacement of personal services with consumer appliances.

service occupations: jobs in sales, marketing, and repairs, not to mention insurance and consumer credit. Much of what we know as the service economy is a direct, complementary, effect of an earlier automation of services: one might even speak of a dialectical pattern, in which a primary term (personal services) passes over into its opposed pole (manufactured good), giving rise to a new, transformed variant of the first term (new or expanded field of services). Only now, these new service occupations are organized, as they were typically not in the nineteenth century, along properly capitalist lines.

It is in this light that we can return to the Oxford Martin School study, cited wide-eyed by so much of the writing around automation today. Cooler heads have parsed this study's methodology and arrived at dramatically different conclusions. Distinguishing between tasks and jobs – that is, between discrete activities and occupations, which are made up of changing groupings of tasks – they have calculated that a mere *nine* percent of current occupations in OECD countries (all of six percent in South Korea) are likely to be automated away in the years to come.[40] The meaning of this crucial distinction elided by the Oxford report is spelled out in another recent study of the effect of introducing ATMs on bank tellers: their numbers rose, if modestly, as their job responsibilities shifted from routine to more 'relational' tasks.[41] In this case – but this pattern is borne out by the history of automation as well – machines displace rather than replace workers as occupations are redefined. Indeed, as the example of the ATMs suggests, the automation of services might expand the market for them: rising productivity may mean falling prices and increased demand, which might require more rather than fewer workers. This 'virtuous cycle' was typical of many industries of the twentieth century, as they became increasingly capital-intensive: up to a certain point in this development, the rising productivity of these lines of production and the corresponding cheaper products widened the market for them. As long as output rises faster than labour is replaced, the effect of automation will paradoxically be to draw in more labour, rather than expelling it. In the case of personal services, in particular, there is another wrinkle to be accounted for. Unlike manufactured goods, the demand for which is said to be income *inelastic* – meaning that a diminishing share of income is spent on these goods, even if incomes rise – the demand for services such as education, healthcare, and entertainment will generally rise in step with the share of income available to purchase them. Even in a world of stagnant real wages, the cheapening of some services through automation

---

40   Arntz, Gregory and Zierahn 2016.
41   Bessen 2015.

would permit more of these services to be consumed, especially as the price of manufactured goods continues to fall in its turn.

What is a service? Most accounts restrict themselves to the most formalistic of responses: a service is a commodity that is produced and consumed in the same instant or interval of time. What this often means is that a service produces no discrete or detachable object that can be kept and, say, sold in turn at a later time. Think of a massage: there is no discrete object exchanged, nor even a visible material change in my body; its consumption is inseparable from its production on the part of the masseur. In the case of a haircut – always the example in the literature on services – we have a material transformation exacted by the haircutter (the removal of hair), but the 'object' thus produced is not detachable from my body: I cannot subsequently sell my haircut to a stranger, or even offer it as a gift to a friend. Because my haircut, even if it conforms to a prevailing style, must adhere to the specific dimensions of my head and its shape, it is difficult to standardize its production. The rationalization of such a service has material limits: it cannot be automated or mass-produced without the quality of the product suffering considerably. More pertinent examples of services that resist such capitalist rationalization – labour processes not subject to intricate divisions of labour, economies of scale, or the substitution of machines for labour – are found in education and health care. Teaching tends to involve one or at most two teachers per classroom, with no complex parsing of the labour process. Productivity gains are hard to achieve without undermining the nature of the service itself. More children can be added to the classroom, but at the expense of the quality of the instruction; the time of teaching cannot be sped up beyond certain rigid limits without a similar, deleterious, effect. Think, alternately, of a nurse specializing in physical therapy: here, too, the quality of the service will be diminished severely once the number of patients reaches a certain threshold, or the time of treatment is reduced beyond a bare minimum.

Karl Marx's account of capitalist development was rooted in the assumption that labour processes would be progressively taken over by capitalist firms and, through organizational and technological innovations, rationalized in the manner I've indicated above (standardization, division of labour, automation). This increasing subordination of labour processes to the demands of capitalist valorisation entailed, in Marx's term of art, a rising organic composition of capital: as capitalists reorganize the production of use-values in order to bend them to the imperatives of profit-making, machinery takes the place of human labour. As labour processes become more machine-intensive, the ratio of what Marx calls surplus labour to necessary labour (i.e., the labour necessary to reproduce the worker's existence) rises in turn, because in society as a whole less labour is required to produce workers' consumption goods. This changing composition of capital means, Marx's theory tells us, that the *rate* of surplus value thrown off by the production process steadily rises as work is increasingly automated. But, and here is the rub, Marx notes that since the total amount of labour incorporated in the production process must necessarily taper off in this scenario, the *mass* of surplus value generated, measured against total capital investment, will in turn decline, even as the rate of surplus value – the amount thrown off per unit of labour – rises. Such a scenario, Marx hypothesized, would mean that as production becomes more and more mechanized, and productivity rises in turn, the general rate of profit for the capitalist economy as a whole would paradoxically – paradoxically, because bourgeois political economists, having no conception of the distinction between surplus value and profit, assumed that rising productivity meant rising profit rates – diminish over some indeterminate amount of time. This declining profit rate would, at a certain point, produce a crisis of accumulation, as capitalist firms find the return on investment too low to initiate new rounds of technological innovation.

As it turns out, the capitalist mode of production has experienced a slowly unfolding crisis of accumulation over the past forty years, punctuated by sudden, and near lethal, collapses: since the early 1970s, we have witnessed diminishing productivity, falling profit rates, and stagnant and even declining real wages. And yet where Marx imagined the rude disciplining of social production to the tune of capital's drive to self-valorisation, and thereby the eventual rationalization – even automation – of labour processes in order to serve the needs of this expanding mass of value, ours is a world in which the vast majority of labour market-dependent proletarians are compelled to perform tasks that resist what Marx called 'real subsumption' under capital, their re-shaping through mechanization to meet the productivity requirements of capital. This is the world of services such as I have defined them above. These labour processes can only be *formally* organized along capitalist lines: personal services that were formerly offered by self-employed domestic servants, or performed without compensation by women in the home, are incorporated into profit-making enterprises, and performed by workers for wages. Fast-growing occupations like the home health aide, or the personal care aide, are particularly refractory to the productivity increases promised by capitalist refinements. These types of labour processes, which are increasingly the norm for the majority of the population of rich countries like the U.S. and Britain, can generate higher output solely by longer working days, or the hiring of more workers. Because they are so labour-intensive, with little capital spent on machinery, plant, or raw materials, capitalist profits in this sector are inversely correlated with wage levels: any rise in the latter squeezes the former. It is for this reason among others that most of the fastest growing occupations in the high-income countries pay so poorly: any rise in wages would either raise prices, eating into demand, or come directly out of capitalists' pockets.[42]

Accordingly, if we return to the rift within the service sector from which we started, we can speculate that it is those occupations in business and professional services – accounting, finance, the treatment of data, etc. – that are most likely to suffer the direct effects of a new wave of automation. Whether these fields can experience sizable gains in productivity is another question entirely. But if we assume that innovations in machine-learning and artificial intelligence will make headway in these lines of work, those whose jobs are usurped by the machines will be forced into the provision of low-paid, precarious consumer and personal services. If the Bureau of Labor Statistics's projections are to be believed, this migration might already have been triggered. Most contemporary speculation concerning the automation of the service sector not only neglects this sector's specific features, they also implicitly assume that the number and type of occupations are finite or fixed. To the contrary: the colonization of human activity by the service sector has most likely only begun. In principle, the entire range of human activity is subject to segmentation; these segments can be transformed into occupations, which in turn can be organized along capitalist lines. Responding to the Oxford Martin School report on the 'computerization' of current occupations, Paul Mason notes that should half of these jobs indeed be wiped out, the result might be less mass unemployment than a vertiginous explosion of the 'human services sector':

> We would have to turn much of what we currently do for free, socially, into paid work. Alongside sex work we might have 'affection work': you can see the beginnings of it now in the hired girlfriend, the commercial dog-walker, the house cleaner, the gardener, the caterer and the personal concierge. Rich people are already surrounded by such post-modern servants, but to replace 47 per cent of all jobs this way would require the

---

42   Some readers will note that much so-called service work might be better called 'circulation labour', whose function is less to produce value (and surplus value) than to secure the exchange of commodities against money in the sphere of circulation. Here the concepts of productive and unproductive labour, such as they are criticized and reworked by Marx in a number of places in his later work, would offer a more coherent categorial framework for understanding what bourgeois economists refer to as the service sector. On this question, see Smith 2020, especially chapters 3 and 4.

mass commercialization of ordinary human life .... [This would] push commercialism into the deep pores of everyday life, [and] make resisting it a crime. You would have to treat people kissing each other for free the way they treated poachers in the nineteenth century.[43]

∴

How does the complexity and fragmentation of the vast, motley service sector affect workers' capacity to organize themselves across occupational types, in view of building anew forms of worker power appropriate to the twenty-first century?

A widely cited paper from the late 1990s on the causes of deindustrialization, written under the auspices of the International Monetary Fund, sizes up in its conclusion the potential effects of the growing concentration of employment in the slow-growth, technologically stagnant service sector of the economy. The co-authors, Robert Rowthorn and Ramana Ramaswamy, emphasize how the fragmentation of this sector, riven by cleavages in skills and wage levels, combined with the material disparity of the concrete labour processes lumped together under this label, will undoubtedly pose insurmountable obstacles to rebuilding powerful trade unions like the UAW of the late 1930s sit-down strikes. 'Trade unions', they warn, 'have traditionally derived their strength from industry, where the modes of production and the standardized nature of the work have made it easier to organize workers'.[44] The historical workers' movement and the industrial unions of the mid-twentieth century endeavoured, through the institution of collective-bargaining agreements, to reduce wage differentials across industries. This objective was formulated not simply on the basis of infra-class solidarity among workers, but on the tendency, driven by competitive pressures among firms, for technological innovations to spread across lines of production and eventually sectors. As firms across the economy adopt similar techniques, the different working conditions of various class segments are smoothed out and over. The rising ratio of machinery and raw materials to labour employed assures a tendential material density of the class. Comparable skill levels, wages, and working conditions prevail in massive plant facilities bringing together thousands of workers at each individual site. The workers' movement itself was at once the product and reflection of this convergent material unity of the capitalist mode of production: if worker struggles

---

43 Mason 2016, pp. 174–5.
44 Rowthorn and Ramaswamy 1997, p. 22.

of the nineteenth century in part impelled the development of the forces of production (cf. the conflicts over the length of the working day), the generalization of this development across lines of production in the early twentieth century shaped the class into a compact and often militant mass. This is what James Boggs had in mind when he spoke of the 'embryo of a socialist society' gestating within this one, 'united, disciplined and organized by capitalist production itself'.

In her magisterial study of the history of the workers' movement, Beverly Silver underlines the way the objective splintering of the service sector outlined by Rowthorn and Ramaswamy is reflected in the isolation of these workers from one another, and their distance from the strategic leverage points enjoyed by workers in fields as different as manufacturing and education. Those who work in the automotive industry – as with Boggs, her key example – are imbricated in a tightly articulated detail division of labour: a work stoppage at one point in the production sequence can bring the entire process to a halt. Teachers, on the other hand, operate with relative autonomy in their classrooms, less affected by a ramified technical division of labour. At the same time, a large-scale strike by educators might reveal their crucial place in the so-called social division of labour, causing widespread disruption at least at the local level, as parents scramble to find someone to care for their children. Workers in the oil sector, however tiny it may be, are able to disrupt the entire functioning of the capitalist economy at least at the national level, as recent struggles in France (in 2010 and 2016) have shown. Workers who find themselves stranded in low-wage service occupations in retail or hospitality (together, one fifth of the workforce) have no such leverage: their workplaces are often dispersed and small in comparison with the great industrial concentrations of the past, and they have little fixed capital to idle. Silver can point to important if modest recent victories by workers in these fields but avers that such successes have come despite the distance of these workplaces – in the case of retail, restaurants, and similar types of work – from the levers of production and social reproduction. They have instead had to 'follow a community-based organizing model rather than a model that relies on the positional power of workers at the point of production'.[45] It is, however, these pre-existing community ties – neighbourhoods, languages, religion – that the ever-expanding ambit of the personal services sector threatens. If these were the foundations of the old workers' movement, whose forms of mutuality and self-aid often relied on affinities derived from ethnic, cultural, and geographical proximity, they are today everywhere in tat-

---

45 Silver 2003, pp. 113–22, 172.

ters, as the social fabric is chewed through by the corrosive effects of money and markets, and communities dissipate into warring, atomized, dysfunction.

In the early nineteen sixties, Boggs foresaw a day when a large number of those expelled from the factories of northern industry would have 'nowhere to go': these were the 'surplus people', 'the expendables of automation'. Today the children and grandchildren of these surplus people remain trapped in collapsing cities, far-flung suburbs, and rural ruins. They scrape by on part-time precarious work and tenuous lines of extortionate credit, commuting to and from work an hour each way, surveilled by heavily armed cops as they make their way home from the bus stop. Some run rackets and hustles, while others sink into depression, or drugs. For many, prison is always near.

Boggs foresaw a world of outsiders, on the margins of the wage relation, yet whose every move was hounded by money. To those who imagined rebuilding the CIO of the two decades prior, he could only say, dream on. The union was lost, he wrote with *sangfroid*, the moment the bosses brought in the computer-controlled machines. The cause of unionism was lost before that: never setting out to attack the bases of capitalist society, it became part of it. 'Historically, workers move ahead', Boggs wrote, in an imaginary retort to those who want to reactivate older figures of organization. '*That is, they bypass existing organizations and form new ones uncorrupted by past habits and customs*'.[46] Boggs was careful not to venture details about what shapes these organs might take; he did not promise they would reconcile the class fractions churned out by changes in the composition of capital. American workers – a term ample enough to envelope his 'surplus people' – would, should they take command again over their own lives, have to launch a 'revolt powerful enough to smash the union, the company, and the state'. But Boggs's accent was less on negation than discovery. Surrounded by 'labor leaders and well-meaning liberals' proposing gimmick upon gimmick in hopes of saving the reigning social order, Boggs wagered on these 'outsiders', who will have to compose, and soon, a 'new way to live'. What he said then is just as true now: 'The means to live without having to work are all around them, before their very eyes. The only question, the trick, is how to take them'.[47]

---

46 Boggs 2009, p. 32; emphasis added.
47 Boggs 2009, p. 52.

PART 2

Key Concepts in the Automation Debate

∵

CHAPTER 3

Time to Automate: The Hidden Labour of Automation

Christina Gratorp

There is no cloud, there are only other people's computers.

∴

The idea that software is intangible and weightless is tenacious. Software, it seems, requires no hard work or maintenance. This notion dovetails nicely with the marketing of that which was formerly deemed automatic as autonomous, a boost in the techno-utopian direction of a future free from work. At the heart of this 'New Deal' of automation is the executing code, engineered by technologists with their eyes on a future in which algorithms and toolchains smarter than humanity itself will tend to our every need. The present is always in a debugging phase.[1] Eventually, we will close the gap of inefficiency – with more automation.

A closer examination of this continuous debugging phase, however, reveals alongside the history of automation a mirrored history of stagnation. Though presented as time-saving and neutral, in reality automation often has more of a resemblance to a Mika Rottenberg video than the clean, standalone process the managerial class claims it to be. Rottenberg's digital art depicts the commodification of the human body labouring alongside peculiar assembly lines, producing items like maraschino-cherries from red fingernails. It shows how, in contrast to its shiny frontend, automation has a backend comprising manual labour and exploitation, coated with ideology. Moreover, embedded in the idea of automatic efficiency is the additional failure to pay for the work required to maintain the production rate as environmentally and socially sustainable. Uncovering the manual work that precedes, builds into, and is absent in automation discourse produces an evidently different picture of automation's alleged time-saving capabilities.

---

1 See Noble 1978.

Leaving behind the much too simplistic notion of automation as an (im)perfect sequence of technological mechanisms, this chapter aims to discuss the broad variety of manual labour involved in automation but left out of its attendance discourse. This silence about manual labour serves to maintain the fantasy of the automatic or 'fauxtomation', as proposed by Astra Taylor.[2] The gaps of inefficiency are closed with the labour time of others than those who enjoy the profits of automation, resulting in a total of time and labour-shifting effects upheld by a scaffolding of money, capital, and social relations. Still, technology is overall seen as an autonomous force, impacting social relations but not itself impacted by the social relations upholding it. History, however, shows that automation is the 'product of a social process, a historically specific activity carried on by some people, and not others, for particular purposes'.[3] With the interlinked digital technologies of today, the manual labour necessitated by automation becomes ever harder to distinguish. Time is fragmented and shifted, and while some labours are accounted for, others are kept invisible, emerging instead as unpaid labour or simply non-technical, thus not worthy of the status given to programmers and designers.

A crucial element in this process is the ability to render labour less expensive in order to make it seem insignificant in cost considerations. Automation then appears as emphatically bound to a capitalist mode of production, where the ability to cheapen labour by displacing it presupposes capitalist class relations. Consequently, automation can be seen as the potential to exploit and reproduce inequality while disguising its character of exploitation from the user, i.e., the individual consumer. The pattern by which automation not only reinforces class relations but also gender relations and racialized hierarchies from which it originates is repeated throughout its development, integration, and maintenance phase, for different locations within the non-capitalist class as well as throughout history. The same applies to the ever-accruing time debts of degraded environmental quality, social damages, and dearth of tools that solve actual problems rather than accumulate capital. These time debts include various forms of labour, where some are postponed (such as decontaminating degraded natural environments), some are performed unpaid in the domestic sphere (as reproductive labour without which there could be no economy), and some only exist as an absence that can actually never be paid back, in the form of shortened lives and lowered life chances (in the sense of Weber's '*Lebenschancen*') caused by precarious conditions. What they all have in com-

---

2 Taylor 2018.
3 See Noble 1978.

mon is the notion that they are not 'technical work', thus they do not seem to be connected to automation. On a larger scale, Raj Patel and Jason W. Moore discuss such debt as 'a kind of ongoing theft' from 'women, nature and colonies'.[4] Nancy Fraser refers to this as 'care deficits'.[5] By highlighting a diversity of work (paid, unpaid and postponed), this chapter wishes to expand the understanding of efficiency in relation to automation, and question the notion that rising productivity equals saved time.

The first part of this chapter is focused on looking at spatial aspects of automation in terms of uneven flows of real resources such as energy and labour time. The second part discusses in more detail the dialectical relation between automation and temporality, while the third part deals with the work of the future and the possibility to automate automation itself.

1   The Extraction of Code

Friedrich Kittler once wrote that 'there is no software', by which he meant to say that writing in the computer age no longer exists outside the hardware preserving it.[6] All code operations are electrical manipulation and information is preserved by signifiers of voltage differences. Where better to begin, then, than with the materiality of the code itself. Code is not nothing and thus has to manifest in something. This something is the result of a complex labour chain, its entry point perhaps a Rwandan tantalum mine, and its last instance somewhere on a factory floor, dictated by Silicon Valley but offshored to China. Although the production of software presupposes large-scale extraction of raw materials and depends on informal workers' substandard working conditions, a discussion about the amount and kind of work required before software-driven automation can even begin is strikingly absent.

In terms of its basic physics, all technology is about converting some form of energy into useful energy.[7] This branch of physics (thermodynamics), dealing with heat, bodies, work, and energy, emerged during the early nineteenth century and was, amongst others, absorbed by Marx.[8] Converting energy also goes for the electronic machines running the code of automated processes. To

---

4   Patel and Moore 2017, p. 95.
5   See Fraser 2016.
6   See Kittler 1992.
7   In thermodynamics called *exergy*.
8   For a thorough discussion of the metabolic-energy perspective on human labour in the works of Karl Marx, see for example Burkett and Foster 2006.

understand the need for energy in software-driven automation, the basic principles of thermodynamics must be expanded upon. Each and every logical one and zero is the result of energy conversions, and each conversion means losing some of the useful energy. Due to the second law of thermodynamics, which has been discussed as a precondition to life itself, we cannot build perpetual motion machines.[9] And as software is information and information is material, at the bottom of the material chain of exchange we arrive at the relation between information and energy. In short, the mass of software consumes its hardware. In this sense, the machines are rather like us: their lifetime is limited.

Based on these unavoidable fundamentals of physics, Alf Hornborg shows how assemblies of technological machinery in the world's industrial 'core' rely on a net import of useful energy from the industrial 'periphery'.[10] Highlighting automation as a spatial rather than a temporal marker moves the perception of progress from a deterministic process of innovation to practices made possible by certain social relations. Each processing of raw material dissipates more energy, while the price of the resulting commodity usually increases. This results in continuous underpayment for the energy derived from natural resources, leading to what is called growth and development at the core of the system and energy depletion in the periphery. Viewing technological development through a lens of a system-independent process like thermodynamics reveals an insidious normative understanding of 'progress', tightly associated with 'uneven flows of real resources such as energy, labour time and hectares of land productivity'.[11] With this insight, Marx argued that the apparent reciprocity of free market exchange is an ideology that obscures material processes of exploitation and accumulation – a view that is today commonly taken up in Eco-Marxist discourse.[12] This is of course not a specific feature of automation, but places automation as part of any technological development arisen from global exchange relations. Different levels of technological development should therefore, argues Hornborg, be understood as a result of differentiated positions within these relations, and not as existing in different periods of history.[13]

---

9   The second law of thermodynamics deals with the direction taken by spontaneous processes. The law can be formulated as an observation that irreversible processes involve dissipative factors, where such irreversible processes are part of the processes of life.
10 Hornborg 2012, p. 42. See also Dorninger et al. 2021. 'Core' and 'periphery' are used analytically and not in a geographical sense.
11 Hornborg 2001, p. 33.
12 See Hornborg 2019a.
13 Hornborg 2015, p. 42.

Capital's aim is then to overcome the limits of space, repeatedly necessitating technological revolutions in transport and communication.[14] But technological development does not necessarily mean that the need for human labour is reduced. As new technology is deployed in the industrial cores, new forms of labour occur in the peripheries. To fulfil its financial purpose and avoid crises of over-accumulation, accumulated capital (stemming from increased productivity as a result of technological development) must also find ways to be absorbed and end up in circulation. According to David Harvey, this instability of capital is an inbuilt contradiction, and 'contradictions have the nasty habit of not being resolved but merely moved around'.[15] Here, the time and labour-shifting properties of automation on a deregulated global financial market suit capital's needs well. While some labour tasks may be taken over by robots and software automation, others are simultaneously created in the fringes; the need for manual labour surfaces elsewhere, alongside automation. Measured over a period of time and from a global perspective, the practice of automation as a whole can be seen as an effect of capital's need to find profitable ways to both produce and absorb capital.

The appropriation of time and space required to achieve the speed of automation is clearly demonstrated by Hornborg, who uses the nineteenth-century railways in England as an example. 'Industrial machines', he writes, '*are* social phenomena. These inorganic structures propelled by mineral fuels and substituting for human work *could not be maintained* but for a specific structure of human exchange'.[16] In the construction of the railways, large amounts of labour/time and nature/space were sacrificed by some for the gain of others. Harvey refers to this phenomenon of modernity as *time-space compression* and Hornborg proposes approaching the concept from a distributive perspective, where time and space are understood as resources available for human exchange.[17] The shortened travel time for passengers in industrialized England must be juxtaposed with the time spent on building the railways – its locomotives, rails, and wagons – and the materials used to manufacture them – timber, iron, coal, and steel. Further, the railways were funded in part by profits from the textile industry, depending on vast *ghost acreages* in southern USA.[18] The import of cotton to England in the year 1850, which required 394 million hours

---

14 Marx 1993, p. 521.
15 Harvey 2015, p. 4.
16 See Hornborg 1992.
17 Harvey 2017. Hornborg 2012.
18 Hornborg 2012, pp. 137–47. The expression ghost acreage is used to explain the consumption of space elsewhere to maintain the productivity in an industrialized country.

of work from mainly slave labour and 1.1 million hectares of arable land in the USA, required only half the working time and a sixtieth of the land area in England to produce the final textile goods.[19] In this way, the railway's time-saving effect for English travellers can be linked to the spatial area of the cotton fields, the working hours of weavers, and the direct consumption of human life under slavery.

The efficiency attributed to modern technology, be it transportation or processors, thus cannot be separated from the social relations of time, space and labour upholding it. Just as the railways during early industrialization established new trade routes for cotton and textiles, the internet as infrastructure has created new trade routes for the distribution of both software and commodities in general. The ghost acreages of software-driven automation facilitated by the digital infrastructure are now found in big economies like China and Brazil, and numerous African and Asian countries with below-average GDP.[20] By 2025, forecasts suggest that the world's mass of software will be contained in over 75 billion devices, connected by the Internet of Things, alongside further billions of industrial devices in the Industrial Internet of Things.[21] These infrastructures require intensive resources – chains of cheap time and space leading up to power centres that can then capitalize on selling software-driven automation on top of these structures. The vastness of the network is crucial for the user's sense of automation; that is, for generating the data on which it also feeds. A network of few connections, few things, and a small amount of accessible data would be an oxymoron, meaning large ghost acreages (equalling labour time) for energy extraction are built into the design.

The sense of automation, though, is reserved for the end user. During the last twenty-five years and the rise of Information and Communication Technology ('ICT'), artisanal and small-scale mining ('ASM') has experienced explosive growth. An estimated twenty-five percent of tantalum, tin, and gold now come from ASM, all listed as conflict minerals, crucial to software-driven automation.[22] Since the early days of ICT, the number of ASM operators worldwide has increased from an estimated six million people to 40.5 million. Some sources estimate a much higher number – up to 100 million miners working without

---

19   Hornborg 2012, p. 144.
20   China and Brazil are two of the world's largest producers of silicon. Africa and Asia are the continents with most informal mine workers.
21   Statistics from Statista, https://www.statista.com/statistics/471264/iot-number-of-connected-devices-worldwide/ (accessed 12 May 2021).
22   Tantalum is used in capacitors, tin as soldering material, and gold in a quantity of components.

supervision or control. These numbers can be compared to the seven million people working in industrial mining.[23] The growth of ASM shows how labour that is necessary to produce automation can be disregarded by building on peripheral, substandard working conditions.[24]

A growing size of the digital infrastructure also means a growing technomass now competing with human and other biomass over living space on our limited planet.[25] The growth of ASM is linked to the increased difficulty of earning a living from agriculture, as a result of water shortages and food crops being crowded out by crops grown for machine fuel and industrial materials. As Hornborg points out, with the relocation of energy to the world's industrial cores, there is an opposite shift of environmental quality, including export of obsolete, toxic technomass to poor or conflict-ridden countries such as Nigeria, Afghanistan, and Syria.[26] Artisanal and small-scale mining is the source of the largest releases of mercury, notoriously difficult to clean up.[27] This type of environmental debt produced by technological development is not only found in the Global South, but also embedded in rich urban environments. Silicon Valley, for example, is one of North America's most polluted sites, a result of four hundred years of hard land and labour exploitation.[28]

Here, social inequality and environmental disruption are intimately linked. This combination of environmental degradation and social inequalities produce what David Naguib Pellow and Lisa Sun-Hee Park refer to as *environmental racism*.[29] This, I argue, also implies large and completely invisible time liabilities, whose financial equivalents are not visible in the cost of automation. If the cost considerations pertaining to software-driven automation were to take into account the effects on the environment and public health, the measure of its efficiency would be quite different. The extremely toxic production of electronics depends heavily on an immigrant workforce and people of

---

23 Intergovernmental Forum on Mining, Minerals, Metals and Sustainable Development, The International Institute for Sustainable Development 2017: *Global trends in Artisanal and Small-scale Mining (ASM): A review of key numbers and issues*, January 2018. The report emphasizes the need to improve national data benchmarking and consistency.
24 For example, during the Corona crisis, the effectiveness of automation was highlighted by showing how white-collar workers were enabled to work from home when the whole world was in lockdown, while in Peru, mine workers were forced to continue working without sufficient infection control measures (Flood 2020).
25 See Hornborg 2008.
26 Swedish Environmental Protection Agency (Naturvårdverket), 'Avfall vid illegala gränsöverskridande transporter' 2015.
27 See García et al. 2015. See also Seccatore et al. 2014.
28 Pellow and Park 2002.
29 Pellow and Park 2002, p. ix.

colour – different groups being variously exploited, restricted, or denied entry to the U.S. throughout the last half century.[30] During the 1990s, up to 80 percent of the production workforce in the Santa Clara region was composed of Asian and Latino immigrants, most of them women.[31] Epidemiological studies show rates of occupational illness among Silicon Valley workers of more than three times that of any other basic industry.[32] Patterns of racialization are a significant part of the history of technology, revealing a contradiction in state policy towards working-class immigrants who are despised and enclosed in different concealment processes made possible by racialized hierarchies while comprising a core part of technology labour.[33] Innovation produces new needs and desires, making parts of the labour force redundant while finding fresh labour forces through immigration or proletarianisation of strata of the population formerly economically protected.[34] Today, much of this production is offshored. The stories of the technical workers in ICT manufacturing are largely untold, but as Christian Fuchs argues they are all part of a bigger story of 'exploitation and imperialism that is inscribed into the phones, computers, screens and laptops that we use every day for talking, writing, listening and watching'.[35]

In a wider discussion surrounding the work that is needed to maintain current levels of automation, the large and hitherto white-collar work force that keeps busy developing, maintaining, and integrating automated processes cannot be ignored either. More than seven percent of the total U.S. workforce are occupied in the developing end of the technology sector, and similar statistics are found in other technology-intensive labour markets.[36] In Sweden, software developer is the eighth most common profession.[37] These are far from the most exploited parts of the working class, but what Erik Olin Wright refers to as a class fraction who sell their labour-power, despite occupying relatively priv-

---

30   Pellow and Park 2002.
31   Pellow and Park 2002, p. 9.
32   See Smith and Woodward 1992.
33   For comparison, see Edna Bonacich et al. 2008. Primary racialization comes from (white) capitalists who seek to exploit workers of colour more than they are able to exploit white workers. Secondary racialization arises among white workers, who fear being undercut and displaced by workers of colour.
34   See Harvey 2008.
35   Fuchs 2014, p. 182.
36   Statistics from https://www.cyberstates.org/. 12.1 million net tech employment in the U.S. workforce of 164.6 million. Accessed in February 2020.
37   Statistics from *Statistics Sweden*, http://www.sverigeisiffror.scb.se/hitta-statistik/sverige-i-siffror/utbildning-jobb-och-pengar/yrken-i-sverige/. Accessed 27 June 2020.

ileged appropriation locations within exploitation relations.[38] Nonetheless, the work is manual and the speed of modern programming, albeit greatly exaggerated, is managed through modern so-called agile workflow management methods, aiming to speed up production in the Fordist engineering tradition.[39] Notions of the transformative significance of engineering labour is persistent in theories of industrial society, but as shown by Peter Meiksins and Chris Smith, engineering ideology, far from having an autonomous scientific rationale, is structured to match the needs of accumulation and profit.[40] That engineering labour is 'determined in the last instance' by the capitalist mode of production is what causes Eddie Conlon to call engineers 'prisoners of the capitalist machine'.[41] Employers commonly favour applicants who spend their free time doing coding work at home, contributing to the big community of free and open source software used to a large extent by commercial companies without cost.[42] This unpaid, hidden labour has successfully been accepted as part of the 'true' coder identity, but however creative the work of the programmer is made out to be, the tasks are often tedious and repetitive, and progress is slow. Coding close to the metal, writing an algorithm that makes an LED flash can literally take days. Having worked as a full-time embedded software developer for fifteen years, I cannot help but wonder how the appeal of my profession would change were it stripped of the economic and status-related privileges so desperately sought for under capitalism.

Thus, the shiny frontend of automation is upheld by a stowed away, unaccounted for, backend. Along with an ever-growing time debt being left out of the equation, it is easy to see how a concept like 'lights-out manufacturing' can gain significance.[43] The darkness on the factory floor demands another kind

---

38   Wright 1997, pp. 20–37. For further discussion, see also Svensson 2019.
39   'Scrum' and 'Kanban' are methods that aim to produce more in less time using different kinds of 'scientific planning'. These so-called agile methods are now implemented widely, and aim to increase management control over production cycles, atomize production tasks and lead to overhead work for employers that continuously need to handle reporting and suggest how to improve work further. If the goal of a 'scrum production sprint' is reached, over time the following sprints will most likely contain more work, to 'push limits'.
40   Meiksins and Smith 1996, p. 9.
41   See Conlon 2019, p. 40.
42   Open source used to be considered underground, but today big companies like Microsoft sometimes contribute to projects of their interest. There is however much more work than coding included in the idea of open source, which is often handled by members of a community without pay.
43   'Lights-out' alludes to such a high level of automation that the factory can operate in the dark; no need for human labour is said to be required. Very few companies claim to be able

of darkness preceding it, a blindness to social, economic, and environmental exploitation. To advocate such a model to achieve equality presupposes a limited view on what technology actually do within society. A frequently recurring idea in socialist utopian visions is to derive freedom from a high degree of productive automation. But these utopian visions are often based on a twofold step of simplified assumptions. First, such ideas import a stabilized and delineated notion of technology from computer science into a social analysis, and thereby fall into an 'epistemic trap' by reducing technology to the result of its technological mechanisms.[44] Second, by adopting this delineated notion of technology, the answer to inequality caused by automation under capitalist conditions becomes an upheaval of those conditions, while keeping – or even accelerating – the technology. This is often referred to as the 'fettering thesis', in which a humane cultivation of the capitalist forces of production is allegedly 'fettered' by capitalist social relations.[45] What this perspective seems to suggest is that property relations under capitalism mainly impact the way technology is used, while being less concerned with how capitalist technology works on a level of design. Thus, capitalist technology can be regarded as having an inherent potential that can be 'set free', if ownership structures were to change. This approach, I argue, is reductionist, and suffers from a lack of interest in both the form and substance of automation as a whole. In contrast, a technosocial perspective should engage with the idea that while technology must be understood within the context of its creation (on all levels), there might be more to 'context' than capitalism. For example, it may be necessary to both analyse an algorithm in a machine learning application based on the conditions for its intended use, and to consider certain aspects as a more particular, embedded policy that is not as easy to renegotiate, such as the energy needs of the application, regardless of its use. A more robust perspective should therefore not depart from treating technology as a standalone phenomenon, but as something that forms *technosocial assemblages*; as something that 'is best conceptualized as an intervening rather than an independent variable'.[46] If there is potential in technology, the focus to realise it should be on making such assemblages a tool for equality. In other words, an analysis should focus on where the possibility for agency, choice, and power resides in these

---

to produce in the dark, and they do not include for example supervision and maintenance workers.

44 See Lee 2021.
45 See for example Bernes 2018, or Huber 2019.
46 Hacker 1979, p. 539.

assemblages, and thus acknowledge their fluidity and complexity.[47] Technology then, is not floating over society, but intertwined with social relations, norms and values, and thus continuously a subject in need of re-evaluation and work.

2 The Production of Labour

The promise to brighten the future – attached to modern conceptions of technological development – allows new automated processes to be introduced as 'solutions'. A solution, after all, presupposes that there is something 'broken' in the present and urges us to abandon that which we already possess. The new, however, is rarely a simple add-on, but requires already existing sociotechnical practices to adapt as well. It stretches backwards, seeking to undo the past while granting a promising shimmer to the future. In the software industry, integration usually means adapting a new product towards an existing set of programs and devices, making it compatible with already deployed interfaces, hardware, and market structures. Meanwhile, our present movements, desires, and thoughts are reshaped, and often are made obsolete. To let go of the notion of technological development as an autonomous force is to admit that this abandonment is not an inevitable consequence of human progress, but a result of struggles between groups with different interests. The old in modernity, it seems, is not always pregnant with the new, but awaits its self-destruction and reappearance as itself, only somewhat warped.[48]

Indeed, studying historic and present examples of the labour that is required for integrating new automated processes in social environments reveals a strong relation between time saved and time appropriated, where the most conspicuous aspect of this relation is the relocation of work. The German Jewish philosopher Adolf Caspary went so far as to express this limited ability of technology to solve social problems as a matter of proportions, saying that while the machine can certainly make the numbers in the economic calculations bigger, the ratio between benefit and necessary mobilization is left unchanged.[49] And so, looking under the hood of automation reveals motives of the ideology of automation itself, tightly associated with class struggle. Although governed by measurability, division, and control, this ideology pre-

---

47 See Lee 2021.
48 I am paraphrasing Karl Marx's notion, 'Force is the midwife of every old society pregnant with a new one.' *Capital Volume 1*, ch. 31.
49 See Caspary 1927.

sents itself as the creator of free and independent subjects. Such liberal views on the subject, however, lean on the labouring body of the other – labour simultaneously upholding the magic and appearing as inferior to it. This structure is underpinned by a trinity of capitalist conditions: the possibility to cheapen, relocate, and thereby conceal labour, since one hour of mineral mining comes cheap (in a monetary sense) compared to one hour of programming. Given the power structures whence the choices of automation emanate, there is little wonder that they tend to reinforce rather than subvert those structures, saving time only for a select few.

From the early days of the telegraph to the present plethora of social media, human integration labour has been key not only to integrate new technologies into society on a technical level, but to perform various kinds of labour tasks integral to make automation work across a wide range of techno-social contexts. Before the twenty-first century, school teachers rarely had to spend time telling pupils to put away their mobile phones. Thus, uncovering integration labour is fundamentally a geographical-temporal question, situation the labouring body within a larger system stretching over time and space.[50] It cannot be separated from capitalist conditions such as the division of technical labour or traditional family patterns. All these forms of labour – ranging from creating social media contexts to performing widely disparate forms of paid and unpaid microlabour – share a crucial aspect: users tend not to see them. That is, they do not show up in the budget. By no means does this imply their lesser importance in making automation technologies work. On the contrary, enclosing labour in different kinds of concealment processes is necessary to uphold perceptions of automation. The telegraph, for example, for one hundred years relied at its core on a teenaged messenger labour force that acted as human links between the post offices, business sites, and homes. Although the telegraph companies did their best to maintain a conception of the messenger boys as being outside the technological system, the messengers actually embodied the automation of the network by connecting its physically separated nodes. As for closing the gap of inefficiency in order to leave the debugging phase behind, the need for messenger boys did not decrease with increased technological sophistication. Instead, it increased.[51]

Integration labour, or *information labour* as Gregory J. Downey calls it, that enables context jumping from the virtual to the physical and from one temporal, organizational, or cultural context to another has since become integral

---

50 See Downey 2014, pp. 141–65.
51 See Downey 2014, p. 151.

to our perception of automation.⁵² It is easy to see parallels in today's digital systems that are seemingly, but far from, interconnected by themselves. Take for example the millions and millions of platform workers engaged in the gig economy, collecting, and charging e-scooters at night, delivering food by bike, or picking up laundry. Or the hundreds of thousands of warehouse workers guided between shelves with *pick to voice* systems attached to their heads, employed to act like human robots while quickly satisfying customers with whatever consumer goods have been brought into existence by capital. Or yet still, the tens of thousands of content moderators clearing the drains of social media, ensuring horror-free feeds to end users.⁵³ 'AI' used in autonomous vehicles depends on hundreds of thousands of people who, through precarious labelling work, manually train computer systems how to interpret huge data sets.⁵⁴ In addition, machine learning systems rely on data creation, that is, the continuous unpaid interaction by humans with said systems.

These labours, although of great variety, share a common notion of not being directly related to any core part of automation. Instead, workers come cheap and are commonly hired by the hour. Automation, then, is mediated by app-distributed user interfaces, concealing the labouring bodies of others who appear to be outside the system. It is not on behalf of technology itself that such automation is cheap, but due to concealed backend integration labour that upholds automated frontends, saving time for those on top of the information food chain.

Capital however has strong reasons for wanting to maintain the idea of this work as being 'outside'. To appear as a software company rather than a labour company has consequences for the financial worth of enterprises.⁵⁵ Hiding labour is key to keeping alive the perception of 'innovation' and 'entrepreneurship' so tightly associated with software-driven automation and the idea of company value. The more visible the workers in human computing become, the less corporate 'automation' looks like automation, affecting stock value and venture capital.⁵⁶ First, making capital appear as a certain kind of capital can be achieved by sequencing and separating labour, hiring by-the-hour instead of

---

52 See Downey 2014, p. 141.
53 Since all software-driven applications automate processes on some level, I choose to include gig workers in this discussion.
54 For comparison, see Marx 1993, The Chapter on Capital: 'The worker's activity, reduced to a mere abstraction of activity, is determined and regulated on all sides by the movement of the machinery, and not the opposite.'
55 See Irani 2015. Also refer to Irani 2015 for line of reasoning in the following section.
56 See Irani 1015.

by employment, and making people work offsite. Second, capital needs workers to be replaceable; depending on a skilled labour force means not having access to an infinite pool of cheap labour. Thus, it is exactly the cheapness of labour, and of course the labour that goes unpaid, which simultaneously enables its required concealment while being instrumental in the creation of value. These forms of labour all have their own implications in regard of precarity, but together they form what Fuchs calls the *international division of digital labour*.[57] Drawing on Marx, Fuchs states that the labour conducted by this collective workforce is necessary for the existence, usage and application of digital technology, while being 'disconnected, isolated ... carried on side by side and ossified into a systematic division'.[58]

As with the telegraph, when more and more people use digital platforms, the need for content moderators increases. The number of poorly paid content workers by far exceeds the number of well-paid software developers employed by platform companies.[59] The psychological impacts from working on the social media shop floor include sleeping disorders, depression, and PTSD, and make for a large time-debt of care labour seemingly delinked from automation.[60] The companies claim to continuously develop better machine learning algorithms to replace the manual work of moderators, but as social, political, and cultural relations change, so do norms and the interpretation of content, limiting the potential of non-human labour.[61] For such socio-material phenomena, quantitative sorting methods fail to add meaning. As Yuk Hui puts it: 'Art and philosophy can't choose science as their point of departure. If they do, they become footnotes to positivism'.[62]

The last decade has taken the concept of context creating 'human robots' even further. While content/context workers are an essential component in making automation in social media technologies useful, requester platforms, such as CrowdFlower, Clickworker, and Amazon Mechanical Turk ('AMT'), base their entire business model on 'artificial artificial intelligence'.[63] The case of AMT, as disclosed by Lilly Irani, serves as an illustrative example of the magic trick of hiding labour. The design of the system allows for massively mediated microlabour, where thousands of workers can be hired for a few hours of

---

[57] Fuchs 2014, p. 5.
[58] Fuchs 2014, p. 5.
[59] Facebook states they had 250 developers in 2017 (Whittaker 2020).
[60] Aschoff 2020.
[61] Wasielewski 2023, ch. 1 (Digitization and Dataset Creation). See also Downey 2014.
[62] Dunker 2020.
[63] See Irani 2015.

work. The hiring does not include a contract or guaranteed payment, enabling a systematic use of surplus and even entirely unpaid labour necessary for the system to be efficient in economic terms. Since managing the workers is done algorithmically, requester employers must find ways to ensure the quality of the work without spending time themselves to manually validate the tasks. This is accomplished with various ways of additional metawork, such as letting the workers perform unpaid tests or letting multiple workers complete the same task. In this fashion, the same work can be performed many times without extra cost, since AMT's participation agreement grants employers full intellectual property rights over submissions, regardless of rejection. As Irani concludes, AMT employers cite the system as enabling them to innovate in new ways by outsourcing tedium (such as labelling huge data sets), which speeds up their experiments and allows them to perform as software companies.[64] In the case of AMT, in the end, to accomplish this apparently without human mediation requires more manual (cheap) labour on the workers' side. Although many of the technologies used in platform services are not at all new, a certain 'language of innovation' that includes phrasing such as 'disruption' and 'creative destruction' is used to approve and justify these new 'solutions'.[65]

Uncovering integration labour thus requires the critical concept of a 'logic of capital' to be central to any analysis of the processes of automation. Such processes should not solely be understood as pure economic competition between capitalists but also as a political strategy, which has implications for how to respond to such movements. This logic means that saving labour time is not the only factor that drives automation. On a political level, the ideology of automation acts on a pattern in which precisely the sequencing and separation of work serves its own purpose. Thus, as a long-term political project, it may be more important to achieve precarious conditions in the labour market in order to create an 'infinite pool of workers', than to maximize profits at any given moment. However, those who do manual work pay the cost of the extra work, as in the case of requester platforms. Swedish postal workers reported on the same occurrence when they had to compensate for the limited functionality of an automated mail sorting device, introduced by the management, which in the end extended delivery working hours. The labouring body 'became an extension of the automation', also causing damages to the workers' fingers, which now had to grip more separate bundles of mail.[66]

---

64 See Irani 2015.
65 See Collier et al. 2017.
66 Nyberg 2020, p. 16.

The most salient features of the relation between management, labour, and automation have maybe never been made as clear as in David Noble's classic study on social choice in machine design.[67] The study of automating machine tools is interesting not least because machine tools are the means by which all machines, including the machines themselves, are made. Noble demonstrates how the technology of production is twice determined by the social relations of production: first by being chosen and designed by the social group in power, and secondly by the realities of the shop floor struggles between classes. Already during the 1940s, the automation of machine tools was accomplished on the shop floor by recording the motion of the machinist on magnetic tape. This allowed for 'record-playback', meaning identical parts could be made automatically by playing back the tape. But record-playback was only a means for obtaining repeatability; it still relied on the skill of the machinist. Hence the alluring prospect of full automation to circumvent completely the individual worker as the source of intelligence behind the production. The transition took more than two decades, required significant amounts of military spending, and caused a 'dramatic transfer' of planning and control from the shop-floor to the office.[68] Due to initially low interest, the state, through the university and the military, mounted a large campaign to interest machine tool companies in the new Numerical Control ('N/C') automation, but the cost was so prohibitively high that it was not until lobbyists got the federal Air Force budget to specifically include N/C that production took off. This created a market for the new software, changing both the horizontal and vertical relations of production.[69] Smaller companies were out-competed, machinists stripped of their pay-grade and methods like record-playback, based on worker-skills, were abandoned. Clearly, the 'defects' of record-playback were conceptual and political, not technical.[70]

Noble's study shows that the relationship between cause and effect is never automatic but mediated by a complex process 'whose outcome depends, in the last analysis, upon the relative strengths of the parties involved'.[71] An important point here is that the software system that became the new industrial standard had been designed specifically with the Air Force in mind. The new technology

---

67   See Noble 1978. A machine tool can be a lathe or a milling machine.
68   See Noble 1978.
69   Horizontal relations of production meaning relations between firms, and the vertical relations of production meaning relations between capital and labour.
70   See Noble 1978. The military spent at least $62 million during the first years alone. In comparison, in 1950 the state budget of California was $1 billion.
71   See Noble 1978.

served the aircraft industry well at the expense of less endowed competitors. Aside from being emblematic for a capitalist course of action, this shows the difficulty in developing general purpose technologies in order to solve general problems, thinking they will have the same effects across a larger system.

That the relation between automation and temporality is not teleological but dialectical is particularly clear when viewed through the prism of the household. While the bulk of the debate on robotics and automation centres on industrial labour, advances in automation have also transformed domestic work, albeit far from unambiguously as regards timesaving.[72] Automated household equipment, now also including various forms of ICT, is commonly described as technology that freed women from housework and allowed them to enter the labour force. Nonetheless, women's workload has not decreased.[73] Rather, the penetration of domestic appliances in households has been accompanied by the social construction of the housewife worker, i.e. the expectation of women to perform double work.[74] Women's resistance to this double labour regime, and the continuous absence of men in the domestic sphere, has been met by capital by introducing more automation into the household.[75] In her study *More Work For Mother: The Ironies Of Household Technology From The Open Hearth To The Microwave*, Ruth Cowan thoroughly dismantles the ideology of automatic household appliances and shows it to be situated within a conservative discourse that confines women to the domestic sphere.[76] Commonly, automated household technologies – not being originally developed for the household but descendants of military technology – attempt to rationalize the single-family household along the lines of industrial production and cause a 'completely irrational use of technology', presenting obvious time and labour-shifting properties.[77]

The failure of household automation to save time can by and large be explained by a failure to recognize how social norms change when technology changes, and their consequential failure to adapt to how a collective unit organizes.[78] Returning again to the argument that rising productivity at some points in production does not equate saved time, it is important to acknowledge that technosocial assemblages also change norms that need not be strictly tied to

---

72 See Fortunati 2018. See also Wajcman 2015.
73 Wajcman, 2015, pp. 111–35.
74 See Fortunati 2018. See also Bittman et al. 2004.
75 See Fortunati 2018.
76 Cowan 1985.
77 Wajcman 2015, p. 115. Here citing R.S. Cowan.
78 See Fortunati 2018. See also Wajcman, 2015, pp. 111–35.

capitalism. Technologies such as the robot vacuum cleaner, for instance, or even installing showers instead of only using bathtubs, change social norms for cleanliness, and the manual cleaning work required to maintain the new norm must follow.[79]

Evaluating the links between technological innovation and time saving is a complex matter. Instead of discussing automation in terms of speed, Judy Wajcman examines how automation in households alters the dynamics and the distribution of time. While offering flexibility for the family to do things at different times – such as the microwave oven offering the family individual settings for dinner – the functionality also enters a feedback loop, whereby it increases problems of achieving collective routines and of scheduling for the family as a whole. This *temporal disorganization* then must be solved, causing more work for the family organizer, usually women.[80] As observed by Leopoldina Fortunati, digital technology, as opposed to household appliance technologies, largely is designed explicitly for the domestic sphere. In a second step, it is exported to the industrial sector.[81] This means that the learning and updating of the knowledge regarding new, digital automation is largely produced at home. Capital heavily relies on such unpaid labour carried out in the domestic sphere and guised as reproductive labour, buying into the notion that it is not contributing to the economy. The so-called automated cashier machines are a telling example, where the low-paid work formerly conducted by cash registers is now replaced with the unpaid labour of women, as they do most of the grocery shopping. It is no coincidence that there is a prevalent feminist critique of techno-utopian thinking.[82]

In relation to previous examples, the microwave oven is particularly interesting in terms of integration of labour and the discussion of inside versus outside automated systems. Being mainly used as a defrosting device, the microwave oven depends on complex food supply chains of unbroken sub-zero temperature, often involving cheap migrant labour and an international workforce invisible to the purchaser.[83] Since frozen food must be transported quickly, long journeys by foot, bicycle, or public transport are made impossible, which

---

79 Wajcman 2015.
80 Wajcman 2015, p. 121.
81 See Fortunati 2018.
82 One of the most prominent critics is Silvia Federici, others include Ariel Salleh, Kate Soper, Astra Taylor and Meredith Whittaker to mention a few. A feminist critique of techno-utopian thinking can also be traced to the absence of the theme technology-as-saviour in the works of visionary feminists in non-fiction as well as fiction, such as Nancy Fraser, Octavia Butler and Marge Piercy.
83 Wajcman 2015, p. 124.

is why grocery shopping for a household reliant on microwave-heated meals reinforces car dependency. Climate impacts from fossil fuels and freezing as well as service time and maintenance labour for cars thus become interrelated to techno-social practices for food preparation.

The promise of automated technologies to help people cope with time shortage is thus not as easily kept as it appears at first sight. Unmasking labour costs in absolute time instead of money discloses the ideology of automation; without the tinkering with time, tediousness, speed and stagnation, capital could not maintain the notion of automation as efficient. It is in this manner, through a faulty perception of efficiency in relation to human time, that automation makes its claim to fame. Fortunati, however, concludes that the domestic sphere also holds great anti-capitalist potential. Since households have become one of the main arenas for capital to launch all kinds of automation technologies, it is also in and around the household that she finds most anti-capitalist strategies. 'It should be evident', she writes, 'that the mother of all battles regarding robotics will not be in the factories and advanced services and professions but in the domestic sphere'.[84]

3 Automating Automation

Should it not be possible to harness the productive force of automation to create wealth with minimal amounts of human labour under a different social order? What is stopping us from automating automation? An obvious first objection is that it would require something essentially different from what is now called automation, which, as highlighted in this chapter, should scarcely be called thus. What is now called automation clearly assumes capitalist class relations. To envision a 'leisure revolution' by incorporating non-existent technologies, whilst believing that building this new social structure would not in its own right require enormous work – of unknown character – is naive at best. Calling the sum of what is unknown 'AI' merely reveals magical thinking. Secondly, hypothesizing about automated automation inevitably closes in on the limits of technological answers to social matters. The notion that everything is accelerating in modern society is not self-evident. With technological acceleration comes stagnation. The golden rule of mechanics – essentially stating that whatever you gain in displacement you lose in power – echoes Harvey's reasoning on merely moving problems around as well as Wajcman's

---

84   See Fortunati 2018.

findings of automation's foremost time-shifting effects.[85] It especially relates to Hornborg's reasoning around technological development as a zero-sum game, upheld by unequal flows of real resources, where some in the periphery do the heavy lifting while the load in the centre seemingly lifts itself. He refers to this belief as machine fetishism, and stresses that while energy cannot be conjured out of thin air, it can be exploited across geographical distances.[86]

For anyone involved in the production of software, the relation between automation and the necessity of real resources is strikingly apparent already at a technical level. The paradox of increasing size and decreasing availability in computer systems is a circle that is hard to square. The larger a system, the less likely it is to be stable, or phrased differently, an increase in speed also increases the potential for gridlocks.[87] Computers in the 1950s required manual repair work that accounted for almost half of the equipment's lifetime. Achieving today's high-availability systems, with at least 99.999 percent up-time, requires that the downtime per year cannot exceed 5.26 minutes. This is not trivial. Approaching the problem of complexity versus fragility involves a range of fault-tolerant design concepts, all essentially a matter of redundancy – of either time, space, or both. Methods to avoid hardware faults include duplexed or triple-module redundancy, that is to say using two or three times as much hardware to keep a super module operating in the event one internal module fails. Design faults are avoided using for example N-version programming, meaning a number of groups of engineers are given the same task to produce code independently of each other.[88] Collaborating between teams to shorten development time is here the opposite of the objective. This single computer view of fault tolerance pales in comparison to faults in multi-computer networks caused by environmental factors such as fire, flood, sabotage, earthquake or power failure, which require mirroring whole systems, using different communications and power grids.[89] Ideally, the systems would have different designs for additional protection against errors. Automation in larger systems thus requires doubling of hardware, development time, testing and operational training, and more maintenance labour. A mode of production relying on automation will be faced with these matters of redundancy. It is not unlikely that

---

85 Harvey 2015, pp. 4–6. Also Wajcman 2015.
86 Hornborg 2015.
87 Virilio 1986.
88 'A and B' code is for example used in trains. The programs are executed simultaneously, evaluating how the train should behave. Disagreements between A and B can lead to an emergency break.
89 See Gray and Siewiorek 1991.

technology created for other purposes than the needs of capital would require even more work, as it would continuously and receptively have to adapt to changing human and environmental needs.

Returning to Adolf Caspary, the position of technological development as a zero-sum game is expressed through a discussion on what justifies the production of machinery. What Caspary points to is the relation between the time-consuming production of machinery itself and its ability to produce goods. For the machine to be profitable it must create a higher demand, because if the demand remained the same, the production of the machine itself would take more time than the non-mechanical production of goods, and thus the production of the machine is not motivated in terms of time-savings.[90] In this view, a machine is not like a tool that satisfies pre-existing needs, but an apparatus created to accumulate capital, that is, to create needs. Machinery, then, is not only the means of production, but the physical motor of the specifically capitalist mode of production of surplus value.[91] Thus, it is the derivative properties of automation – the rate of change of production, not rate of production – that is automation's driving force. If the demand remained the same, the need for automation would dissolve, since the real costs of development, integration, and maintenance would exceed the benefits. Scrutinizing the 'dark factory' as an expression for capitalistic production, automation palpably appears as a method for making profit, not to lighten the load of labour. Production in the dark is only cost efficient for large production volumes of the exact same goods over a long period of time, for the sake of gaining market share. As argued by Caspary, this can only be fulfilled for consumer goods when the goods are produced for the market and not to fulfil needs.[92] That is to say, when the individual is dependent on the market and not the other way around. For Caspary, argues Mårten Björk, the general tendency of capitalist machinery is to –

> ... relocate the saving of labour time in the production process where machines have been introduced to another part, for example to the miners who gather the materials needed for the production of the machines. Machines must be produced and reproduced by workers and this implies work, and thereby machines not only posit the possibility of an economy based on surplus labour. They also reproduce the necessity of the labour that reproduces the machines that, in some production

---

90   See Björk 2020.
91   See Björk 2020.
92   See Caspary 1927.

processes makes labour superfluous, but which as instruments for capital reproduces an economy based on surplus labour.[93]

Thus, Björk continues, the relation between 'justice and machine' for Caspary is utopian, and therefore cannot liberate humanity from unnecessary work.[94] By this, Caspary repudiates the 'fettering thesis'. Based on his readings of Marx, Caspary contends that capitalism is a process of industrialization stratifying humanity into divergent and antagonistic classes, rendering automation as practices specifically tied to capitalist property relations.[95] In *Grundrisse*, Marx argues that the specific way of industrial capitalism to produce goods by introducing machines into production processes enters a kind of self-oscillating mode, in which an increasingly large part of the production time must be spent on producing the means of production.[96] This is only possible when a certain degree of overabundance has already been reached, or, as Marx writes, '*surplus population* (from this standpoint), as well a *surplus production*, is a condition for this'.[97] Thus, 'the employment of machinery itself historically presupposes ... superfluous hands'.[98] And so, the machine according to Caspary emerges because of the 'free disposal' of an 'army of free workers'.[99] Applying this perspective on automation as an infrastructure to produce surplus labour as well as creating the need for work in other and new industries accentuates the contradictory argument made on software automation as being overall efficient. For example, that modern production processes now rely on complex labour chains that include digital machines has equally brought forth a global workforce of maintenance workers in the cyber security sector, a sector now predicted to exceed one trillion U.S. dollars in spending in 2021.[100]

One does not need to go as far as Caspary does in fencing off automation from time saving to see how a Casparian/Marxian perusal of the examples in this chapter elucidates the recursive relationship between automation and surplus labour. Across the field of development, integration, and maintenance, the creation of new forms of labour as well as the production of surplus labour appear as an all-pervading leitmotif. Moreover, automation's inability to reduce

---

93   Björk 2020, p. 349.
94   Björk 2020, p. 349.
95   See Björk 2020.
96   Marx 1993, The Chapter on Capital.
97   See Marx 1993.
98   See Marx 1993.
99   See Björk 2020.
100 Morgan 2019.

overall labour time does not stand out as a bug but as a feature. The concealment processes necessitated to exclude this work from the idea of automation take on different forms, exploiting and reinforcing class and gender relations and racialized hierarchies within the global labour force, as well as reproducing gendered relations within the domestic sphere.[101] The common theme is not a reduction of total labour time, but a relocation of labour time to other parts of the production process by tapping into these inequalities. Relocation is made from centre to periphery, where the periphery is not only found in countries with below-average GDP, often in the Global South, but also as embedded inequalities in rich urban environments. On a global scale, software-driven automation requires cheap manual 'offset' labour (performed by for example informal mine workers), manual production labour (performed by immigrant manufacturing workers) and a diversity of nestled micro tasks (performed by an 'infinite' pool of micro labourers and by unpaid women), all working under different detrimental conditions. This work is then simply not accounted for as labour that is necessary to uphold the present level of automation. Further, integrating automated processes into daily life demands of us new forms of integration labour that must be considered necessary in making technology work within a techno-social context. On top of this, automation has a tendency to appropriate subsequent time, its inbuilt condition of speed combined with yesterday's data carrying the structures of the past into the future – be it prejudices of race, class, or gender hierarchies. Such processes, then, are rigid, while the humans surrounding it must submit to its inflexibilities. To find a truer measure of equality, an unknown amount of 'shadow work' completely left out of the picture must therefore be added to any calculus of automation efficiency. Those views are conspicuous by their absence in the present forms of automation discourse, which do not originate from a working collective with agency but focus on individuals as consumers.

Considering this, it is clear that focusing on automation's time saving properties as a means to achieve equality risks reproducing the same capitalist class structures it was alleged to upheave. William Morris once said that every utopia betrays the personality of its author, perhaps giving us a clue as to how to relate critically to current liberal hegemony and techno-utopian narratives. Linking an increased level of automation to a post-capitalist future free from work, as in the recent rise of some socialist visions of automation, particularly stands out as a utopian position disconnected from the deeply unequal, labour-intensive,

---

101   As well as racialized hierarchies in the domestic sphere, since childcare (nannies), home cleaning services and so on, is often based on immigrant labour. This falls outside the scope of this chapter but is well worth noting.

and even labour-producing conditions upon which automation is based. To conclude, none of the arguments put forward in this essay intend to oppose automation as playing a possible role in production, but to question the notion of automation itself as liberating. Even if some automated processes can save time, focusing on this distracts from the proper goal of reconstituting technosocial relations. Treating everyone's time as equally important – which in the proper sense of the word would be radical – instead harks back questions about how work should be organized and how it can be sustainably, equally, and fairly distributed. The focus then, must be (e)quality, not quantity, of labour.

CHAPTER 4

Capitalism without Workers: On the Impossibility of Automation and Its Relation to the Question of Value

Benjamin Ferschli

1   Work in the Age of Its Supposed Automatic Reproduction

We truly live in interesting times, but as the saying goes, this may be more of a curse than a blessing. For some years now, the world has seemed haunted by a spectre of revolution in the forces of production with, often contradictory, publications having become asymptotically infinite. This *'automation discourse'*, which we are currently in the 'late stage' of according to Benanav is fractured along the lines of –
1. proclamations of new 'hyphen'-capitalisms.
2. (Supposedly) revolutionary technologies.
3. National economic projects framed as technological paradigms.[1]

Building on this abundance and arguably chaos of engagement, more strategically oriented political proposals have emerged, particularly in the United Kingdom, where notions of 'Post-Work' and 'Post-Capitalism', often through the mechanic of an unconditional basic income (UBI), have even come to feature in some degree in electoral political programs.[2] The political formula for these forays, primarily on the left but not exclusively, is summarized by Dyer-Witheford, Kjøsen and Steinhoff as the rallying call of: 'AI+UBI,' demanding the full automation of production and the severance of the wage relation.[3] It is for these two reasons, the ordering of proclamations of new capitalisms through new technologies as well as their suggested overcoming through concrete political proposals, that we must (once again) consider the question of automation and value in greater detail. A clear understanding of the theoretical functions and

---

[1] Digital-, Platform-, Surveillance-, Extractive-, Gig- etc.; advanced robotics and AI ('Machine Learning'); Industry 4.0; AI made in Germany; China's 'New Generation Artificial Intelligence Development Plan'; Benanav 2020, p. 7.
[2] Srnicek and Williams 2015; Bastani 2019; Mason 2016; Cruddas and Pitts 2020; Pitts and Dinerstein 2017.
[3] Dyer-Witheford, Kjøsen and Steinhoff 2019, p. 6.

contradictions of automation is often eschewed in exclamations of imminent and revolutionary breaks within or beyond the capitalist mode of production. The goal of this paper is to provide the basic structure for such a viewpoint based on Marxian considerations.[4]

'Automation' connects to Marxian debates on at least two obvious sides: *value* and *competition*. In this nexus, *value* certainly takes the fundamental role. All terms, propositions and theories of automation must offer answers to the questions how, why, and where value is produced, or not, and particularly why this may or may not be changing.[5] This 'Fundamental Question of Value', as it may be called, has in some form recently begun to '(re-) surface'.[6] The intimately connected, yet distinct, side of 'competition', which I take to contain accumulation and crisis, has been an equally prominent issue of debate in understanding contemporary automation projects.[7] In order to specify the interdependence of these two sides, I propose an 'inversion' of the view of automation to the perspective of capital in this contribution. Rather than considering automation as a lever in the further dispossession of workers or for the easing of their toil,[8] I suggest focussing on capital's fundamentally contradictory desire for the radical expulsion of living labour from production, while crucially relying on it. This view is useful because understanding this contradictory character of automation for capital, also allows for an understanding or 'unification' of the contradictory assurances contained within the 'automation discourse':

1. Robots will do more and more work, but employment will remain high. Work will become less toilsome, yet more 'interesting' (intensified) and workers need to be up skilled for this.
2. Automation doesn't eliminate jobs, but also it does, however only the 'bad' ones, but also, we are progressing to a fully automated society without work in general but there are still a lot of jobs to do.
3. Wages will increase for everyone, but if they go any higher competition will force our hand on automation and lay-offs will become inevitable, so maybe it is better to keep wages down.
4. Everyone benefits, yet automation is essential for competitiveness and maintaining profits, and thus primarily the interests of capital.

---

4 'Marxian' here meaning thought relying on or in reference to Marx, as opposed to the absolutisations in historical political programs which may in this sense be thought of as 'Marxist'.
5 Specifically, this often means answering the question: what is the differentia specifica of '*digital* capitalism'?
6 Pitts 2021; Briken et al. 2021; Saad-Filho 2020. As *the* central category of Marxian political economy, value was, of course, never 'gone'. Publications dedicated to the analysis of value appear to be on the rise once more, however.
7 See Benanav 2020.
8 As in 'Post-Capitalism' and 'Post-Work'.

It is not surprising that contradictory assurances follow from contradictory ambitions. There is thus an internal split of automation between being an abstract fantasy as well as concrete current and historical project. This split must be constantly navigated by capital and may be summarised as the project of 'Capitalism without Workers'. The contradictory nature of the above assurances contained within that project, even where they may formally not be maintained as 'contradictions proper', thus lies in automation's internally fractured nature. This does not mean, however, that arguments could not be found (and are constantly being found) for why these propositions are in fact not mutually exclusive. As is evident with the last assurance at the very least, however, a fundamental opposition between capital and labour cannot be relegated to a mere 'it depends on the circumstance' or 'everyone will benefit'. Importantly, these contradictory propositions are also unified in that they abstract and distract from the work that is incessantly being done, and continues to be done, under the capitalist mode of production. The vision of 'Capitalism without Workers' and the automation projects at its heart, thus obscures just this, the continuing and unbroken subordination of most of the world to wage-labour. Correspondingly, diagnoses in the economic mainstream as well as Marxian literature that production has lost its material base in favour of the production of knowledge, are in this sense at best western-centric, and even then, hardly to the point. Material production subsumes and relies on labour to an unprecedented extent globally.[9] It is here, however, that inversing our view of projects of automation to that of capital becomes fruitful again, as it precisely sheds light on this 'hidden' abode of production, the work performed there, and its systemic centrality. It is the focus on the logics of capital which reveals the centrality of workers. More than anything, therefore, the arguments set out in this paper, aim to give a reason for why capital's desire for the expulsion of living labour from production remains unfulfilled even in the sincerest automation projects, and despite continuously changing configurations of global labour.

The point of this contribution thus lies in thinking through the relation of automation to value, via the fantasy of 'Capitalism without Workers'. It thereby provides arguments for understanding limitations of contributions and political strategies in the present automation discourse, which do *not* consider automation's contradictory nature. To address these points, I first explore the historicity of the contradictory project of capital's expulsion of living labour. Second, I argue that the technological fetishism of hegemonic understandings of auto-

---

9 Pfeiffer 2021.

mation in neoclassical economics must be traced to its tautological discussion of utility. Finally, in rehearsing the basic architecture of Marxian thought on automation and value I present a classical account on the ultimate impossibility of the expulsion of living labour, or 'Capitalism without Workers', and thus suggest concrete limits to automation. These three arguments may be used as points of reference for the critique and dismissal of many contributions in the present debate on automation as well as their political conclusions, hopefully providing a basis for alternative proposals seeking to '(de)-automate' the future, and to render it a little less 'interesting'.

2 Capitalism Without Workers: History and Fetishisation of Machinery in the Production of Value

To begin, I would like to advance two arguments for two reasons. First, the relevance of the vision of 'Capitalism without Workers' by illustrating its contradictory and cyclical permanence in the long history of automation. Secondly, to motivate its engagement through Marxian literature, by illustrating its insufficient engagement in mainstream economic theory due to its fetishisation of technology.

3 Permanently Cyclical Resurgence of Automation as Societal Boundary

> *'Taking its periodicity into account, automation theory may be described as a spontaneous discourse of capitalist societies that, for a mixture of structural and contingent reasons, reappears in those societies time and again as a way of thinking through their limits.'*[10]

∴

Despite a truly extensive matrix of visions of where the automation of work *will* lead, at this junction, the world still spins around the sun, that is, around living labour.[11] In this persistence we can see a disconnect between the vis-

---

10 Benanav 2020, p. 8.
11 Prominent visions range from Neo-Feudalism (Dean, 2020; Ford, 2015); machine-led hyper

ions of automation and the future of work that are currently being put forth and work's drab continuing mundanity. The fantasy of automated work or the 'factory devoid of humans' itself, reaches far back, of course, as do its disappointments.[12] Aristotle's dreams of self-weaving looms and Rabbi Löw's kabbalistic Golem are echoed in Karel Čapek's 1921 play R.U.R., introducing the term 'robot' to the modern social imaginary.[13] Automation's decidedly economic-theoretical treatment emerged with the industrial revolution, or rather the inception of classical Political Economy itself in the works of Smith, Ricardo and Marx. An equally old socio-political relevance is of course illustrated by movements such as the Luddites forming an unbroken chain with more recent fears of 'technological unemployment' particularly in the 1930s and 60s as well as our present debates.[14] Automation and its debate thereby appear as cyclical invariance in the historical development of capitalism, grafting onto ancient fears and fantasies of worklessness. Most importantly, however, automation has always represented an essential part of the arsenal of capital in its desire for independence from, or at least control of, living labour. It has thus served as both, concrete weapon in competition with other capitals through the expropriation of worker-knowledge and the reduction of (labour)-costs, but also to threaten workers into submission without being in fact implemented.[15] If the worldviews of complete control of arch-industrialists Ure and Babbage, which live on in not just a few industrial engineers, speak to anything, then to capital's fundamental desire for the expulsion of antagonistic living labour from an otherwise harmonious sphere of production (a vision of 'Capitalism without Workers'). This expulsion of Cain from paradise has, of course, never fully taken place and thus the crucial element lies in correctly judging the relative historical importance of this contradictory double nature: when does automation appear as threatening fantasy, and when as threatening concrete project. As

---

capitalist dystopia (Dyer-Witheford, Kjøsen, and Steinhoff 2019); to the end of capitalism itself (Wark, 2019); nothing new at all at (Wajcman 2017; Taylor, 2019) or substantial societal reform (Srnicek and Williams, 2018).

12   Uhl 2019.
13   Taylor 2018; "R.U.R" had its premiere almost exactly one hundred years ago. The word 'robot', having its origin in Čapek's work, derives from the Czech word for feudal compulsory labour.
14   This is of course also the meaning of Grey and Suri's paradox of automations 'last mile' (2019): true automation is always just around the corner and that 'this time' it is truly different. On the Luddites see Hobsbawm 1952.
15   Marx: *'It would be possible to write a whole history of the inventions made since 1830 for the sole purpose of supplying capital with weapons against the revolts of the working-class.'* Marx 2013, p. 300.

Dyer-Witheford, Kjøsen, and Steinhoff point out in critique of Taylor's 'fauxtomation', of merit in its own right, automation cannot entirely be reduced to the first or relegated to a charade.¹⁶ While the laws of motion of capital may often have been lofty in words, they have just as often been hard as steel in deeds – and vice versa. Hence, while it is obvious that previous visions of automation may have been defeated *vis-à-vis* their ultimate outcome, they have equally succeeded in different degrees in concrete cases: as the horse superseded human muscle, so did the machine the horse, all with very real consequences for different types of work and categories of workers. The question remains, however, why despite all such material strides, capital's independence from living labour has so far always been thwarted. A reason for this is explored below. What can be maintained here is the long and cyclical history of fantasies and projects of 'Capitalism without Workers'.

4 Value, Utility and Automation

> *'Utility is the quality in commodities that makes individuals want to buy them, and the fact that individuals want to buy commodities shows that they have utility.'*¹⁷

∴

As a second point, I want to motivate the engagement with automation based on Marxian literature, by arguing that most contributions in the automation discourse are unhelpful, since they rely on a fetishization of technology. This encompasses the twin certainties of: 'robots *will* take your job', but to believe that this will lead to unemployment is a 'Luddite fallacy'.¹⁸ I argue that it was the turn to utility as placeholder of value, during the marginalist revolution that has necessitated the fetishisation of technology and thus automation, contained within these dominant views.

The ambitions of the seemingly similar labour theories of value offered by Smith, Ricardo, and Marx were, of course, not congruent. While Smith and Ricardo mostly viewed their labour theories of value as a foundation for prices,

---

16   Dyer-Witheford, Kjøsen, and Steinhoff 2019.
17   Robinson 1962, p. 48.
18   As Kevin Drum 2017 asserts: 'You *will* lose your job to a robot'; emphasis added.

Marx sought to delineate the conditions of possibility of the capitalist mode of production itself. This fundamental difference is also reflected in their respective theories of industrial development, the division of labour and the use of technology therein. For example: the early British arch-entrepreneurs Babbage and Ure, opposed the propositions of Smith, that division of labour is solely a technical question of efficiency, which is still an assumption in modern economics. For them it was obvious that a further division of labour meant the *breaking down* of existing skills, rather than their increase as Smith argued. The overall function of automation was also clear for Ure, namely cheapening labour. He even extrapolated that the entire tendency of manufacturing was to constantly cheapen the cost of labour, either by substituting artisans with workers, men with women and children, or everyone with machinery. The points of Ure and Babbage were directly received in Marx (and later escalated by others such as Braverman in his degradation thesis).

From the very beginning of Political Economy therefore a schism can be identified between Smithian and Marxian forms of thought, arguably in terms of an opposition between abstract categories and real abstractions.[19] Smith's views still constitute the central axis in present mainstream economic thought, even if parts, such as his labour theory of value, have been dismissed. During the 'marginalist revolution' and subsequent dominance of neoclassical economics, 'value' was replaced with 'utility' as the basis of economic theory. This introduction of 'utility' as an abstract placeholder of economic value in combination with Leibniz's infinitesimal calculation, and given assumptions of maximization, allowed neoclassical economic theory to derive 'stringent' formal models still representing the core of economics.[20] The primary purpose of this development was to explain that which it itself was meant to replace: 'value' and more specifically prices and their movement in a market. However,

---

19 Most notably defined of course in the work of Alfred Sohn-Rethel and meaning that the central abstractions of the capitalist mode of production are not mere mental exercises or formal categories but concretely enacted and constitutive of reality behind appearance. It is in the acts of exchange, production or the application and development of the natural sciences that abstractions are necessitated and assume this reality: 'they know not what they do but they do it'. Centrally a critique of ideology would attempt to uncover precisely what is in fact being done, rather than what is merely thought that is being done based on how matters seem to appear.

20 See Ferschli and Kapeller 2019; How fundamental utility is, is particularly evident in microeconomics, where analyses of production and consumption only differ semantically. In this sense indifference curves (consumption) and isoquants (production) are direct mirrors of each other as well as their 'utility' basis, which from the perspective of the firm is of course 'profit'.

'utility' or 'preference', as the value of pleasure derived from consumption, are naturally not directly observable. The solution of neoclassical economic scholasticism was to refer to 'indirect observational mechanisms' or 'revealed preference', meaning that the prices, which are observed in a market, are the result of 'given preferences' underlying exchange.[21] Hence, while attempting to explain prices, the solution was to claim that prices are what they are because of an unobserved 'unmoved mover' of utility. The consequence being, that utility equals the price a consumer is willing to pay for a commodity and vice versa. Price thus becomes the 'objective' value in a 'subjective' value theory: one pays as much as it is valuable to oneself, and it is as valuable as much as one pays. This represents the opposite interest of Marxian value theory, of course, which arguably seeks to constitute an 'objective' value theory, exceeding the mere appearance of subjective, albeit socially constituted, (use) values. Utility theory is easily identifiable as tautological, since ex post any behaviour can be integrated into the theory based on unobserved preference and utility.[22] Every possible price can thus be argued to derive from an unobserved underlying preference structure. As Ferschli and Kapeller show, the law of demand and utility theory are kindred animals which have the same problem: if the structure of preferences or their stability is not known, counterfactuals can always be dismissed by recourse to changes in preferences. While utility can therefore explain every behaviour, it does not *understand* anything (it is void of informational value).

How does all this relate to a specific understanding of technology? As utility functions as an unseen mover of prices, so does technology in the case of production. Excluding considerations of value and how it is constituted in production, only leaves commodities with prices, the costs of production of which are determined by technology. This is evident in standard economic modelling where technological change is assumed to be an exogenous shock which serves as the most important driving force of economic development.[23] It was in this sense the turn of utility in the replacement of value which produced this need for exogenous shocks to market movements, or put differently, it is the lack of

---

21   Samuelson 1948; A meaningful distinction between preference and utility cannot be made. Utility designates economic value resulting out of the consumption of a commodity and preference the ordering of commodities on the basis of their largest subjective utility. The largest utility thus always is the highest preference, and since the highest utility must always be pursued, the ordering of preference must equate the ordering of utility and vice versa.

22   Ferschli and Kapeller 2019.

23   See the Solow (1956) model for economic growth, the pivotal role of technology growth therein, and the models derived from it.

non-tautological discussion of economic value which necessitates the fetishisation of technology as *vis maior*.[24] Following Bastani, it is thereby also not surprising that neoclassical economic theory comes into contradiction with its own assumptions when confronting actual technological change such as in the breakdown of the relation of price to marginal cost with the physical 'reproduction' of software incurring almost 0 cost.[25]

5 The Fetishisation of Technology

Pointing to the fetishistic treatment of technology in economic theory is no novelty of course.[26] Such fetishism, constitutive of many contributions in the present 'automation discourse', ultimately represents a 'sub-form' of capital-

---

24 At this point it must be revealed that the charge of tautology has of course also been levelled against Marxian value theory, arguably most famously by Böhm-Bawerk and Joan Robinson. While it is not possible to detail the many discussions on this, or the question whether it even matters for the functional analysis of capitalism, some thoughts are in order. Formally, a tautology is constituted in that an expression is always true even if parts of this expression are not, put differently, a necessary conclusion is already contained in one or more parts of an argument or: it is always true. Drawing on Robinson's tautological equivalence of utility theory in the introduction of this chapter, we may reformulate her point: labour is the quality in commodities which gives them value, and the fact that they have value means that they are based on labour. This certainly seems like it would be always true. It does, however, neglect the duality of value (use and exchange) contained within commodities and thus prices. (Abstract) labour may be what allows exchange value and the exchange of commodities, but it is precisely not the exchange of commodities which gives them use-value or exchange value which gives them abstract labour. More importantly, however, even if we were to assert that the labour theory of value is in fact logically tautological, there would nonetheless be a crucial difference to utility theory: it is not historically tautological. It always operates under radical historical contingency, and it is thus precisely not always true, but only ever under the historical delimiters of the capitalist mode of production (the production of use-values through labour as panhistorical constant is a different question). It is thus in the movement between modes of production that means value is not always and necessarily based on abstract labour and in this sense not necessarily tautological. No matter how tautological the relation of labour to value may thus be identified under the capitalist mode of production, its effect is clearly demarcated in its historicity. Such a barrier is missing from marginal utility theory, explaining its imperialistic use in all social spheres and disciplines. This being said, the question whether Marx's labour theory of value is tautological and whether it truly matters for the relevance of its conclusions, is a fundamentally different conversation than that on the tautology of marginalist utility theory. More than anything, this is evident in their respective treatment of technology, returning us to the purpose of this paper.
25 Bastani 2018, p. 64.
26 Harvey 2003 and Hornborg 2013.

fetish. I follow Butollo and Nuss in locating such fetishisation where the physical appearance of social relations is 'absolutised', barring an understanding of these relations. For example, where the necessity of UBI is derived out of automation, or economic development depends on exogenous technology shocks.[27] The charge most often levelled in critical engagements with the automation discourse, however, is rather 'technological determinism' than a critique of fetish, prompting the question how they interrelate. Indeed, a large part of the debate on automation can in fact be judged as technology deterministic, as it views the process of automation implicitly or explicitly as unproblematic, linear, homogenous, and exogenous.[28] This does not necessarily mean the same as technological fetishism, however. Discussions and critiques of vulgar materialisms certainly have a long tradition in Marxian debates, leading at points to a decided focus on class as force in the shaping of technology.[29] In turn, some have submitted to technological determinism based on some form of materialism and determinism rejecting any constructivist claims.[30] It would, however, hardly seem likely that such proponents would also knowingly submit to a fetish of technology. Even in Lenin's 'soviets plus electrification' formulation, the deciding element are the soviets. Whereas the critique of technological determinism arguably proposes to consider social factors in shaping automation, a critique of the technology-fetish must point to neglected contradictions, chief of which is of course the persistence of work outlined in the introduction. While the concrete relation of technological determinism to the technology fetishism must be left open at this point, we can maintain that what is important is that wherever an 'absolutisation' of the physical appearance of social relations forbids an understanding of these relations (How does capital produce profit? How do robots produce value?) we are in the realm of fetish.[31] A concrete example of this is given by Butollo and Nuss who cite Brynjolfsson and McAfee, and their view that the polarisation of the labour market in the 1980s was a direct consequence of technological change, rather than, for example, rampant de-regulation.[32] The point being that the consequences of technology are never 'exogenous' as assumed by dominant economic theory and do not represent an 'outside shock' of a *deus ex machina*.

---

27 Butollo and Nuss 2019 and 2022.
28 Judging alone by the exorbitant numbers of citations on the automatability of jobs: Frey and Osborne 2013; Arntz et al. 2016; for a critique: Spencer and Slater 2020.
29 Noble 2011.
30 See the examples provided by Adler and Borys 1989 and Sayers 2007.
31 Butollo and Nuss 2019, p. 12. See also Butollo and Nuss 2022 for the english translation.
32 Brynjolfsson and McAffee 2014, cited in Butollo and Nuss 2019 (2022), p. 11.

I have stressed this point, in the particular context of this volume, because I believe it central to understanding that automation can never be a mere technical question, as assumed by many of the most prominent contributions. As Marcuse writes in his critique of Weber, it is precisely Weber's devoted attempt at 'value-lessness' which reveals him as bourgeois and opens the doors to distorting external valuations in his analysis of industrial capitalism: the 'ought' has revealed itself in the 'is'.[33] This is a fitting charge against much of the automation discourse as well. Following the desire, therefore, to move beyond a fetishised treatment of automation, I suggest a turn to Marx. This will make it possible to further specify the relation of automation to value, the dismissal of which runs parallel to technological fetishisation in neoclassical economic theory, and thus provide an explanation for what may have so consequently disappointed the projects of radical expulsion of living labour from production in the past.

6   The Fundamental Question of Value

> *'The true barrier to capitalist production is capital itself.'*[34]

∴

Marx was famously concerned with industrial machinery and its internal and external logics.[35] Considering his 'three cardinal facts' alone illustrates the centrality he accorded to it in understanding the capitalist mode of production.[36]

- *'Concentration of the means of production in few hands'* relates, among other things, to the process of real subsumption of labour in the factory and its technological apparatus and the competitive drive for technologisation and accumulation.
- *'Organisation of labour into social labour'* summarises the development of an increasing division of labour, based and acting back upon science, and its dependence on the 'general intellect' (although of course not included

---

33 Marcuse 1965, p. 2.
34 Marx 1991, p. 358.
35 Meyer (2019) goes as far as retracing much of the interest of German industrial sociology in technology centrally to Marx's heritage.
36 Marx 1991, from the third part of *Capital* volume 3.

in *Capital* but brought up here due its significance for the development of Marxian currents, in particular post-operaismo).
- '*Creation of the World-market*' surmises the immense productivity increases, based on competitive compulsion, the expansion of capital and its accumulative struggles, ultimately leading to crisis and not just technologically uneven development.

If machinery is so fundamental and constantly revolutionary, however, why then has automation not led to factories devoid of humans, but rather the (partial) de-industrialisation of developed economies, whose super-market self-checkouts are in the same supply chain as 'labour intensive' cobalt mines in the global south? Why has capital not radically emancipated itself from living labour? After all, from a purely technological viewpoint much work presently done could arguably have already been automated some time ago, in particular when 'financial feasibility' and considerations of returns on investment are put aside.[37] A possible answer lies in the building-block readily eschewed by neoclassical economic theory: value.

The central intuition when thinking through the architecture of Marx's thought on *value* and automation is quickly told: given the validity of the labour theory of value, an increasing organic composition of capital, through increasing automation in the form of mechanised fixed capital, in the course of real subsumption and thus the extension of relative surplus value extraction, forced by the competitive need to reduce labour costs and increase control over the labour process, must reduce the mass of extracted living labour power (ceteris paribus), thus surplus value and thereby profits, forcing capitalism into crisis and prompting its potentially final and impossible contradiction. At the heart of this logic of a 'final contradiction' of autonomous machinery thus lies the fundamental question of (surplus) value, and its reliance on living labour. As quickly as the overall story is told, as deep and old are debates on its individual parts. The contradiction at its base remains the same, however: the fetishisation of machinery and the seeming 'equivalence' of capital and labour in the production of value.[38]

---

37   Benanav 2020.
38   The goal of this text is not to defend this position or its parts against well-known charges such as 'the law of the tendency of the rate of profit to fall is obviously wrong' or 'why does capitalism then still persist if it isn't wrong' or also more detailed accounts such as Michael Heinrich's dismissal. Rather, I think it is worthwhile to consider this position in light of its historical significance (for example in the formulations of Mandel and Pollock) and more importantly as perspective on our current fragmentations of globally labour-intensive production. Again, the goal is not a defence of the labour theory of value or the law of value, but rather the attempt for a potential explanation

Caffentzis has specified this contradiction, reinforcing that the automation of production was indeed seen by Marx as 'last metamorphosis of labour' with the visions of replacing living labour with the automatic machine system defeating themselves in the end.[39] In a dialectical twist, however, it is workers protesting their own replacement, unemployment, and worsening working conditions, which provides a central motor for the vision of 'Capitalism without Workers' in the first place. Assuming workers as 'docile factors of production' carrying their own skin to market expecting and *accepting* the tannery, the great need and haste in technological automation would arguably be reduced. As Caffentzis put it, the capitalist class (and in a mirror image also the working class) is thus locked in a permanent contradiction it must navigate: on the one side a moment of necessity of automation as measure against the collective knowledge and action of workers, on the other, the existential gamble of the reduction of the mass of surplus value. In Caffentzis's own words: 'Hence the capitalist class faces a permanent contradiction it must finesse: (a) the desire to eliminate recalcitrant, demanding workers from production, (b) the desire to exploit the largest mass of workers possible'.[40] Or put differently: to reduce the extent of required living labour but drawing on its expense as foundation of value.

It is within this tension that the empirical phenomenon of automation must play out. It is also at the junction of this contradiction that a detailed return to value theory serves as useful basis for understanding our present technological dilemmas, as they have been summarised in the contradictions constituting the notion of 'Capitalism without Workers'. The goal of this section is to rehearse several relevant key parameters of Marxian thought relevant to this notion. Since it is impossible to even begin to sensibly outline the reception of value in Marxian thought at this point, or even the controversies on value and automation, the following text must eschew more detailed debates and controversies in favour of the outlines of an interconnected web of basic considerations.[41]

---

of the partial and specific automation of developed economies as it is embedded in labour-intensive global value-chains. I view this ambition as akin to that of Morris-Suzuki 1984.

39 Caffentzis 2013.
40 Caffentzis 2013, p. 72.
41 As the lines of thought here traced largely represent common knowledge within Marxian debates, despite massive controversies at particular points, I have abstained from directly referencing Marx himself for the most part. It may be useful, however, to mention that in this overview I have turned to chapter 5 of *Capital*, Volume 1, for relative surplus value production, chapter 6 for use value, exchange value, and labour power, chapter 7 for my reading of the labour and valorisation process; chapter 9 for the rate of surplus value,

7 Productive Forces and Their Countervailing Pressures

Beginning with the most general point of departure already provides insights into supposed 'productive revolutions' presently underway. As is well known, the capitalist mode(s) of production are constituted by the *forces of production* (or productive forces consisting of the means of production [equipment, technology, raw materials] and productive capabilities of labour [skill]) and the *relations of production* (the social distribution organizational structure within and between classes).[42] A closer look at the Marx's notion of 'productive forces' is already revealing here, as also suggested by Butollo and Nuss.[43]

First, it must not be forgotten that the development of the productive forces is not an end in itself but serves capital accumulation. This means, in line with the critique of the technology fetishism outlined above, that they are not an exogenous factor of economic organization, but subject to the very contradictions of the capital employing them. This means that all technology must be understood in terms of the uses and strategies of capital, shaping its form and function, relativising conclusions of the automation discourse in which specific constraints of historical and geographical accumulation strategies are not considered.

Recognising these constraints forces a very specific distinction to be made between the constant laws of the motions of capital, allowing the designation of an economy as 'capitalist', and their transfigurations. As Butollo and Nuss put it: 'the methodological mistake of most visions of full automation lies in not thinking through the necessary mediation between abstract technological potential and labour market developments' … Assumed endless possibilities of automation must be contrasted by the 'complexity of processes which have constantly be adapted to environmental conditions'.[44] The example of Butollo and Nuss here is that had the automobile industry been constrained to only producing Ford's Model T, full automation could have already been achieved a long time ago.

A further point to consider is the interplay of the forces with the relations of production, vividly illustrated in the current overwhelming push of representations of interest, consultancies, and private research institutes for the use of robots and AI as the only and inevitable solution for slowing growth and

---

chapter 10 for the absolute and relative surplus value production. And finally, volume 3 chapter 13 for the tendential fall in the rate of profit.
42 Adler 2009.
43 Butollo and Nuss 2019 and 2022.
44 Butollo and Nuss 2019, p. 13; my translation.

stagnating productivity growth rates.[45] More than anything this promise of technological growth is likely to remain 'false', merely representing a facet of competition among capitals (those selling and those buying robots).

Moving further down into the *relations* of production, two more specific sub-relations are contained, which are relevant here: 1. competition among commodity-producers and 2. the subordination of workers to managerial authority in the firm. It is here that the specific question of value fundamentally comes into play.

8   Value and Labour

Why should only living labour be a source of value? After all, would a car built solely by machines not drive just as well? Or a pot not boil water at the same temperature? Deriving the insufficiency of exchange based in the 'utility' and desires of those involved in the exchange of commodities, Marx argues that commodities must take recourse to a commonality contained but not reducible to either one in order to be exchanged. This 'vanishing mediator' of 'abstract labour' allows for the values of commodities to be related, by representing the social form of work under the capitalist mode of production. This commonality can never be escaped without tautology or break with said mode of production. The concrete quantity of value of a commodity is therefore derived from the (socially necessary) labour time required for its production. The dual nature of the labour contained in a commodity (concrete and abstract) and its relation to value, as well as the understanding of the abstract commonality of commodities in exchange, this 'fundamental question of value', is in Marx the lynchpin for the understanding of political economy itself.[46] The goal is not merely a theory of pricing but relating labour and the wealth it has produced through the intermediary of 'value'. Thus, as Tsogas details, all commodities carry the 'social character' of their producing labours and thus the entirety of the 'capitalist social exchange relations'. Commodity fetishism here then means the obscuring of these relations and the commodity as their most basic form.[47] Thus, the 'recognition' (meaning the social reality of abstract labour under capitalism as real abstraction) of the underlying social relations of the commodity, rather than merely the labour contained within it, allows exchange and thus value.

---

45   Butollo and Nuss 2019, p. 17.
46   Lange 2019.
47   Tsogas 2012, p. 380.

The 'mystery' of labour does, of course, not conclude here but merely begin. Digging deeper into the concrete side of labour, the distinction between labour and labour power becomes central. Originating in the ironic double 'freedom' of the worker, free from capital and free from feudal servitude, she is 'free' to present her labour power as commodity to the capitalist. However, it is not a concrete result that is being sold, but the capacity to work during a certain amount of time during the day. This presents the capitalist with the problem and desire of extracting as much concrete labour from the purchased labour power as possible, as the extent of this extraction determines the surplus value, constituting a central part of profit. The origin of surplus value in production is then the transformation of labour power into concrete labour within the confines of labour time in factory and office.

9 Surplus Value and the Law of Value

Expressing the above relationship more formally, surplus value is constituted in the difference of the labour time required for a worker to produce the value of commodities for her own reproduction, and the actual time she spends working. It is in this sense is a 'gift of labour time', or surplus labour time, to the capitalist. The notion of a correlate of the working time (social necessary average) which goes into the production of a commodity and its measure of value is also summarised as the 'law of value'. What is important at this stage is that the capitalist seeks to extract as much surplus value and surplus labour time as possible, as it stands in an intimate, though not congruent, relationship with profit.

We are thus quickly approaching the pivotal dynamics in Marx's thought on the contradictions of automation. Marx has outlined two central strategies in increasing surplus value extraction. 1. Absolute: extending and intensifying the working day in order to increase the absolute mass of surplus value, or 2. reducing the part of the day the worker labours for her own reproduction through technological and organizational improvements of labour productivity, increasing the relative share of surplus value in work. It is the second strategy which is most relevant for our present purposes. By virtue of lacking 'natural limitations', such as exist in the extension of the working day, relative surplus value extraction allows for an escalation to the point of contradiction: on the one hand, if less time is required for the worker to reproduce herself with her work, and the part of the working day where the worker works for the capitalist 'for free' is increased, this should increase surplus labour for the capitalist, on the other hand, however, this also means that the surplus value embodied in each commodity is reduced. Thus, through the use of machinery more commodities may be produced with less application of living labour power, however reducing the value of the commodity of labour power. As Lange summarises: strategies of relative surplus value extension build on further automation and the increase of constant capital relative to labour or variable capital and thus represent the seed of capitalist crisis inherent in the development of the productive forces.[48] The removal of living labour from production and its replacement in a relative-surplus value strategy through machinery, robots and automation, or the development of the productive forces more generally, thus should spell out crisis, or rather an impossibility of (full) automation and 'Capitalism without Workers'.

10 Machinery and Control

The prominence and importance of relative surplus value extraction coincides with developmental stages of capitalism, being representative of large-scale industry and thus necessarily the real subsumption of labour under capital in Marx.[49] This means that there is a historical axis on which the likelihood of crisis must be seen. In extending this (partially historical) development scheme, Dyer-Witheford, Kjøsen, and Steinhoff[50] have recently suggest a stage of 'hyper'-subsumption as next logical step of capitalist development: the progressive subsumption of labour *into* machinery, were the autonomous subject of production manifests fully in artificial intelligence. While absolute and relative strategies are not mutually exclusive or strictly historically distinct and often pursued in tandem (see for example recent legislation in Austria on the extension of the working day next to projects of 'Industry 4.0'), there is a certain historicity to absolute surplus extraction in relation to labour laws.[51]

Notwithstanding such questions of the historicity of capitalist development, the central moment of contradiction for automation lies in relative surplus value extraction, as outlined above. Machines thus cannot produce or increase

---

48 Lange 2019, p. 50.
49 Schmidt 2019.
50 Dyer-Witheford, Kjøsen, and Steinhoff 2019, p. 21.
51 Since September 2018 extensions of the working day to 12 hours are permitted without permission from work councils. This was made possible by Austria's ultra right wing (neo-fascist) government. At the same time a focus on 'Standortwettbewerb' (location-competition) has intensified around subsidizing and developing technological advantages in particular in manufacturing (https://plattformindustrie40.at).

the mass of surplus value themselves. Labour power is the only commodity capable of producing value upon consumption (by the 'lucky' capitalist), explaining the centrality of the desire for its control. While the entire 'secret of value and surplus value' is thus buried in the difference of time between necessary and surplus labour, it is precisely this distinction which breaks down when considering automation through robots.[52] Robots do not reproduce
themselves, do not receive a wage and do not produce value. There can be no distinction between necessity and surplus, pivotal in Marx's categories. The valorisation of capital results only from the variable part of capital, more specifically the relation of surplus labour and variable capital, or 'rate of surplus value'. This means that the development of the productive forces has as its goal the effective valorisation of work, meaning fully automated 'Capitalism without Workers', again, represents a fantasy and a fetish.

Another point must be made here, namely while machinery represents the central strategy of relative surplus value extraction, its 'labour saving' functions seldom coincide with the easing of work for the worker, as is often argued and assumed under the capitalist mode of production. As Schmidt reminds us of Marx's very clear reply to Mill's question whether 'any machine ever made the work of anyone easier', was that this was never the purpose of machinery.[53] Its only objective is to decrease the part of the working day the worker labours for her own reproduction. This point thus also explains the use of present digital technologies and their retention of living labour leading to the intensification of their work rather than their replacement.[54] Meyer details a dichotomy connected to this in the debates of German Marxism in the 1980s still relevant in Marxian schools today, namely a 'subsumption model', which saw the point of technology in domination and control over labour and a 'production model' assuming genuine concerns of productivity.[55] The relationship between the two, however, is certainly not exclusionary and more complete control of the worker in fact in a sense must come with her increased productivity and vice versa. Abstracting from the final purpose, the immediate interest of mechanising and automating production, following Marx, certainly lay to a large degree in breaking the power and resistance of crafts workers in whom the monopoly of productive knowledge has resided (picked up by Braverman as a general tendential law of the capitalist mode of production). The extracted skill, formed into machinery would of course be able to control production

---

52   Lange 2019.
53   Schmidt 2019, p. 57; Marx 2013, p. 391.
54   Huws 2015.
55   Meyer 2019.

more readily and increase its productivity. The consequences for workers Marx put as *'even the easing of the burden of work becomes an instrument of torture, in that the machine does not liberate the worker from labour but his labour from purpose'*.[56] In this sense, automation does not lighten the load but worsens conditions of work, prompting Harvey to identify a 'dialectical inversion' in the instrument of reducing labour time, becoming a means of torture and integrating the entire life of the worker into labour time.[57]

11 Competition and Crisis

Now, if the outcome is designated as crisis, this begs the question why a capitalist would pursue a strategy leading to her own demise. Arguably, because it is not her demise that is at stake in the imminent drive to automate. Schumpeter's theory of 'creative destruction', often falsely presented, derives at this point from his reading of Marx. An individual capital seeks to 'undercut' competitors by going below the socially necessary labour time, while selling at socially average prices, equalling monopoly profits. What follows, however, is destruction indeed: labour saving technology becomes generally used, levelling the difference to the socially necessary labour time and 'setting the scene for the next wave of automation'.[58] With the wider adaption of technology necessitated precisely by the pressures of competition, the rate of profit is diminished in the long run. The seeming rationality of the single capital becomes irrationality for its entire class of capital. What I have put above in the brackets of ceteris paribus is of course that all this depends on production outputs remaining the same. If production extends overall, the drop in the mass of surplus value should be mediated. This also means, however, interestingly that monopolies have a stabilising, countervailing function on the profit rate in that they, with all their might, stifle precisely those competitive pressures which would spur the forces of production and spell out crisis. Hence, we can surmise that the goals of automation lie not just in saving labor costs (and not primarily in displacing labour), but crucially in the promise of monopoly profits.

Limits to automation other than the law of value, are also given by configurations in the organisation of labour such as the relative difference of machine costs and the costs of the replaced worker, which is why, Marx emphasised that

---

56 Marx 2013, pp. 445–6.
57 Harvey 2005.
58 Dyer-Witheford, Kjøsen, and Steinhoff 2019, p. 17.

there were machines developed in Germany in the sixteenth and seventeenth century which were only used in Holland, and French innovations of the eighteenth century, which were widespread only in England.[59] The reason being that if the reserve army, or surplus population, is accessible enough for certain parts of production in certain localities, the use of machinery may be more expensive and thus unnecessary. The wider political economy of industry and occupational structure centrally determines the progression behind or beyond the static law of value. Consequently, automation does not follow a universally uniform law, capital may for example be interested in retaining skilled workers and rationalisation can and has been achieved by other means, such as Taylorism and lean production methods etc.[60] Finally, capital's drive to ensure surplus value may also lead to a fettering tendency in the use of productive technology for the further automation of production itself, stifling its own technological prerequisites.

In spite of the above relativizations through countervailing powers to automation, its main drive cannot be dismissed. The motor of implementing automating machinery in relation to variable capital, is thus given as twofold by Dyer-Witheford, Kjøsen, and Steinhoff: class conflict, in desiring to expand control over living labour (see Braverman) but also to undercut competition and establish monopoly-profits and rents (see also Baran and Sweezy).[61] The two-pronged drive for the increasing mechanisation of production leads to capitalist crisis on two fronts:

> 1. A crisis of overproduction (as Benanav recently identified in the context of global deindustrialization, based on Brenner's 'long downturn' thesis): increasing production while reducing wages, gives imbalance between the increasing volumes of commodities and purchasing power available to buy them.[62]

> 2. The tendency of the rate of profit to fall, as outlined above, wherein capital '[runs]against its own value-decreasing machinic momentum'.[63]

The different crisis theories, their compatibility or rational have been a central point of contention (see for example the confrontation of Heinrich and Mose-

---

59 Schmidt 2019, p. 71.
60 Schmidt 2019, p. 72.
61 Dyer-Witheford, Kjøsen, and Steinhoff 2019, p. 17.
62 Benanav 2020.
63 Dyer-Witheford, Kjøsen, and Steinhoff 2019, p. 18.

ley on the fall of the profit rate), which cannot be engaged here.[64] The unifying feature is the inherent drive of capitalist accumulation, and the substitution of labour through machinery, resulting in crisis. The point is that where automation seems like the 'final triumph' of capital, it is its 'final downfall' by subverting the logic of wage-labour and wage-based consumption, undermining the social basis of value, and thus capitalism. This then is the final contradiction of the capitalist mode of production, undermining its own conditions of possibility. The question thereby becomes, how it manages to survive despite its contradictions and/or what this may mean for political proposals.

12 Conclusion: The Impossibility of Captalism Without Workers, Its Meaning for the Debate on Automation and Where to Go From Here?

In this paper I have suggested to view ongoing projects of automation from the perspective of capital in order to better understand their contradictory development. I have outlined how the split between fantasy and concrete action in the project of 'Capitalism without Workers' rests on a fetishisation of technology and the mystification of social relations. I have outlined how the dismissal of value in favour of utility is connected to this fetishisation, on which the majority of contributions in the automation discourse relies. Rudimentarily thinking through the basic Marxian architecture on automation and value I have suggested that 'Capitalism without Workers' appears as impossibility, suggesting, in turn, fundamental limits to currents automation projects. As put by Tomšič in recourse to Moishe Postone, we can surmise that Marx's essential point was that capitalism could exist without capitalists since the capitalist drive to self-valorisation is 'structural, systemic, autonomous', but capitalism without the proletariat represents a structural impossibility.[65] However, the line of thought presented above is based on theoretical commitment, such as to *a* labour theory of value and the law of value, which has been drawn into doubt within Marxian debates in the past decades. Nonetheless, the simple analysis provides us with an argument for why decades upon decades of automation have not resulted in the expulsion of living labour often envisioned before, as well as why new digital technologies have led to the intensification of work and a greater reliance on precarious work instead of factories devoid of humans. As

---

64 In *Monthly Review Press*: https://monthlyreview.org/commentary/heinrich-answers-critics/.
65 Tomšič 2015, p. 66.

a concept 'Capitalism without Workers' thus appears relevant to our present debate on automation by illustrating automation's limits but also in capturing the fundamentally contradictory ambitions, fantasies and concrete projects of automation from the perspective of capital.

I have glossed over much in this text. Fundamental debates on the tautology of the labour theory of value, the transformation problem and the tendency of the rate of profit to fall, have not been engaged and what was presented was a static view, when in fact the inner-Marxian debate has significantly progressed, or at least shifted at points, on these questions. While this progression cannot be engaged here in detail, I would like to point to three relevant moments of debate, as signposts for further engagement. First, I would suggest the reception of Mandel's 'absolute inner limit' by Morris-Suzuki as central.[66] This limit, essentially a reformulation of what was said above, postulates quite simply that full automation would be incompatible with capitalism. Morris-Suzuki attempted a circumvention of Mandel in arguing that capital can and has shifted away from the extraction of value in production to extraction in the production of innovation, hence a commodification of the innovation process itself. The result being a constant struggle over innovation monopoly (see Google, Facebook etc.). While Morris-Suzuki maintains that Mandel's inner limit may thus recede it does not disappear. The second moment of debate I would suggest as pivotal development, does in fact argue for its dissolution: Negri's postulation of the end of the law of value and the resulting post-operaist concepts such as immaterial labour and 'cognitive capitalism'. As described earlier, surplus value is fundamentally constituted in the difference of labour time required for a worker to produce the value of commodities for her own reproduction, and the actual time she spends working. The correlate of the socially necessary working time which goes into the production of a commodity and its measure of value is also summarised as the 'law of value'. It was Negri (with Hardt) who pronounced this law as 'dead'.[67] The reason for this 'death' of the law of value in the widest sense is rooted in the shifted role of knowledge in production, and the importance of the 'general intellect'. While Negri maintains that the source of value still lies in labour in an extended sense, the distinction between working time and leisure has become blurred under the regime of cognitive capitalism, meaning the law of value has outlived its usefulness. The only invariant of capitalism thus is the law of antagonistic exploitation, and not the law of value. Finally, third, in the negation of the negation, and as more recent

---

66 Morris-Suzuki 1984.
67 Begun in the 1970s in *Marx beyond Marx* and culminated in *Labor of Dionysus* in the 1990s.

development, I would like to point to what may be called a recent 'British turn' against post-operaist thought. While this is arguably a dramatised formulation, as obviously there is much agreement and disagreement with post-operaist thought independent of the country of institution, a certain pushback in and around UK-institutions seems discernible.[68] This front for the defence of the law of value, and its diverse arguments, strikes me as relevant and recent point of engagement. These three moments of debate, more than anything, leave us with a schism on the fate of the law of value and corresponding metamorphosis of capitalism *vis-à-vis* new automation, which requires more detailed attention than can be provided here.

Despite the fragmentary and basic character of what I have laid out above, I believe the arguments I have presented suffice for critiquing much of the present automation discourse. Naturally, the question remains regarding the workers' horizon for struggle against capital's concrete and abstract automation projects. While general points about political proposals can also not be made here, it is important to consider functional limitations in capital's capacity of renouncing living labour. Whether 'full automation' may or may not present a fundamental problem for value and thus capitalism ultimately, given the continuation of work and employment it does not seem to be the present ambition anyway, or rather 'Capitalism without Workers' appears to have been disappointed *vis-à-vis* its vision of radical 'liberation' from living labour once again.

Even where such goals are formally or informally upheld, we must consider the global areas of extraction which function in tandem with the partial automation of developed economies. It seems to me that as ever, intervention at the point of production is essential, regarding the ownership of the means of production and who extracts value, why and how. As Benanav has put it, it is important to remember that post-scarcity is not a technological project but one of political struggle.[69] Finally, in connecting such struggles beyond spheres, sectors and regions of production I would suggest to keep David Harvey's dictum in mind that all work produces value, and we all struggle against it.[70]

---

68   See for example Pitts 2018.
69   Benanav 2020.
70   Harvey 2005.

CHAPTER 5

Deskilling: Automation and Alienation

Amy Wendling

In *Capital*, Marx writes that modern labour processes 'presuppose labour in a form that stamps it as exclusively human'.[1] In the famous passage that follows, he defines this exclusivity as the faculty of imagination: the ability to forecast the project before bringing it to fruition. Two points of contrast highlight this faculty's supposed exclusivity. The first is animal life, and the second machines that require external design and direction. In this sketch, the imagination marks the precious attributes of will and consciousness.

Why would Marx, inheritor of the critical tradition from Kant, dissolver of all prejudices, have insisted so dogmatically upon the distinctiveness of human labour? The latter was already eroding in the nineteenth century. Darwin is at sea looking into the realities of animal life and, in many ways, German chemists have already beat him to his realizations. Babbage imagines computational machines that not only can add and subtract as well as humans, but far outstrip human mathematical powers. From the vantage point of the twenty-first century, it is harder to imagine what exactly a labour that is exclusively human might be. Even though it reflected the best natural science of his time, the Romantic view of animal life Marx espouses appears flat and overly simple, and has been replaced by better biological accounts that minimize the split between human and animal life. As for machines, as Nick Dyer-Witheford, Atle Kjøsen, and James Steinhoff demonstrate, capital is distributing the cognitive and perceptive tasks that traditionally have made up the faculty of imagination to machines, intelligences in particular.[2] The pillars that shore up the exclusively human labour that is Marx's fundamental premise have given way.[3]

---

1 Marx 1990b, p. 154; 1997, p. 192. The author wishes to thank Anne Ozar and Claire Shinners for their editorial work on this essay.
2 Dyer-Witheford, Kjøsen, and Steinhoff 2019, p. 62.
3 Compare Engels in the *Dialectics of Nature*, especially the passages on the influence of labour in the becoming human of humanity (Engels 2012, pp. 172–86). Here he argues that labour shaped human evolution. Engels sets a boundary between humans and other non-human animals, but also shows how this boundary is unstable, sometimes on the very same page! Non-human animals lack intentionality, and yet they plan (p. 181). Only humans 'stamp their will on the earth', and yet this is hubris (p. 182).

But Marx's texts, too, are already resources for thinking through the labour process even once the dogma of exclusively human labour has been put aside. Automation is a threat to human labour when this labour is elevated and aspirational, and so the kind of labour Marx thematized as skilled labour. For it is skill that machines threaten to absorb. Still, automation is a relief to human labour when this labour is drudgery, and no one worries over giving up drudgery. For this reason, a deeper understanding of the skill concept will be fundamental, and this essay undertakes its analysis. Ultimately, I argue that the skill concept is mobile and can be deployed in capitalism's interests. Capitalism portrays the skilled as non-bodily, the unskilled as skilled, and the skilled as unskilled. So too, the legibility and illegibility of various skills are determined by capital's requirements.

1   On Automation

'Automation' is basically a twentieth-century concept; 'automatic' a nineteenth. Both are rooted in the older automaton concept, whose seventeenth- and eighteenth-century versions rely routinely on the analogy between human and mechanical bodies to explain each other. In the wake of this analogy, it may not be possible to think either 'human' or 'machine' without recourse to the other term, as Anson Rabinbach demonstrates.[4]

According to the *Oxford English Dictionary*, the automaton engages in 'action that is not accompanied by volition or consciousness'.[5] To speak more precisely, the automaton engages in action unaccompanied by a volition or consciousness internal to the automaton itself. And while an automaton may appear to be self-possessed of will and consciousness because the behaviour we traditionally associate with these attributes is on display, this is an illusion. Descartes writes about how the humans outside his window may just be hats and coats concealing automata, and thus the will of another.[6]

For Aristotle, the set of political distinctions built on the importance of volition, consciousness, and their true location was the basis of the classical subdivision of the human concept into free and enslaved – and, albeit somewhat differently, male and female. If a whiff of slavery can be detected in the nineteenth- and twentieth-century automation debates, this is hardly acci-

---

4   Rabinbach 1990.
5   'Automatic-Automaton' 1989, p. 805.
6   Descartes 1993, p. 22.

dental. Computer identification systems demand declarations that 'I am not a robot'. Nor is it accidental that gendered divisions of labour are central to discussions of skill, and especially to the concept's conventional definition, where skilled labour is a male provenance.

We rightly think of Marx as an origin of the critique of automation. However, he is also automation's apologist. He celebrated the free time that automation of tasks would produce, once this was liberated from the capitalist social form. It is probably helpful simply to think of Marx more globally as one of the origins of the twentieth-century concept of automation, a concept that relies on both political economy and its critics.

According to the *Oxford English Dictionary*, automation is the 'automatic control of the manufacture of a product through a number of successive stages; the application of automatic control to any branch of industry or science; by extension, the use of electronic or mechanical devices to replace human labour'.[7] The definition has a particular context; Marx calls it 'modern industry'. This industry occurs in a reasonably advanced moment of the division of labour, thus the mention of successive stages. The definition also clearly requires the replacement of human labour. Marx is careful to emphasize, though, that this is not a full replacement. There is a shift from variable to constant capital, but working humans are still necessary to operate machines. According to Marx, there is also a shift in the direction from the skilled labour of manufacture to the unskilled labour of tending machines.

Both a promise and a threat, the role of replacement in the labour process is usually misunderstood. Human labour is certainly replaced, but not in a one-to-one fashion, and not without significant changes in what labour means, accomplishes, and does – and, perhaps also, what the human is and should be. There are both hopeful and terrifying iterations of the concept of labour and the concept of the human. Maybe Descartes's fears about misjudging machines are really fears about what humans will become.

2   On Alienation

Marx's theory of alienation relates to the issue of automation, as this theory bears directly upon the changes to the labour process characteristic of the 18th and 19th centuries. This theory predates *Capital*, and aspects of it actually predate Marx.

---

7   'Automatic-Automaton' 1989, pp. 805–6.

Georg Lukács draws our attention to Hegel's assessment of the changes to the labour process in his Jena manuscripts of 1805–6:

> [Man] can hand over some work to the machine; but his own actions become correspondingly more formal. His dull labour limits him to a single point and the work becomes ... more and more one-sided [...]. The individual's skill [*Geschicklichkeit*] is his method of preserving his own existence [...]. Thus a vast number of people are condemned to utterly brutalizing, unhealthy, and unreliable labour in workshops, factories, and mines, labour which narrows and reduces their skill.[8]

Later in his *Aesthetics*, Hegel develops the theme:

> [T]he long and complicated connection between needs and work, interests and their satisfaction, is completely developed in all its ramifications, and every individual, losing his independence, is tied down in an endless series of dependences on others. His own requirements are either not at all, or only to a very small extent, his own work, and, apart from this, every one of his activities proceeds not in an individual living way but more and more mechanically according to universal norms. Therefore there now enters into the midst of this industrial civilization, with its mutual exploitation and with people elbowing other people aside, the harshest cruelty of poverty on the one hand; on the other hand, if distress is to be removed [i.e. if the standard of living is to be raised], this can happen only by the wealth of individuals who are freed from working to satisfy their needs and can now devote themselves to higher interests.[9] In that event of course, in this superfluity, the constant reflection of endless dependence is removed, and man is all the more withdrawn from all the accidents of business as he is no longer stuck in the sordidness of gain. But for this reason the individual is not at home even in his immediate environment, because it does not appear as his own work. What he surrounds himself with here has not been brought about by himself; it has been taken from the supply of what was already available, produced by others, and acquired by him only through a long chain of efforts and needs foreign [*fremder Anstrengungen*] to himself.[10]

---

8   Quoted in Lukàcs 1975, p. 331; Hegel 1969, p. 232; second omission in the original.
9   Compare Marx's famous passage from the *German Ideology* about the post-prandial critical critic who hunts and fishes in the morning (Marx 1978b, p. 160) and the realm of freedom passage from the third volume of *Capital* (Marx 1991, pp. 958–9).
10 Hegel 1975, p. 260.

Marx's 1844 *Manuscripts* take up the charge of describing the alienation Hegel begins to chart here, adopting the German vocabulary of *fremd*: the more virulent term from the *Manuscripts*, which comes into English as alienation, is *Entfremdung*. The 1844 *Manuscripts* track Hegel's insight that alienation is not a general psychological or social phenomenon. Instead, alienation is tied to the way work unfolds in industrial modernity. Interestingly, already in Hegel, alienation is a problem both for those engaging in unreliable labour in workshops, factories, and mines and for those devoted to higher interests.

Two mistakes of interpretation, both rooted in Althusser, begin in this connection between Marx and Hegel. The first is to regard the early Marx solely or exclusively as a romantic humanist. In fact, already in the 1844 texts, major cracks can be seen in the notion of an authentic human essence from which one is alienated. The idea of a human species that becomes what it is only by virtue of its historical circumstances, a rehearsal of the Aristotelian idea of second nature, is the biggest one. The second mistake of interpretation is to regard the early Marx as split off from the late Marx, with the earlier Marx as a kind of humanist and the late Marx as a fully scientific thinker who has put his romantic notions aside. In fact, the early and late Marx are best considered together and, once this is done, the subtle shifts around the notion of a labour that is exclusively human can be reckoned with more acumen. Even Althusser ultimately came to reject the split he had posited.[11]

In the 1844 *Manuscripts*, alienation has four moments. Humans are alienated from the product of their labour, the activity of labour itself, the species, and as a consequence of this species-alienation, from other humans and themselves. Modern life ensures material poverty; boring, repetitive, and dangerous work for the majority of people; a degraded and overly fixed notion of human nature; and fractious societies with false concepts of what the human is and does. Marx develops his account of second nature in the *German Ideology*. He develops his account of alienation, progressively, into the accounts of the labour process that we find in the *Grundrisse* and *Capital*.

Automation both develops all four moments of alienation and, paradoxically, lays the groundwork for the overcoming of alienation, both as an experience and as a concept driving the definition of the human. As for alienation's first moment, material poverty can be reckoned in many different registers. Massive wealth inequality is only the most obvious. A second dimension of material poverty is the standardization of the consumption object and, indeed, after material poverty, this is the overriding feature of automation on the con-

---

11 Althusser 1994, p. 27; Balibar, Cohen, and Robbins 1994, pp. 170–88.

sumption object. A third dimension is the one Hegel notes: automation guarantees a very partial participation in production. Automation's effect on the other three moments of alienation, and on alienation's overcoming, require more sustained attention.

3   Automation and the Labour Activity: The Skill Concept

Already in Hegel, we can see the central term around which the changes to labour characteristic of the modern period have been theorized: skill. This is unfortunate for two reasons. The first is that the overemphasis on the intellectual dimension of skill neglects a second important term, namely, strength. The second is that the skill concept, considered in general, is structured not only by this reduction but also by other biases particular to capitalism's interests, including a set of issues related to how capitalism metabolizes scientific literacy. The pressures produced are so intense that the skill concept borders on incoherence.

In the account of alienation from the 1844 *Manuscripts*, Marx writes that the labourer does not develop his mental *and physical* energies freely.[12] In the *Grundrisse*, Marx writes, 'it is the machine which possesses skill [*Geschick*] and strength [*Kraft*] in the place of the worker, is itself the virtuoso'.[13] And in chapter XV of *Capital*, Marx still writes of skill in a bodily way that emphasizes the handling of tools: 'Along with the tool, the skill [*Virtuosität*] of the workman in handling it passes over to the machine'.[14] Subsequently, Marx will even call light labour a torture, because not only intellectual but also all bodily interest in work is annulled.[15]

But in other passages Marx has dropped strength or bodily habit from his account of alienated labour and substituted the intellect. He writes in the *Grundrisse* that 'in machinery, *knowledge* appears as alien, external to [the worker]'.[16] Though Marx himself largely avoids thematizing deskilling primarily or even exclusively as knowledge, this opens the door for others who will. Particularly when deskilling is thematized as knowledge, and then knowledge is thematized primarily or exclusively as cognitive, the bodily element to skill is

---

12   Marx 1975b, pp. 275–6; 1982, pp. 368–9; emphasis added.
13   Marx 1993, p. 693; 1987a, p. 29; 1980, p. 572; emphasis added.
14   Marx 1990b, p. 366.
15   Marx 1996, 35, p. 426; 1990b, p. 369; 1989b, p. 411.
16   Marx 1993, p. 695; 1987a, p. 84; 1980, p. 574; emphasis added.

lost. For better and for worse, the category around which much of the critique of automation unfolds, from this point on, is cognitive deskilling.

Once the bodily element of the skill concept is lost, it opens the door to the bourgeois apologist, who can always argue that there will be new skills to develop as others are eclipsed. Since one kind of knowledge can be learned just as well as another, knowledge is unlike the slow acquisition of bodily habit, one form of which is strength. The limitation to the domain of knowledge covers over the fact that relatively few workers will need to gain these new skills compared to those who are displaced. It also covers over the fact that the new knowledges, at least for the vast majority who do not become a part of the managerial class, are thinner and thinner in content and so speedily acquired. As Harry Braverman writes:

> [S]hort term trends opening the way of advancement of some workers in rapidly growing industries [...] simply mask the secular trend toward the incessant lowering of the working class as a whole below its previous conditions of skill and labour. As this continues over several generations, the very standards by which the trend is judged become imperceptibly altered, and the meaning of 'skill' itself becomes degraded.[17]

In light of this, Carl Frey's distinction between replacing and augmenting technologies from his 2019 book *The Technology Trap* may not exhaustively explain the tensions of developing automation. Frey bifurcates technologies into replacing technologies, his name for those technologies that eclipse skills and cancel jobs, and augmenting technologies, his name for those technologies that require new skills and enable new job prospects.[18] Frey then argues that Marx and Engels were writing at a time when replacing technologies were the norm. Conversely, augmenting technologies characterized the twentieth century until about 1970 and enabled a rosier view of the relationship between technology, job creation, and capitalism.

However, skill is not a fixed term in the way Frey's distinction requires. If skill is progressively degraded in the way Braverman outlines, augmenting technologies themselves are increasingly alienating or even themselves replacing because they degrade the skill concept, even as they enable new jobs. The automobile, arguably the twentieth-century augmenting technology par excellence, does indeed employ many, but it does so on assembly lines that attempt

---

17   Braverman 1998, pp. 89–90.
18   Frey 2019, p. 131.

to restrict worker movement and curtail all but the most repetitive and limited motions. If I am temporarily well paid for swivelling in my chair and typing answers to my email, this alone does not mean that the clerical environment is very different from the Taylorist factory, as Braverman also argues.[19]

It will be much harder for the bourgeois apologist who considers the loss of strength and the general eclipse of bodies in late capitalist society. The best argument for the apologist – that machines can do hard and harmful physical labour, a proposition to which even Marx would agree – is not a sufficient alibi for removing all physical movement from work or for constraining workers from natural or spontaneous physical movements. And, indeed, it is exactly the complete removal of physical work of any type that we witness in clerical situations and in the industrial situation that becomes increasingly clerical, where the worker is reduced, in the words of Marx, to 'watchman and regulator'.[20]

Braverman describes the 1958 Bright scale, which charts twelve kinds of worker contribution, such as physical effort, mental effort, dexterity, general skill, and education, through seventeen levels of increasing mechanization.[21] There are complex shifts in most of the kinds, though as the mechanization levels reach 12–17, all ultimately devolve to non-existent worker contribution. Among the twelve kinds of worker contribution, physical effort is distinguished by its utter lack of complexity. It reaches the level of non-existent contribution already at level 5, from which point sedentary workers are assured. As Braverman relates, water fountains are deliberately relocated to ensure that workers do not have to walk too far to access a drink, ensuring their sedentariness, as every step is considered a loss to capital.[22]

In the artificial intelligence debates that characterize our own era, we see the eclipse of the body in Moravec's paradox: sophisticated computers can play chess, do taxes, and diagnose your cancer but lack the perceptive and mobility skills of most humans one year of age or older.[23] Not having a human body, including perhaps even its weaknesses like lapse of attention and fatigue, likely will remain a significant block for machine learning. However, in an environment where the skills of perception and mobility are not required by the labour process, those same computers are not unskilled. It does not matter if I cannot mimic human perception if my task is simply to transcribe or tabulate numerical data.

---

19   Braverman 1998, pp. 203–47.
20   Marx 1993, p. 705.
21   Braverman 1998, pp. 152–3.
22   Braverman 1998, p. 214.
23   Dyer-Witheford, Kjøsen, and Steinhoff 2019, p. 12.

In addition to the issues raised when we restore strength to its rightful place among the alienations brought on by automation, there is another difficult question: What is skill exactly? Not even the vocabulary is consistent. In fact, Hegel and Marx already use several different German terms to try to capture the skill concept in its relationship to the changes of modern labour. Hegel's term, *Geschicklichkeit*, is richer in content than *Fertigkeit*. He thus draws attention to what is being narrowed and reduced in factory labour. Marx moves us in a still richer direction of the skill concept, using the Latin: the labourer whose skills are being given over to the machine is a virtuoso. This term invokes the highly complex skill set, both bodily and cognitive, we associate with performing music.

We know with some precision what Marx meant by deskilling as he considered the onset of factory labour, with giant steam engines replacing the manufacturing workshop. He writes in *Capital*: 'from the moment that the tool proper is taken from man, and fitted into a mechanism, a machine takes the place of a mere implement'.[24] Here deskilling is a specific term: Marx is describing the function of controlling the pathway and pacing of single tools operated by hand, both the knowledge and the specific bodily habit of operating them, including the strength acquired and required to do so. But once the skill concept is generalized beyond this specific historical example, and limited to the domain of a knowledge that excludes bodily strength and habit, it becomes much less clear what actually it is describing. Unclear concepts are easy targets for occupation by capitalist norms.

Consider farm and ranch labour, as Braverman invites us to do at the end of his book.[25] Considered unskilled by the knowledge-based educational metrics of the twentieth century, farm and ranch labour is actually a peculiar combination of highly evolved skills, the transmission of which is based in an apprenticeship system. This apprenticeship system really begins in human infancy. The cumulative abilities are the acquisition of a lifetime and of an environment, from the rhythms of meals to the physical strength needed to lift a forty-pound calf. The ability to grow grain and also mix it to feed cattle when grass is scarce – or, alternatively, to move cattle to grass feed when grain prices are high – is essential. So are basic carpentry, masonry, irrigation technology, and veterinary medicine, to which one might also add meteorology. Long-range timing and planning are crucial, as the micro-tasks of each day must be matched to longer-range, and changing, patterns of climate and mar-

---

24   Marx 1990b, p. 326.
25   Braverman 1998, pp. 300–1.

ket. Finally, there is reading to keep abreast of how artificial intelligence will shortly change the operations of the tractor. This is 'unskilled' work.

Consider domestic labour. As Martha Giménez reminds us, 'domestic labour, like all forms of labour, is an acquired skill'.[26] The propensity to see domestic labour as unskilled, however, comes from a variety of factors. The first is a confusion between the unskilled and the unwaged. The second are the gender, race, and class markings of domestic labour. The third is a disproportionate use of the skill concept, in sociological literature, only in association with non-domestic tasks.[27] To thwart our prejudices, consider the following highly skilled but unwaged domestic consumption labour of a male accounting professor. He organized his family grocery shopping according to community commodity price circulars, searching them for market oversupplies. He also conducted the shopping, often traveling to multiple stores, and sometimes pursued several shopping tasks each day. He then calculated the marginal savings to his family over decades: tens of thousands of dollars less for precisely the same items they would have purchased anyway.

Capital does tend to reduce all human work to unskilled work. However, it simultaneously portrays various data entry tasks as 'skilled' and their acquisition as 'upskilling'. By contrast with the consumption labour of the accounting professor, consider the 'highly skilled' work that lands me at a sophisticated computer monitor. The labours are nearly all cognitive and the lifestyle sedentary. Programmatic features of computer software and hardware condition the pacing and timing of work: from bodily posture, to an awareness of abstract time, to the multitasking and interruption that are features of office suite programs, telephone and text messages, and email. Much of my work undertaken at the monitor shares a common essence: data entry or programming. I program a calendar so a program can access my availability without having to engage me in real time. I convert documents from one format to another. So, too, my leisure hours are devoted to data entry on increasingly smaller devices in the format of social media, accepting or rejecting notifications, downloading an app, or checking in for medical appointments. In connection with this, I will examine the function of calendar programs in the final section of this essay. Thus the category of skill itself is essentially malleable to capital's purposes. Marx himself was the first to recognize this. In the third volume of *Capital*, he calls attention to two basic instabilities in the skill concept under capitalism. The first of these – I will address the second in the next section – is Marx's

---

26 Giménez 2019, p. 265.
27 Steinberg 1990.

observation that the division of labour makes skills more and more basic and so easy and quick to acquire.[28]

Capital calls unskilled those actions that are either illegible to its purposes or whose importance it wishes to deny. It celebrates as skilled those actions, however degraded they may become, that advance its interests. Just as there is no technology per se, outside of a social form, *there is also no skill beyond what capital names*. And similarly with educational attainment, as Braverman also points out.[29] What capital calls and celebrates as education will be thin and poor: an education for data entry and political docility.

4 Automation and the Human Species

A third problem with the skill concept is that, when predicated of humans, it is often individualistic – or, in a variation of this, regarded as a talent of a particular subset of individuals. Though skill nearly always comes about in human community, once possessed it becomes the private property of the human being or subset of human beings who have, in the traditional vocabulary, 'acquired' it. This falsely shapes our understanding of skill and what the losses are when skills become obsolete. It is not merely a question of dispossession in the manner of a pair of shoes or an iPhone, or even the dispossession of a set of guild skills and secrets. Skills of individuals and small groups also neglect the kind of skill that may be generalized to a larger group, whether the group is humans or non-humans. Marx's *Grundrisse* even names this: the general intellect.

The passage that names the general intellect is a famous one from the section of the *Grundrisse* that is customarily called the 'Fragment on Machines'. The passage is frequently parsed in Marxist literature, especially the Italian autonomist traditions and in contributions to the journal *Futur Antérieur*.[30] Unlike in *Capital*, which is explicitly circumscribed by the social form of capitalism in many of its discussions of machinery, and so argues primarily about the negative aspects of machines within that social form, the *Grundrisse* is not exclusively bound by capital as a social form, and so ranges out beyond to revolutionary developments and uses of machines. This section of the *Grundrisse* is additionally remarkable because it still uses the 'alienation' vocabulary alongside some of the discussions of machines, especially those that advance capital's

---

28 Marx 1991, p. 414.
29 Braverman 1998, pp. 294–310.
30 Dyer-Witheford 1999, pp. 219–40; Smith 2013; Negri 1992.

interests. The alienation vocabulary will largely disappear in *Capital*, absorbed by concepts like real subsumption.

In the *Grundrisse*, Marx uses the category of fixed capital to discuss technology. This is the context for his discussion of the general intellect:

> Nature builds no machines, no locomotives, railways, electric telegraphs, self-acting mules etc. These are products of human industry; natural material transformed into organs of the human will over nature, or of human participation in nature. They are *organs of the human brain, created by the human hand*; the power of knowledge, objectified. The development of fixed capital indicates to what degree general social knowledge has become a *direct force of production*, and to what degree, hence, the conditions of the process of social life itself has come under the control of the general intellect and been transformed in accordance with it.[31]

Note that the exclusive dichotomy between nature and humans with which the passage begins migrates, imperceptibly, into the more permeable and no doubt also more accurate 'human participation in nature'. Note too that the power of knowledge includes both the brain and the hand: that is, knowledge is not defined exclusively as a cognitive property, but includes some dimension of strength or bodily habit as discussed above.

One thing Marx himself ties to the general intellect is the practice of science. Tony Smith revisits the issue of science at the time of the Industrial Revolution. He argues that science was already a matter of daily practice of skilled craftspeople. He writes:

> An exclusive focus on 'deskilling' [...] oversimplifies Marx's position. Such an exclusive focus understates the extent to which the general intellect was already 'diffused' at the time of the Industrial Revolution: that is, not monopolized by a small group of scientific-technological experts.[32]

Braverman also supplies evidence for the role of the general intellect for working-class craftspeople in the Industrial Revolution's scientific and technological practices, and even in the development of science, full stop. Braverman thereby links the skill concept to the practice of science, citing Elton Mayo: science is a 'reflective moment attempt[ing] to make explicit the assump-

---

31 Marx 1993, p. 706; original emphases.
32 Smith 2013, p. 227.

tions that are implicit in the skill itself [...]. Science is rooted deep in skill and can only expand by the experimental and systematic development of an achieved skill. The successful sciences consequently are all of humble origin'.[33] This observation is borne out by the history of biology in medicine, which was a trade; chemistry in metallurgy, also a trade; and even physics in optics and mechanics. Da Vinci, the illegitimate son of a notary, was apprenticed to a craftsman.[34] The history of Florence is a history of masonry, among other things.[35]

If we cannot easily see how the general intellect was already diffuse during the Industrial Revolution, this might partly be conditioned by what came after. Braverman cites an actual moment of transition: Bernal's observation that the Royal Institution in London had to brick up its back entrance to 'keep out the mechanics who stole into the gallery'.[36] Such a wall attempts to sever the connection, in both directions, between skilled working persons and scientific knowledge. If working persons are the losers here, so too science.

Motivated by the kind of control this wall emblematizes, a mythology of the separation of concept from execution separates us off from actual histories of science and technology. Indeed, according to Smith and Vercellone, it also separates us from the actual histories of factory labour, since the Taylorist idea of separating conception and execution in the workplace is more of a control fantasy than an exhaustive description of what actually goes on in capitalist workplaces, still today as in the industrial revolution.[37] On this view, capitalist workplaces have always relied on a set of 'skills' appropriated without recognition from 'unskilled' working populations: capitalism just does not wish to recognize them, and so inculcates them while denying that they exist. Marx points this out frequently in the *Grundrisse*, using the vocabulary of 'free of charge' to designate such unrecognized appropriations.

Scholars have speculated about how best to imagine the universal social knowledge that makes up the general intellect. Marx's vision for the role of the general intellect in the society he anticipates is, in fact, very clear. Recall that universal education was one of the demands in the list at the end of the *Communist Manifesto*. While capitalism appropriates the substance of the general intellect free of charge from working persons, in an inheritor society, the general intellect would play a very different role. The dream of free time, defined

---

33   Quoted in Braverman 1998, p. 91.
34   Isaacson 2018.
35   Goldthwaite 1982.
36   Quoted in Braverman 1998, p. 92.
37   Smith 2013, p. 227.

not as an empty bracket to fill with consumer objects, but as time for self-development, would be realized. Through these measures, Marx envisioned a developing general intellect in both social and productive life, including continuing technological development.

At stake is whether the general intellect can really develop in the capitalist social form, or not. Tony Smith argues that Marx's theory is less helpful here than on other issues, as his meditations on the future of the general intellect were accompanied by the assumption that the capitalist social form would be overcome.[38] Still, Marx's diagnosis of the contradictions of the general intellect is likely most useful to us while the capitalist social form endures. Indeed, the question of whether and how the general intellect is able to develop within the capitalist social form is exactly the problem Marx himself investigated.

We can define the capitalist social form as one in which everything is measured by profit and then developed according to the profit it enables. For both education and technology, this has meant a very selective development, as Marx also emphasizes, albeit with reference to a different century. Consider technologies that did not develop, or even those that were actively fettered, such as biofuels, in light of the profit demands of the fossil fuel industry. Consider also educational possibilities that did not develop, as workers were trained instead to occupy existing Taylorist jobs.

Still, Dyer-Witheford follows Marx most closely when he argues that, with the general intellect, capital calls into being forces it cannot control.[39] In this sense, capitalism itself, at different times, actively develops the general intellect, contributes to its development, or, at minimum, does not hinder it, all the while attempting to discipline and constrain it into desired channels that may themselves be contradictory. First, the general intellect must be fettered to profit-maximizing strategies. Second, capitalism relies on appropriating the things that the general intellect produces free of charge, as Smith also points out.[40]

We can think of the general intellect as embodied in any of several institutions. Science and technology are one such institution. A second is education systems, especially the universal education systems of the capitalist welfare states.[41] A third, the one Dyer-Witheford calls the general intellect's 'quintessential institution' is, of course, the internet.[42] Subsequently, Dyer-Witheford

---

38   Smith 2013.
39   Dyer-Witheford 1999, p. 237.
40   Smith 2013, p. 220.
41   Vercellone, quoted in Smith 2013, p. 216.
42   Dyer-Witheford 1999, p. 228.

also calls our attention to a fundamental truth: while we are in the capitalist social form, there will always be a battle over the general intellect.[43]

The wall at the Royal Institute was itself such a battle. When we turn to the Bernal text, we learn that the Institute was put into place in 1799 precisely to cultivate scientific literacy among the working population: Count Rumford was no democrat but needed scientifically trained mechanics to shore up the productivity of the Industrial Revolution.[44] The wall went up swiftly, however, under its first director, sometime in the first third of the nineteenth century. Here we see the battle over the general intellect is one that is interior to capital: it both needs and fetters the general intellect's development in accordance with contradictory tendencies. The working classes both must have, and cannot have, science.

We see this play out in debates around education, for example. This is the second instability in the capitalist skill concept that Marx himself calls our attention to in the third volume of *Capital*, where he writes that 'basic skills ... are reproduced ever more quickly, easily, generally and cheaply, the more the capitalist mode of productions adapts [...] popular education'.[45] When popular education universalizes a skill – Marx's examples are commerce and languages – the result is deskilling, since the supply of the skill is increased, decreasing its value. Skill is legible only relative to context and determined by relative scarcity. The mass acquisition of skills through education is itself a deskilling.

Today, through the education of the capitalist welfare state, workers possess a set of abilities that we might loosely define as a kind of background skill: among them basic literacy, mathematics, and, at the end of the century, facility with computer technologies. However, Braverman, following Marx, is not sanguine about the prospects for universal education to develop legitimate skill. Tasks become more basic, and more and more workers can perform these basic tasks, which are thus worth less and less. A progressive dumbing down of education, and so of the general intellect, occurs within the capitalist social form. To Braverman's examples we might add the issue of bodily reduction in the notion of skill, and the subsequent curtailment of physical education, art, and music programs. Such examples show that the contest for the general intellect is alive and well in the twenty-first century.

If the pre-eminent function of automation, in the capitalist social form, has been control over workers, its companion concept is humans who are seen, and who may see themselves, as needing to be controlled. Illustrating at once

---

43 Dyer-Witheford 1999, pp. 230–1.
44 Bernal 1969, p. 538.
45 Marx 1991, p. 415.

capital's selective development of technology and the contest for the general intellect, Marx writes famously in *Capital* that 'it would be possible to write quite a history of inventions, made since 1830, for the sole purpose of supplying capital with weapons against the working-class'.[46]

In order to imagine inventions with a different teleology, two shifts are required: to remove the exclusive control function from automation, and to restore autonomy to humans, and perhaps also other actors. And this brings us back to the third and fourth moments of alienation from the 1844 *Manuscripts*. When Marx says in 1844 that humans are alienated from their species being he does not mean that they are alienated from a fixed and eternal essence into degraded forms. Instead, Marx inherits and then redeploys the Aristotelian notion of a second nature.[47] In this idea, natures are shaped in society, but are nonetheless truly natures by virtue of that shaping. What remains of the human essence is simply the vanishing point of a capacity to become. There is nothing to prevent such a vanishing point from adopting good, non-alienated relationships with sophisticated production technologies – including, ultimately, artificial intelligences.

Alongside education and the internet, the artificial general intelligence or AGI is yet another institution in which we might see the general intellect embodied. Dyer-Witheford, Kjøsen, and Steinhoff write, speculatively, that –

> [W]ith the advent of proletarian AGI, [Marx's surplus population] would become absolute, co-extensive with a human species rendered obsolete to the valorisation of value. Humanity would become a 'legacy system', outdated hardware unsuitable for running the inverted world of capital. The status of humans in such a situation might be comparable to the current status of wild animals, tolerated on the fringes of capital so long as they do not detract from valorisation, or so long as they are not usable as raw material in the production process. In contrast to the malice of the

---

46   Marx 1990b, p. 411.
47   The clearest link between the species being idea and Aristotle occurs in the *Grundrisse*, where Marx writes: 'But human beings become individuals only through the process of history. He appears originally as a *species-being* [*Gattungswesen*], *clan being, herd animal* – although in no way whatever as a ζῷον πολιτικόν in the political sense' (Marx 1993, p. 496). Do not be fooled by the negation in the passage: Marx is not asserting that the *Gattungswesen* and ζῷον πολιτικόν concepts are dissimilar, but that they are similar, with a crucial qualification. Marx prefers the species-being concept to the ζῷον πολιτικόν because he means to get at the deeper structures that precede the sort of determination Aristotle has in mind, asserting that the very essence of what it is to be human is to take on a second nature in history. For broader context, see Wendling 2013.

machines in the *Terminator* series, in this scenario humans would simply no longer be of interest to capital. According to this view, we have not seen capitalism yet.[48]

In this passage we see the three poles, humans, animals, and machines, anchoring one another, as they always do in modernity, though with a different hierarchy than the one proposed by Marx in the passages cited at the beginning of this essay.

Allow me to pose an alternative speculative conclusion. What if the human, machine, and animal concepts do not proceed unchanged from those we received from the nineteenth century? Indeed, the notion of the human animal is already with us, and as Rabinbach and others have argued, the concepts of human and machine have relied on one another for definition since at least the seventeenth century. What if, instead of the human animal as a legacy system, the notion of the human animal expands to include AGIs in symbiotic co-evolution? This is the direction that Donna Haraway pointed to decades ago with her famous cyborg.[49] More recently, the work of Galit Wellner on the face-like features of the cell phone, and its functions as a part of human embodiment in the form of what she calls 'everyday carry', points in the same direction.[50]

In such a symbiosis, the concept of the human species itself will undergo change, rather than lingering as a superseded Romantic residue alongside the operation of capitalist AGIs. If AGIs only became possible by studying large data sets of human behaviour, they will only improve by continuing such studies, and perhaps by expanding them to include issues of human embodiment, mobility, and perception. In the end, we are already talking about ourselves when we talk about them.

5 Automation and the Organization of Time: The Calendar Program

To bring together the threads of this argument, let us consider an extended contemporary example of automation, namely, the calendar program. Where a person or set of persons was responsible for coordinating the significant affective and social dimensions of setting human activities into the space-time continuum, now calendar programs use cloud technologies to proliferate sched-

---

48 Dyer-Witheford, Kjøsen, and Steinhoff 2019, p. 144.
49 Haraway 1991.
50 Wellner 2016.

ules through various devices. Calendar programs attempt to automate abstract time in late capitalist societies. Indeed, they display features of all four categories of alienation, features of the deskilling-reskilling debate, and features of the struggle for the general intellect.

Abstract time, as Marx argues, was one of the achievements of capitalism. Once labour could no longer be measured by useable objects, as every contribution was very partial, we began to measure it with time. Following Marx, E.P. Thompson and Moishe Postone have charted the consequences of abstract time for the twentieth century, including and especially struggles around time.[51]

Calendar programs continue the intensification of temporal domination in our own era. They are tailored to an age where concentration of persons in a workplace, the factory or office model, has begun to erode, or even, especially when combined with video, where labour is dispersed from direct human observation, such as in janitorial work. In the unprecedented dispersion of clerical labour inaugurated by the coronavirus, calendar programs are a part of the arsenal of monitoring software that functions to observe workers and standardize time without the need for workers to gather in a single physical space.[52]

There is even a specific skill that the calendar program seeks to supplant: time management. Competent clerical workers who manage schedules lose jobs as they are replaced by other kinds of 'schedulers'. Individuals conditioned to manage their own work time in other ways, such as with a clock or watch not yet outfitted by the new calendar functionalities, also lose the internal time consciousness required to manage such a skill. Instead, if very autonomous, a worker alternates between programming his or her calendar and then jumping to its set of bells and whistles: she is both commander and executor, serially, with these functions separated. If less autonomous, there is only the jumping to the pre-programmed increments, with the calendar acting as a time card.

The internal time consciousness that was the specific achievement of the abstract time era now consolidates, in one way, even as it begins to erode, in another. As with so many features of capitalist technology, it is ambivalent. On the one hand, calendar programs enable an unprecedented domination of time, segmented into micro-units. However, on the other, this frees the operator from some of the bounds of internal time consciousness. Especially with longer increments, his or her mind can drift in passive or spontaneous reflec-

---

51 Thompson 1967; Postone 1993.
52 Hern 2020.

tion that is not accompanied by a consciousness of the abstract time that has passed, because a programmed interruption will notify him or her that it is time to return to a task.

But these longer increments are increasingly under assault by the notifications themselves. One feature of the calendar program is its fixed increments and start times. In Starfish, an appointment scheduler developed by the educational technology company Hobsons, every meeting is divided into precise 15-minute units. While these can be stacked for a longer meeting, their boundaries cannot be flexed. Artificial intelligences can calculate not only the start time to an appointment or task, but also the transit time needed to get to it. Time to leave for work, says the phone. While at work a buzzer sounds fifteen, five, and two minutes in advance of the start time for the next task.

In addition, the control function of the calendar can overwhelm its productivity function. Some dead time that could have been allocated to other tasks, were the skilled operator still able to flex the temporal continuum, is now instead spent waiting for the next increment or buzzer. Passive time and spontaneous time are thereby converted into waiting time, in which the next task is anticipated. Not only are passive time and spontaneous time each important to the worker's well-being and sense of autonomy over labour, they are arguably important to the labour process. Passive and spontaneous time is precisely the time in which a skill may be considered reflectively, in the way Mayo outlines, and improvements to process or technology undertaken. Once waiting time supplants passive and spontaneous time, all that is left are tasks. And, indeed, the vocabulary of task replaces that of skill.

Calendar programs are, moreover, embedded within other programs: email programs and teleconferencing programs. Multitasking, another offender against passive and spontaneous time, is thereby built into their functionality. So is surveillance, as supervisory technologies can look into your environment at any time, including from a remote location; can take screen shots of your activity in random, frequent, increments; and can compile productivity data about task completion. For this reason, the electronic stamp of logging into a computer monitor or smart phone is as reliable as a punch card or identity badge, and the electronic surveillance systems that ensue subsequently far outstrip human powers to physically observe workers.

Humorously, calendar programs even attempt to speak to other calendar programs, though with only mixed success so far. An appointment calendar allows students to book advising appointments in one asynchronous system. Still, this has to be both pre-programmed by trained advisors and then integrated into their individual calendars. At present the translation of one calendar program to another requires a great deal of human labour. So, while the programs claim efficiency and self-sufficiency, human labour is again pressed into service, free of charge, to sync them up.

As I book a medical appointment or train journey, a program asks if I want to send it automatically to my calendar. In fact, all of life's activities are subject to unprecedented synchronization. As a feature of proletarianized life, the smart phone becomes an object of what Wellner calls 'everyday carry'.[53] In the process, calendar programs are part of the ongoing erosion of the thin distinction that may have remained between working and non-working times.

As for relations with other humans, the calendar program largely makes them disappear. Instead of calling one another, each of us programs a schedule and then asks the other to access it, addressing any gaps that the programs may have interfacing with still more programming. The great achievement of calendar programs, asynchronous scheduling, is thus also their liability. It is also a fetishisation of connection with other humans, embodied in an object, albeit one with quasi-human properties.

The calendar program thereby brings to our attention another boundary concept that shapes the notion of what skill is and what it is not: namely, programming. And it is precisely the difference between skill and programming that undergirds the worries about what makes human labour distinctive from that of machines, the very worries with which this essay began.

If, to capture the distinction between skill and programming, we reintroduce a concern not only for volition, imagination, and consciousness, but also for strength, habit, and mobility – in whatever kinds of entities may come to have such properties – this illustrates a general truth. Just as there is a struggle for the general intellect, so too there is a struggle for skill. The ability to name skill is a weapon in capital's arsenal. If it can call unskilled work skilled for workers to save face, it will do that. If it can determine skill as non-bodily, it will do that. If it can avoid remunerating skilled care work by calling it unskilled, it will do that.

This argument has limits, of course. I can either play the violin, or I cannot. Skills also admit of degree: my Arabic can be advanced, intermediate, or novice, and I may be able to speak it but not read or write. And, indeed, some skills are acquired only at great pains, while others are acquired swiftly. Still, while recourse to the skill concept may indeed be essential to highlight Marx's critique of automation, critics should use it only with great specificity and deliberation. Otherwise, it itself may come to harbour the very social condition of alienation they wish to diagnose.

---

53   Wellner 2016.

CHAPTER 6

De-alienated Labour, Technology, and the Social Heart of Socialism

Jeff Noonan

More than thirty years after criticizing Marxism as essentialist and economistic, the Covid-19 crisis has awoken Chantal Mouffe to the important role that meaningful work plays in a good life. Writing in *The Guardian*, she rejects the capitalist reduction of workers to 'human resources' and calls for the democratization of work. 'If we leave these things solely to the market', she argues, 'we run the risk of exacerbating inequalities to the point of forfeiting the very lives of the least advantaged. How to avoid this unacceptable situation? By involving employees in decisions relating to their lives and futures in the workplace – by democratising firms. By decommodifying work – by collectively guaranteeing useful employment to all'.[1] I agree, but her position was central to the Marxism she once repudiated. This chapter will return to Marx's argument that non-alienated labour would become central to meaningful human lives in a socialist society. My goal is to deploy Marx's still living philosophical-political understanding of non-alienated labour as an intervention in contemporary debates about socialism as a post-work society. I will argue that his understanding of the nature and importance of non-alienated labour in valuable and valued human lives remains both unsurpassed and misunderstood by the advocates of full automation and socialism as a post-work society.

The connection between socialism and post-work is not new. Marx himself had an ambivalent relationship with the idea of technological progress, sometimes suggesting that the elimination of labour was one of the material conditions of socialism.[2] His son-in-law Paul Lafargue developed this technotopian side of Marxism in his minor classic *The Right to be Lazy*.[3] Post-work demands returned in the 1980's as responses to the economy transforming power of computing technology. Andre Gorz in Europe and Jeremy Rifkin, Stanley Aronowitz, and Jonathan Cutler in the United States argued in their own ways that job-

---

1 Mouffe 2020.
2 I will briefly discuss this ambivalence below, but I have examined it in detail elsewhere. See Noonan 2017, pp. 1–20.
3 Lafargue 2011.

destroying technological development was inevitable.⁴ Consequently, the Left would have to drop its demand for full employment and instead demand the freedom from work made possible by the spectacular productivity gains new computing technologies would generate. Those arguments are the template for contemporary demands for full automation and freedom from labour that define the technotopian wing of the socialist movement. Thinkers like Kathi Weeks, Nick Srnicek and Alex Williams, and Aaron Bastani have all criticized Marxism for a lingering attachment to a repressive work ethic and urged Marxist to embrace a labour-free future. Time now spent working would become leisure time for citizens to consume, connect, and play.

I do not disagree with the argument that socialism is incompatible with the capitalist form of *alienated* labour. I also do not disagree with the post-work claim that socialism would expand the amount of time available for self-realizing activities. Under capitalism, technology reduces socially necessary labour time, but not the amount of time workers must work to earn wages sufficient to satisfy their needs. Under any form of socialist society, the productivity gains that intensify exploitation and alienation would be realized as a reduction in the amount of time workers would be compelled to devote to work. However, overcoming the capitalist form of economic compulsion to work is not, I will argue, identical to eliminating the need to work. Our human need to work is a function of the social nature of individuality. I claim that socialist post-work advocates do not understand the full implications of Marx's claim that 'the individual *is the social being*'.⁵ Defenders of socialism as a post-work society tend (like liberals) to valorise individual choice over and against social commitments. However, if individuals are social beings, then our commitments to each other are the ethical foundation of socialism.⁶ Thus, I will argue that freedom from the imposed necessity of wage labour will be experienced as *freedom to* realize our talents and interests in life-valuable and socially valued ways.

The argument will be developed in three steps. In the first, I will review Marx's analysis of the violent processes by which human labour is subsumed under ruling class power and reified market forces. In the second part, I will evaluate the claims of post-work theorists and conclude that while much of

---

4 Gorz 1989; Rifkin 1995; Aronowitz and Cutler 1998.
5 Marx 1975b, p. 299.
6 I use 'ethical' not in the sense of abstract rules to imposed upon individual behaviour but in its original sense of ethos, or way of life. Ethics concerns better or worse ways of living. Ethical foundations therefore refer to the mode of living that will characterise life in a socialist society.

their vision of the future is ethically sound, it suffers at the deepest level from a failure to understand Marx's conception of the individual as a social being. Consequently, they fail to understand that the need to contribute to collective well-being is a cornerstone of meaningful individual lives. In the final section, I will return to Marx to substantiate that argument and conclude with some brief remarks about the future of the struggle to free human labour from capitalist exploitation and alienation.

1 Subsumption, Alienation, and the Negation of Social Individuality

Before I turn to the problem of the relationship between the real subsumption of labour and alienation, I need to clarify the meaning of the term 'life-value.' The term was developed by philosopher John McMurtry as an attempt to both understand the universal foundation of value and as a basis for social criticism. McMurry argues that all value whatsoever presupposes the existence of sentient life-forms, human beings above all. Conscious, valuing beings are the subjective pole of objective value relations. Particular values can be decided by the application of what McMurtry calls the "primary axiom of value": *"X is value if and only if, and to the extent that, x consists in or enables a more coherently inclusive range of thought/feeling/action than without it."*[7] Life-values are therefore either instrumental (resources, relationships, or institutions which satisfy a fundamental need) or intrinsic (the enjoyed expression of sensuous, intellectual, creative, and affective capacities).[8] This conception of value is more comprehensive than Marx's understanding of value, use-value, and exchange-value and enables us to integrate Marx's categories to a deeper understanding of the social conditions of good human lives.

For Marx, the value of a commodity is a function of the socially average labour-time required to produce it.[9] Value is the social substance that allows different commodities to be priced and exchanged. Exchange-value is expressed as the market price of a commodity, while use-value is determined relative to the needs or desires of the purchaser.[10] These definitions are fine in the realm of political economy, but they are too limited as justifications for the

---

7   McMurtry 2011b, p. 213, original emphasis. This paper is not focused on value-theory but the life-value of work. Hence, I must set aside a detailed defence of this position from criticisms. The relevant arguments may be found in Noonan 2012, pp. 3–89.
8   McMurtry 2011a, p. 14.
9   Marx 1986a, p. 188.
10 Marx 1986a, p. 44.

superiority of a socialist society. In order to justify the claim that one form of life is better than another, we require a more general understanding of what makes human life valuable. Socialism cannot be understood simply as the inversion of the capitalist relationship between exchange value and use-value, because not all use values are life-valuable. A hydrogen bomb has use value, but its use is to threaten or destroy life. Definitions of socialism in terms of the prioritization of use-value over exchange-value fail to distinguish goods, relationships, and institutions that improve life by satisfying needs and enabling the realization of our life-capacities from those which are harmful. Capitalism produces an extraordinary array of use-values, but many of them are either harmful to life or ultimately meaningless.

The struggle for socialism is a struggle to free the life-value of labour from the exploitative and alienating dynamics of capitalist class structure. Once the life-value of labour has been freed from class power and reified market forces, labour will become, as Marx said, life's primary need.[11] Non-alienated labour would indeed be a source of *self*-fulfilment, but of social selves that need and want to develop their capacities in ways which are valuable and valued by others. We want to be free from capitalist labour not because it is labour, but because it is alienated, exploited by the ruling class, and subsumed under reified market forces.

In Marx's use and my own, 'alienation' has a socio-economic foundation but pervasive psychological and ethical effects. Labour is alienated under capitalism because workers have no choice but to exchange their labour-power for a wage.[12] 'Freed' from control over the means of production, they must sell their capacity to labour to a capitalist who will pay them, not because they have a particular talent, but because their labour-power can help produce saleable commodities. This alienation (lack of control) over the object of labour and the labour process transforms all aspects of social and individual life. It turns essentially cooperative, life-serving work into competitive struggles between workers for scarce jobs. It makes nature appear as a mere object for profitable exploitation. Finally, it turns our 'species being' – our capacity for world-transforming and self-realizing creative activity – into the bare capacity to exert labour power for a fixed period of time.[13] While Marx would soon drop the Feuerbachian idea of 'species being' from his theoretical vocabulary, the fundamental ethical contrast between a life dominated by reified forces and class

---

11   Marx 1978a, p. 531.
12   Marx 1975b, p. 272.
13   Marx 1975b, p. 275.

power and a life of free cooperation and self-realization was central to his critique of capitalism until the end of his life.[14]

Defenders of a post-work society correctly identify the alienating effects that wage labour has on our sense of ourselves and the value of work. However, they tend to see the capitalist causes of alienation as second order expressions of a deeper problem: externally imposed demands on our time. Capitalism is to them the social expression of a deeper problem that originates in the natural fact that human beings, like all life forms, are dependent on the natural world for survival. Up until this point in history this natural dependence has compelled us to work. If thinkers like Srnicek and Williams are correct, then the deepest problem is dependence on labour for survival, not dependence on wages.[15] Full freedom requires that we allow automated systems to free us from the need to labour at all. Overcoming capitalism is a necessary but not sufficient condition of the highest goal: to free our lifetime form all forms of external dependence and establish the social conditions for free choice in every sphere of activity and relationship. If they are correct, then socialism must be a post-work society. On balance, I do not think that Marx thought of socialism as a post-work society, but rather as a post-alienated work society in which workers have freed themselves *and their life-activity* from *capitalist* productivism and its psychological effects.

Marx always stressed that labour was a two-sided process. On the one hand, it is the means by which human beings produce and reproduce our lives. In this sense it is an 'ever lasting, nature-imposed necessity'.[16] This side is the material basis of labour's life-value across time. Its forms change, but its essential connection to life-creation and life-support remains. Unless we want to conclude that keeping ourselves alive is by definition an alienating waste of time – which would be an absurd conclusion for mortal human beings to draw – we ought to distinguish this natural necessity from oppressive forms of social necessity. That one needs to work in order to live is – in principle – no more an oppressive imposition on lifetime than that we need to sleep. However, labour is not simply

---

14   This position was famously denied by Althusser, who claimed that there was an 'epistemological break' between the *Manuscripts* and *The German Ideology*, and that the later work in no way presupposed the humanist categories of his early philosophy. See Althusser 1977, pp. 21–41. I disagree with this argument, but fully justifying my critique would take me too far afield. In any case, I believe that Althusser's position has been definitively textually refuted by Istvan Meszaros, who shows how analogues of the idea of alienation and species being occur throughout Marx's career, even if the words themselves do not. See Meszaros 1986.
15   Srnicek and Williams 2015, pp. 85–6.
16   Marx 1986a, p. 179.

a natural requirement of life, it is also a social activity. The specific forms that it takes are shaped by the class structure and economic, political, and cultural institutions that define different societies.[17] These social forces are responsible for oppressive forms of necessity. The alienation of workers from the product of their labour is the hallmark of capitalist society. Unique amongst human societies, labour under capitalism is not directly productive of the means of life. Instead, workers work for a wage with which they then purchase from the market commodified need-satisfiers. The need to work thus appears as the social need for a wage. Wages must be earned by subordinating one's goals and desires to the discipline of the labour market and the boss. *That* form of self-discipline is oppressive, but it is a function of capitalism, not labour as such.

The process by which the natural and social life-value of labour is negated by the development of capitalist forces was and remains essentially violent. The original subsumption of labour under emergent market forces mediates but does not destroy the connection between human labour, nature, and life-support. In Europe, peasant labour was torn from the land through the enclosure of the commons. In Africa, indigenous persons were murderously removed from their homelands and enslaved. In the colonized world, indigenous peoples were violently expelled from their lands, killed in wars or by disease, and forced to serve the growth of colonial agriculture and industry. Across the world, these processes of 'primitive accumulation' were written, as Marx says, 'in letters of blood and fire'.[18] David Harvey has reminded everyone that the fires of 'accumulation by dispossession' continue to burn and blood continues to be spilled.[19]

These processes of expropriation and dispossession are the violent means by which human labour is subsumed under capitalist economic dynamics. Marx distinguishes the formal subsumption of labour from the real subsumption of labour. Labour is formally subsumed when it is converted from pre-capitalist to capitalist wage labour.[20] This process occurs at uneven rates in different regions of the earth. The formal subsumption might leave earlier traditions and non-market values intact. The real subsumption of labour occurs when individual workers become part of a global capitalist division of labour and every aspect of their work activity is shaped by market forces. The real subsumption of labour is a social process, but Marx also notes that it has psychological and political effects on the consciousness of workers. The dispositions that emerge

---

17   Marx and Engels 1976a, pp. 31–2.
18   Marx 1986a, p. 669.
19   Harvey 2003b, pp. 137–82.
20   Marx 1986a, p. 313.

once labour has been really subsumed under capitalist dynamics inure people to the repressive demands that the temporal rhythms and spatial distributions of human activity under capitalism require. 'The advance of capitalist production develops a working class, which by education, tradition, habit, looks upon the conditions of that mode of production as self-evident laws of Nature'.[21]

Although Marx does not explicitly make this point, it follows from his argument that the capitalist work ethic is one such psychological effect of treating capitalist norms as laws of nature. Their internalization is aided by the bosses' sermons that hard work is good for the soul. However, when workers preach the value of hard work *under capitalism*, they are giving voice to a necessity that is forcibly imposed on them. They are, we might say, adapting their preferences to an imposed and oppressive reality.[22] They accept capitalist constraints on their horizons and potentialities in periods of suppressed class struggle because there appears to be no alternative. In order to overthrow capitalism as a social reality, workers also have to free themselves from this mindset. However, I maintain that while the desire to work hard under capitalism is pathological, the desire to work hard under changed social conditions would not be.

Overcoming beliefs about the goodness of hard work in its capitalist form is necessary because it keeps workers subordinated to life-destructive forms of productive activity. Marx devotes careful attention to the meaning of 'productive' labour under capitalism. His work as a whole is radically misunderstood if we think that his definition of 'productive' labour under capitalism is applicable to all forms of labour. Indeed, the very definition of productive labour under capitalism distinguishes it from all past (and, I argue, future) forms of human labour by the fact that it formally severs productive activity from direct service to life-support. A portion of the product of peasant labour was taxed to support Church and State, but the primary product of peasant labour *was food*. The primary product of capitalist labour *is not* food, or clothes, or computers, but surplus value that is appropriated as profit when consumers pay the purchase price. Productive labour under capitalism creates surplus value, not need-satisfying goods. Of course, the realization of surplus value depends upon the production of objects with use value, but these will not be produced just because people need them, but only when it is profitable to do so. 'This productivity is based on relative productivity – that the worker not only replaces an old value, but creates a new one; that he objectifies more labour time in

---

21 Marx 1986a, p. 689.
22 See Cholobi 2018, pp. 2–17.

his product than is objectified in the product that keeps him in existence as a worker. It is this kind of productive labour that is the basis for the existence of capital'.[23] Capitalist productivity will let people starve if there is no profit to be made and it will destroy any human talents that cannot be formally subsumed under the capitalist labour process.[24]

Marx himself notes that capitalist productivity is a perversion of true productivity. The true productivity of labour is a function of its life-serving necessity. 'Assuming, however, that no capital exists, but that the worker appropriates his surplus labour himself – the excess of values that he has created over the values that he consumes. Then one could say only of this labour that it is truly productive'.[25] Labour in capitalist society is also truly productive to the extent that it does in fact produce life-values that everyone consumes in order to live. The species would die out if it ignored the 'nature-imposed necessity' of working for a living. True productivity is nevertheless subordinated to the production of surplus value and profit. The difference between true productivity and capitalist productivity explains why abject misery and staggering wealth exist side by side in the capitalist world. Marx makes the link between life-value and true productivity explicit in *The 1844 Manuscripts* when he asserts that labour is 'life engendering life'.[26] Once freed from its capitalist form, people would be motivated to engage in truly productive non-alienated labour because it would be experienced as the production of meaningful, life-valuable, and socially valued life itself.

It is true that Marx argues that capitalist productivity is a material condition for the development of socialism.[27] This argument is the source of the ambivalence towards automation that I mentioned above. In the *Grundrisse* he argues that the 'historical mission' of capitalism is to reduce socially necessary labour time and the historical mission of the proletariat to liberate the

---

23 Marx 1989a, p. 9.
24 Capitalist productivity is thus the process that 'degrades' labour, in Harry Braverman's famous phrase. See Braverman 1998. One of the grotesque ironies of the Covid-19 pandemic has been that at a time when people were worrying about food shortages, capitalist farmers were destroying food. Yaffe-Bellany and Corkery 2020.
25 Marx 1989a, p. 9.
26 Marx 1975b, p. 276.
27 Marx does speculate in a late letter to Vera Zassulich that there might be roads towards socialism that do not run through capitalist industrialization. In particular, he saw the Russian peasant commune as a possible basis of socialist transformation that did not have to pass through the crucible of expropriation and industrialization. Nevertheless, these are comments made in passing in response to a request, not a full blown alternative transitional theory. See Marx 1979, pp. 335–336.

resulting surplus time as free time.[28] For the later Marx, liberation of surplus time as free time involves allowing 'machines to do for us what we formerly had to do for ourselves'.[29] I do not doubt that liberating human beings from imposed drudgery in the service of capital is liberatory. However, the development of technology under capitalism has reached levels Marx could never have imagined. While there remain grounds for scepticism about just how perfectly Artificial Intelligence can simulate human creativity and intellect, it would be scientifically, philosophically, and politically naive for socialist to argue – as the post-work theorists do – that AI and other advanced technological systems are in themselves benign and only their capitalist use is to be feared.[30] If AI does develop in the way serious and sober minded scientists think that it can, it would be able to replace human labour across the spectrum of labouring activity, from the most menial to the highest flights of speculation and creativity.[31] All that would be left for humans to do is play, and a life of pure play, I will argue in Section Three, is a liberal, not a socialist ideal.

Capitalism not only alienates workers from the product and process of production, it also alienates us from nature and each other. At the same time as it alienates us from each other, it also develops the social nature of labour to the highest degree. Capitalist exchange relations are motivated by selfish interests, but within them a more extensive sociality grows. 'The social interest which appears as the motive of the act as a whole', Marx argues, 'is certainly recognized as a fact on both sides, but goes on, as it were, behind the back of the self-reflected, particular interests'.[32] While the socialization of labour takes place unnoticed both by workers, who are struggling to survive, and capitalists, who are reaping profits, capitalist markets draw ever greater numbers of people into an unconsciously but nevertheless materially effective cooperation. 'Materially effective cooperation' means that even the humblest forms of work help produce the shared conditions of life. Capitalism is driven by the goal of developing surplus value, but in the process, by socializing labour, increasing its productivity, and bringing the whole world into communication, it also develops the material and social conditions for the future satisfaction of the full range of human needs and all-round realization of life-capacities. These

---

28 Marx 1986c, p. 250.
29 Marx 1986c, p. 250.
30 I have myself voiced doubts elsewhere and it would take me too far afield to repeat those arguments here. See Noonan 2018, pp. 72–3.
31 In the worst case scenario sketched out by Nick Dyer-Witheford, Atle Mikkola Kjøsen, and James Steinhoff, technological development could produce a "capitalism without human beings." See Dyer-Witheford, Kjøsen and Steinhoff 2019, p. 111.
32 Marx 1986c, p. 176.

needs and capacities are, Marx maintains, real wealth in contrast to the money value, which measures wealth in its alienating, capitalist form: 'if the narrow bourgeois form [of productive labour] is peeled off, what is wealth if not the universality of the individual's needs, capacities, enjoyments, productive forces etc.?'[33] The real subsumption of labour by the forces of capitalist productivity is thus, on the one hand, alienating and exploiting, but, on the other, viewed from the standpoint of materially effective cooperation, the definitive refutation of 'great man' theories of history and the integration of wider and wider circles of people into life-valuable, material effective cooperation and social labour.

The human historical project is not maintained and advanced by singular talents and virtuosi, but by the daily labour of millions of human beings. They are the condition without which life could not continue: socialism liberates the life-value of this cooperative enterprise because real, living human beings decide to work together to liberate the social life-value of work from its 'narrow' bourgeois form. Enrique Dussel argues correctly that the real basis of Marx's critique of capitalism are living humans who can never be completely integrated by the forces of capitalist productivity. 'When the "social relations of production" are mentioned in the Marxist tradition, it is frequently forgotten that the social "relationship" is, first, a "relationship" among persons (practical, political, ethical relationship: namely, it can be just or unjust, perverse or correct) and second, the "social" moment of the relationship indicates the already perverse nature of capitalism ("isolated" worker, non-communitarian labour, etc.)'.[34] The liberation of the totality of needs, capacities, and enjoyments of these real people must include their capacity to freely commit to serving each other and the common good through meaningful, life-valuable labour. Freeing the social life-value of labour means that no one would be forced to toil without cease at meaningless detail work, but it also means, I will argue in Part Three, that all sorts of work which are today alienating could become fulfilling, once their real social life-value is freed in a socialist future. As we will now see, the defenders of full automation and socialism as a post-work society fail to appreciate the social life-value of labour.

---

33 Marx 1986c, p. 411.
34 Dussel 2001, p. 350.

2 Full Automation and Socialism as a Post-Work Society

The post-work camp is unified by the claim that a socialist society should fully embrace automation for the sake of freeing human beings from the need to work. However, they diverge on their interpretation of Marx's position. One side argues that Marx did not go far enough towards re-thinking the role of labour in a socialist society. Kathi Weeks argues that Marxism (and some versions of feminism) remain wed to a 'productivist' ethic that overvalues productive labour.[35] For Weeks, the overvaluation of productive labor would perpetuate repressive psychological complexes like the work ethic within a socialist society. 'Rather than a revised version of the work ethic', she argues, 'Marxists and feminists must make critique of the work ethic a priority of the struggle'.[36] Weeks' argument can be read as a feminist extension of an older critique of Marx's productivism. The older criticism argued that Marx was too much in thrall to capitalist industry, technology, and material wealth. Ted Benton, for example, while broadly supportive of Marx's critique of capitalism, warned that there was a 'Promethean-productivist' strain in Marxism that would perpetuate the destruction of nature under socialism if left unchecked.[37]

The other side of the post-work camp celebrates the Promethean-productivist side of Marx. They add that he could not foresee the full liberatory effects of the technological power developed under capitalism. Aaron Bastani argues that 'despite repeated calls for the working class to liberate itself, Marx did not believe that work makes us free – nor that the society of work expands the scope of human possibility. To the contrary, his view was that communism was only possible when our labour – how we mix our cognitive and physical effort with the world – becomes a road to self-development rather than to survival'.[38] There is something correct in what Bastani argues. Marx believed that communism would free our cognitive and physical efforts to transform the world from *capitalist forms of wage labour*. However, Marx also believed that *non-alienated* labour was at the very heart of human freedom and would remain central to a socialist society as both a felt need for self-development and a willingly accepted obligation of membership in the socialist community. Although he argued that technological development pushed back the bound-

---

35 Weeks 2011, p. 4.
36 Weeks 2011, p. 110.
37 Benton 1996, p. 83. For a critical discussion of this line of thinking see Foster 2000, pp. 134–136.
38 Bastani 2019, p. 55.

aries of the sphere of necessity, he never maintained that it would or could disappear entirely.[39] Hence, dependence on nature and interdependence with others would always remain, and therefore too the natural and social need for non-alienated, democratically distributed work.

Post-work advocates miss this essential dimension of Marx's understanding of the relationship between non-alienated labour and human sociality for three inter-connected reasons. First, they do not regularly, explicitly, and rigorously distinguish between alienated and non-alienated aspects of labour under capitalism and alienated and non-alienated labour as such. As we saw above, Marx made explicit both distinctions. Even within capitalism, labour produces life-value. Socialism thus liberates the life-value of labour submerged under capitalist class relationships. Post-work theorists see only the burdens of labour under capitalism and not the life-value. Believing that socialism is the simple opposite of capitalism, they conclude that that which exist under capitalism will be abolished under socialism. Hence, they see socialism as a post-work society rather than a society that has freed the life-value of labour from exploitation and alienation.

Consider Nick Srnicek and Alex William's defence of a post-work society. They maintain that the belief that work is an essential component of a meaningful life derives from an archaic, religious view of life as suffering: 'This thinking', they argue, 'has an obvious theological basis – where suffering is thought to be not only meaningful, but in fact the very condition of meaning. A life without suffering is seen as frivolous and meaningless. This thinking must be regarded as a holdover for a now transcended stage of human history'. In their view, the contemporary left must embrace a resolutely technophilic understanding of the future and demand 'full automation'.[40] They regard work merely

---

39   Marx 1986b, p. 820.
40   Srnicek and Williams 2015, p. 125. There are at least two problems with this claim, which I cannot pursue here, but they are germane to the argument so I will mention them. First, they implicitly identify suffering with harm. While suffering *can* imply harm, its more fundamental meaning is 'to undergo'. In that broader sense, suffering stems from the passive and receptive side of human being. Humans, like all life forms, are material organisms that depend upon nature and each other. Our lives necessarily involve undergoing experiences that we do not choose. Marx, following Feuerbach, thought that suffering in this sense was part and parcel of being human. Not only is this conception not necessarily religious, to the extent that religions comprehend the passive and receptive side of human existence, they actually contain deep insights into the human condition. For the Feuerbachian foundations of Marx's position, see Feuerbach 2013, pp. 48, 52. For Marx's appreciation of the passive-receptive side of human beings see, Marx 1975b, pp. 299–300. I explore the deeper onto-ethical implications of human finitude in Noonan 2018. The second problem follows straightaway from the first. One could argue that Srnicek and Williams also argue

as 'our jobs, our wage labour'.⁴¹ Marx regarded jobs and wage labour as alienating, but they alienated us not only from our own individual talents and potentials, but also from the 'truly' productive, life-valuable aspects of labour under capitalism: our *social* individuality. The capitalist is only concerned with the production of surplus value, but Marx never forgot that beneath this capitalist appearance lay a deeper reality: the collective satisfaction of shared needs. The distinctions between alienated and non-alienated aspects of labour under capitalism and alienated and non-alienated labour as such are both implicit in the post-work argument, but failure to make them explicit introduces a confusion into their political and ethical conclusions.

The second reason why the defenders of post-work do not understand the crucial link between the life-value of non-alienated labour and human sociality at the heart of the socialist alternative is that they focus one-sidedly on individual self-realization as the essential good of human life. Srnicek and Williams argue that liberation *from* work means liberation from all external constraints on the way that individuals use their time to satisfy their desires. Full automation means-

> liberating ourselves from the decrepit economic image of humanity that capitalist modernity has installed, and in making a new humanity. Emancipation in this vision would therefore mean increasing the power of humanity to act according to whatever its desires might become. And universal emancipation would be the [...] maximal extension of this good to the entire species. It is in this sense that universal emancipation lies at the heart of a modern left.⁴²

Bastani likewise defines his 'Fully Automated Luxury Communism' in terms of *individual* self-realization and desire-satisfaction. He chastises the environmental movement for its 'small is beautiful ethic'. In contrast to environmental asceticism, fully automated luxury communism 'rallies against that command, distinguishing consumption under fossil capitalism [...] from pursuing the good life under conditions of extreme supply. Under FALC we will see more of the world than ever before, eat varieties of food we have never heard of, and lead lives equivalent – if we so wish – to those of today's billionaires. Luxury

---

from a theological perspective: a technological theodicy in which science takes the place of God as redeemer of humanity.

41   Srnicek and Williams 2015, p. 86.
42   Srnicek and Williams 2015, p. 83.

pervades everything'.⁴³ The life of billionaires is certainly an odd model for citizens of a new *communist* society to emulate, but it is consistent with Bastani's consumerist conception of the good life. Luxury is premised on an absolute abundance of consumer goods that eliminates the need for all effort to acquire them. His conception of the good is rooted in the philosophical principles of liberal capitalism that one would expect a communist critic to criticize and go beyond.

Srnicek and Williams do not go all the way down the yellow brick road of luxury communism, but they do argue that the emancipated future must be defined in terms of leisure rather than work. 'Work can be defined in contrast to leisure [...] But leisure should not be confused with idleness, as many things that we enjoy doing involve immense effort. Learning a musical instrument, reading literature, socialising with friends and playing sports all require immense effort'.⁴⁴ Unlike Bastani, who thinks of communism primarily in terms of consumption, or, on the other hand, Brian O'Connor, who has recently defended idleness as the antidote to capitalist and Marxist productivism, Srnicek and William see emancipation in terms of the liberation from all constraints on *self*-realizing activity.⁴⁵ They recognize the value of the effort that learning, relating, and creating pose. Their definition of leisure activity seems to be synonymous with what Marx meant by non-alienated labour under socialism.

On one level this identification of leisure activity and non-alienated labour is true. However, it is one-sided and remains as wed to the same liberal understanding of individuality as Bastani. Marx conceived of socialism as the liberation of the human individual from the dominating power of the ruling class and reified market forces. However, as I noted, for Marx, the individual is a social being. The liberation of social individuals means that the veil of competitive appearances is pulled back and the cooperative reality and life-value of labour becomes apparent. The examples that Srnicek and Williams give of non-alienated leisure activity are – with the exception of socializing with friends – examples of activities that are good because the solitary individual finds them interesting. One might love playing guitar, but not be very good and thus unwilling to play in public, or even for one's friends. One edifies the self by reading literature, but this personal cultivation might make no difference to how one behaves in public – there are plenty of well-educated egomaniacs

---

43   Bastani 2019, p. 189.
44   Srnicek and Williams 2015, p. 86.
45   O'Connor 2018, pp. 61–2.

in the world. The problem is not that individual freedom does not involve personal cultivation and pursuit of one's own interests, it is that there is another side of freedom essential to Marx's understanding of social individuality that is overlooked by the post-work camp. Failure to grasp the social dimension of individual freedom is the third problem with the conception of socialism as a post-work society.

Liberal philosophy conceives of freedom in opposition to external constraints. One is free when material or political circumstances do not impede the satisfaction of whatever desires one feels. However, desires are formed within social contexts, and hence it matters what sort of desires we pursue. Shopping satisfies our desires, but also helps perpetuate capitalist consumer industries. While alienation is rooted in capitalist class structure and private control over the means of life, it also has psychological and ethical effects. Cut off from collective control over what we need and set in atomized opposition to each other in competition for scarce labour and resources, capitalism sets people in opposition to each other. The reign of private property makes 'direct, physical possession' of things and people 'the sole purpose of life and existence'.[46] In this world, people who willingly contribute to the good of the whole look like suckers. The best life is to free ride and consume. Alienation, therefore, makes us think of ourselves as isolated desire-machines pursuing our self-interest. Socialism, as the collective overcoming of the social causes of alienation, would also allow us to revise our self-understanding and conception of our own good. Once we see the truth of social individuality and the social life-value of labour, we will cease to think of other people as barriers to our freedom and externally imposed demands on our time as oppressive burdens. The good of self-realization will remain, but as a good of social individuals for whom contribution to overall well-being is a component part of their private good.

Negatively, the de-alienation of individual consciousness exposes as illegitimate any desires whose satisfaction would use up natural resources in unsustainable ways or harm other people. Positively, the de-alienation of individual consciousness would motivate us to think of our own good in terms of contribution to the whole. That means that the good for individuals in a socialist society cannot be understood exclusively in terms of leisure. Leisure activity is *personally* valuable; non-alienated labour is socially life-valuable, both to the individual labourer as a social being, and to their fellow citizens. Bringing these two sides together under the conception of life-value defined above, socialist society would promote and cultivate the desire in each citizen to realize their

---

46   Marx 1975b, p. 294.

goals and talents in ways which are both personally satisfying and socially valuable and valued. Of course, people would have a right to privacy and their own pursuits, but the socialist conception of individual freedom would overcome the forms of opposition between private and public good typical of capitalism. If freedom is self-realizing activity, and self-realizing activity presupposes the appropriation of resources produced through collective labour, then individual freedom is always a social achievement. The post-work argument does not coherently integrate the social nature of individual freedom.

I say that they do not coherently integrate this argument rather than overlook it, because the role that the obligation to contribute plays in a free life is ironically recognized. Srnicek and Williams, for example, make one exception to their demand for full automation: care labour (including childcare and medical care). They except caring labour because of what they call its 'moral value'. 'A final limit to full automation is the moral status we give to certain jobs, including care work'.[47] I agree, but ask: what does this 'moral status' consist of? My answer is that it consists in its life-valuable content: the material contribution that our labour makes to the satisfaction of the needs of our fellow citizens. To find value in working for the sake of satisfying social needs is to recognize the equal value of other people's interests and needs. When we devote some portion of our life-time to forms of activity that are not only individually interesting but valued by other people as positive contributions to their well-being and the world that we share, we are not altruistically sacrificing ourselves but realizing our individual talents in socially valuable ways.

Socialism, as I argued elsewhere, is a society that is beyond egoism and altruism.[48] Capitalism makes it appear as if good lives are normally a private affair that we might choose to disrupt if we happen to feel the pull of the needs of others on our sympathies. Once we understand the social nature of individuality, we realize that if no one contributes then there will be nothing to appropriate for our own use. To realize ourselves in personally interesting ways thus means that we must also devote some time to producing *common wealth*. The decision to use our time in that way is not a subtraction from free individual activity but one form free individual activity takes. The freedom, goodness, and meaning of individual lives are bound up with recognition of our social nature. We need rest, repose, leisure, fun, frivolity, and hobbies, yes, but we also need to work in the service of the well-being of others. The Covid pandemic has not disrupted the ability of those with wealth to consume, but it has disrupted

---

47   Srnicek and Williams 2015, p. 113.
48   Noonan 2004, pp. 68–86.

our ability to be together in public space. The rush to gather when restrictions have been lifted and riots against further lockdowns suggest that people feel the unbearable emptiness of being isolated consumers waiting alone for Amazon to deliver packages. We need to be together in public space, but we also need to act together to produce and distribute the things that we all need to survive and flourish. Mutual need-satisfaction, and not one-sided self-realization, is the ethical core of socialism. I will conclude by demonstrating that Marx recognized this truth, and that his conception of socialism is not premised upon luxury consumption, individual leisure activity, or freedom from non-alienated labour, but the free *obligation* to contribute our labour to the well-being of others and the society as a whole.

3 Sociality and Socialism

Socialism as a post-work society would fully automate work and maximize leisure-time. As Srnicek and Williams argued, automation potentially frees time for individual projects that we enjoy doing, and the time in which we enjoy doing whatever it is that we are doing is leisure time. The implication of their argument is that work is never enjoyable, because it involves constraints on our activity: while working, we do what we must, while at leisure, we do what we want. The problem is not that they affirm the importance of enjoyment, but rather that they accept the relationship between leisure and work as it is established under capitalism and project it into the socialist future. However, as Sean Sayers pointed out in his defence of the need to work, the capitalist configuration of the work-leisure relationship constricts the potential of *both* work and leisure. 'The need for work and the need for leisure […] is ultimately an […] expression of the development of modern industry […]. These have been defined within the framework of capitalist relations of production'.[49] If those relations of production were transformed, then it follows that we would think of work *and* leisure differently.

Srnicek and Williams reserve enjoyment for leisure activities and see the post-capitalist future in terms of automating all activity that lacks 'moral value'. But if this moral value is, as I argued above, the life-value that our work has for others and ourselves as social individuals, then it could be freely chosen too, and thus enjoyable – but in a different sense – than leisure activities. The contrast between non-alienated labour and leisure is drawn in terms of

---

49   Sayers 1987, p. 24.

how social individuals evaluate demands on their time that arise from other peoples' needs. Recognizing and responding to the needs of others is the substance of meaningful human relationships. Freed from capitalist alienation, people would experience the work that they still choose to do as a life-valuable response to others' needs that gives purpose and direction of their life. Leisure is valuable because it engages our personal interests. Leisure activities are fun. But fun is not the only form of human enjoyment. We can also enjoy doing that which is necessary for collective well-being. Personal interests are necessary but not sufficient for a good life. We are also linked to others through various forms of obligation, which only the most self-enclosed narcissist wishes to live without. Life-valuable labour is one of those obligations.

Post-work defenders have a too narrow conception of individual enjoyment. They defend full automation because it will free our time as completely as possible from necessary activity. From their perspective, freedom means self-realization, and self-realization means the cultivation of our talents and capacities to the highest possible degree. This conception of enjoyment privileges creative pursuits over the mundane – think of Srnicek and Williams' examples of learning a musical instrument and reading literature. These are valuable pursuits to be sure, but artistic self-creation is not the only valuable or enjoyable way to live. As Ursula Huws reports on her thirty years of empirical study of workers' complex struggles with their work under capitalism, workers enjoy work when they feel that it is valued by others. She does not use the term life-value, but I think it explains clearly her ethical position. Like Srnicek and Williams, she uses the example of care workers:

> Care workers, for example, often express strong satisfaction with the feedback that they get from grateful patients, even when their pay and working conditions are appalling. In other words, the experience of labour mixes the positive and the negative in complicated and fluctuating ways. The most mundane tasks may bring some sense of satisfaction to the worker who has carried them out well, while the most exciting may involve an element of drudgery.[50]

Why does the grateful patient enable the caregiver to feel satisfied with their work? Because the gratitude tells the worker that they have done something life-valuable for another person: they have made their life better, if only for a moment. That experience, I argue, is the experience of the positive contribu-

---

50   Huws 2019, p. 16.

tion that fulfilling obligations to others makes to the good of our lives. Freedom is self-realization, including all those forms of self-realization that involve willing response to the needs of others.

Socialism is incompatible with workers toiling in 'appalling conditions'. It is not incompatible with people choosing to devote some portion of their lifetime to life-valuable but physically and emotionally demanding labour. Jasper Bernes understands the value of hard work from a socialist perspective when he criticizes technotopian socialists for their assumption that a revolutionary society can simply appropriate technologies developed under capitalism and put them to alternative purposes. Agriculture is one example where socialist citizens might have to work harder in revived labour-intensive traditional methods. However, since everyone would be conscious of their contribution to the community's healthy food supply, the life-value of the labour, though demanding, would be a source of meaning and enjoyment. Bernes roots his hope for a socialist future in people awakening to the threat that capitalist agriculture poses to our food supply and activating and expanding the bonds of care that link human beings together. 'I take as my baseline an assumption that people organize their lives with an eye to their own survival and well-being and the survival and well-being of those they care about, where the radius of care can be as small as the family nucleus or "friend group" but far more expansive as well'.[51] Provided no one was dragooned into the fields and there was time for other forms of work and self-cultivation, the physical demands of the job could be enjoyable because people would feel that they were working to produce healthy food for each other, not money for the boss.

In a different vein, but leading to a similar conclusion, Sean Sayers' turns the critique of Marx's 'productivism' upside down and concludes that those who reject work as a necessary part of good lives under socialism confuse his critique of *capitalist* productivism with the 'true' meaning of productivism that I discussed in Part One.

> Socialism is a productivist philosophy [...] in the sense that it recommends production simply for the sake of production, but in the sense that it regards production as "man's essential activity" and as a primary human and social value [...]. [W]hat socialism demands [...] is not the liberation of people from work but rather the liberation of work [...] from the stultifying confines of the capitalist system.[52]

---

51   Bernes 2018, p. 363.
52   Sayers 1987, p. 24.

Freeing work from alienation and exploitation will be experienced as liberation and growth of the meaning and value of our lives because it is tantamount to freeing the social nature of our individuality from the competitive egoism of capitalist life. As Peter Hudis correctly argues:

> a society is unfree not because labour-time is a measure, but because socially necessary labour-time is a measure. And socially necessary labour-time ceases to be a measure once individuals *as a social entity* organise social production [...] in accordance with their natural and acquired talents and capabilities.[53]

The precondition of socialism is not that life is freed from the need to labour, but that the organization of labour is freed from the capitalist imperative to produce more of that which is profitable in ever less time. This in-built drive is the deep cause of the environmental crisis confronting us on multiple planes of ecological breakdown. Unless the capitalist imperative to expand production regardless of the material costs to the world, the extinction and climate crises cannot be solved. We cannot move past the capitalist growth dynamic unless we also move beyond a conception of the individual good defined by accumulating possessions. This paper will not solve environmental crisis, but its conception of the good life as *enjoyed, life-valuable* contribution to others' well-being and the world we share does point us beyond the destructive egoism of the consumer desires strategically cultivated in us by capitalist culture. Socialism will not be so much a world of absolute abundance of commodities as it will an abundance of time. The concrete goal is to appropriate the productivity gains that technology makes possible as free time. People can choose how they will realize that time, but if they act as social individuals they will desire to contribute to overall well-being, not just cultivate their private talents and satisfy their own desires. They will need, and create opportunities for, meaningful, life-valuable labour.

I cannot specify in advance every particular job that ought to be preserved from automation because its life-value is a source of meaning for those who perform it. I do want to stress that the list might well-include jobs that intellectuals tend to look on with horror. I live next to the busiest commercial border crossing in North America, criss-crossed by trucks 24/7 under normal conditions. At the height of the pandemic, the flow was reduced to only essential

---

53 Hudis 2013, p. 203, original emphasis. Hudis goes on to criticise Marcuse's version of socialism as a post-work society. I have also examined the ambiguities of Marcuse's understanding of labour's role in a future socialist society. See Noonan 2020.

goods. Supply chains have become so tightly integrated since the passage of the first North American Free Trade Agreement (now the United States, Mexico, Canada Trade Agreement) that the production of food and life-serving goods depends upon the uninterrupted flow of material between Canada, the United States, and Mexico. On the one hand, this fact reveals the dangers of capitalist globalisation; on the other, it puts the work of truckers in a new light. It is no exaggeration to say that their labour has been every bit as important as frontline health care workers during the crisis. When my neighbours and I banged pots and pans every evening at 7:30 to recognize the courage of commitment of essential workers, truckers were amongst those we celebrated.

The point is not say that trucking will necessarily be preserved once artificial intelligence perfects the science of self-driving trucks. It is to say that people might continue to find truck driving valuable and choose to preserve it, even once it becomes possible to do away with it. Moreover – and again, as Huws points out – technological progress has not simply eliminated jobs, it creates new ones, many of which are life-valuable as means of satisfying others' needs and intrinsically valuable.[54] This fact complicates the post-work narrative of a linear relationship between technological change and decline in the demand for labour. The empirical complexities involved in predicting capitalist demand for labour are daunting and I cannot venture into them here. The impossibility of seeing even the capitalist future clearly means that the future configurations of socialist labour practices is *a fortiori* beyond my reach. However, there is enough evidence to support my *principled* conclusion: labour is a source of meaning for individuals who understand themselves as social beings. As Marx argues in one of his earliest attempts to understand alienated labour, capitalism alienates us from human community. 'The community from which the worker is isolated by his own labour is life itself, physical and mental life; human morality, human activity, human enjoyment, human nature. Human nature is the true community of men …'[55] Choosing non-alienated labour is not a choice to perpetuate suffering in isolation; it is a choice – under changed social conditions – to reconnect to life, the human community, activity, and *enjoyment*. It is a choice to devote some of our life-time to realizing our creative capacities in ways that meet others' needs. Life-valuable labour allows us to escape the narrow confines of egocentric self-maximization and do things that are valuable and valued by others. What else could socialism be?

---

54   Huws 2019, p. 14.
55   Marx 1975a, pp. 204–5.

Lest it be objected that I am too wedded to the young Marx's romance of the worker, I will conclude with a principle he espoused in one of his last works. Although it is the most famous aphorism about socialism that he ever uttered, it is never mentioned by the post-work camp. In 'Critique of the Gotha Program', Marx rejects mathematical equality of rights as the principle governing the distribution of wealth under socialism. Instead, Marx argues, in the highest phase of development of the new society, goods will be distributed according to our concrete needs. However, need-satisfaction presupposes that there are goods to satisfy them. Thus, Marx adopts a principle of reciprocity to determine who gets what in the new society: 'after the productive forces have also increased with the all-around development of the individual, and all the springs of co-operative wealth flow more abundantly – only then can the narrow horizon of bourgeois right be crossed in its entirety and society inscribe on its banners: From each according to his ability, to each according to his needs!'[56] Socialism will ensure that everyone gets what they require to live and pursue their individual projects, but people must also contribute their individual talents to the production of need-satisfying social wealth.

Since everyone is working willingly for the good of the whole, of which their own good is a part, the duty to contribute is not an externally imposed burden. It is true that for some it may still feel like an imposed burden, so long as they think of their own good as separate from and opposed to the good of others and the whole. In those cases, education and argument will be necessary to get people to more carefully reflect upon the real conditions of their individuality. The good of the whole is not a reified abstraction apart from the good of each individual to which our private lives can be sacrificed. There is a dialectical relationship between individual and social good. Individuals depend upon relationships with nature and each other in order to survive, and they individuate themselves by channelling shared symbolic resources (language, gesture, inscription, dress, etc.) in unique ways, using their creative intelligence and imagination to produce novel artefacts and shape their personality. Once we learn to see that our self-creative acts have social and natural conditions, we can further understand that to contribute to the social conditions (and not destroy the natural conditions) is not altruistic self-sacrifice but simply what is required in order to become a unique individual. Assuming that we enjoy the process by which we make ourselves into the person we want to be, the contributions that we must make to society as a whole through non-alienated and life-valuable labour will cease to be self-negating and come to feel life-affirming.

---

56   Marx 1978a, p. 531.

A similar argument can be made in regard to the objection that a strict interpretation of Marx's aphorism that those who cannot contribute to society could justly be deprived of the resources that they need. This interpretation perhaps draws the bounds of ability too narrowly and ignores the bonds of solidarity that would motivate production and distribution in a fully formed socialist society. As the disability rights movement has shown over the past three decades, 'ability' and 'disability' are not fixed natural kinds but socially determined.[57] The belief that there is a large category of people unable to contribute to society is rooted in ableist assumptions. Once socially imposed barriers to workplace participation were lifted, people with disabilities proved those assumptions wrong. Incidentally, their struggles *to* work provide very strong evidence in support of my claim that performing non-alienated labour is a fundamental element of a meaningful and enjoyable life. That does not mean that their work is less alienated under capitalism, but they do prove that underneath the alienated forms of work under capitalism is the value of work as meaningful contribution to the social whole of which one is an individual member. That said, even after removing every conceivable social barrier to non-alienated labour, there may be a very small subset of citizens totally unable to work. Those people nevertheless remain human beings and citizens whose needs can be satisfied by drawing upon surplus wealth that the other citizens generate.

The future society will be wealthy in Marx's sense of being rich in human needs. However, it is called *social*ism and not consumerism for a reason: liberated from the tyranny of the boss, the reified forces of the market, and their alienating effects, we will be able to understand clearly the true conditions of our lives, in their biological, social, economic, and ethical dimensions. The most important of these conditions is that we need one another. Mutual need is not a tie that binds us to a repressive work ethic: understood properly, it is an ethical relationship which frees life from both meaningless drudgery and meaningless accumulation of things *for* labour that is meaningful because valuable and valued by our fellow citizens.

---

57 See for example Carpenter 2018, pp. 229–41.

PART 3

*Automation, Labour, and Resistance:
Production, Distribution, Representation*

∴

CHAPTER 7

The Transformation of the Retail Sector: Automation and the Warehouse

Larry Liu

Increasing concerns about automation have pervaded academic and public discourse.[1] Within the capitalist economy the technological content (constant capital) usually increases over time, while the labour content (variable capital) decreases relative to constant capital. This shift occurs to retain competitive advantage against other capitalists and realise more surplus value. The competition among capitalists, in turn, is producing a concentration of capital or a tendency to monopoly capitalism.[2] Marxist labour process research led by Harry Braverman noted that within the context of monopoly capitalism, managers use technologies to standardise output, increase surveillance, and deskill workers, which makes them more replaceable cogs in the wheel and undermines worker organisation.[3] However, before the current titans of technology in Silicon Valley are capable of automating all kinds of work tasks (routine and non-routine) newly created job opportunities in highly robotised workplaces remain precarious, highly surveilled, and deskilled. This finding corresponds to Marx's description of machinery in the labour process: 'In proportion as the use of machinery spreads, and the experience of a special class of workmen habituated to machinery accumulates, the rapidity and intensity of labour increase as a natural consequence'.[4]

Taking the retail sector as a case study, I argue that the automation in the retail industry led by Amazon Inc. reveals a substantial transformation through technology, auguring a shift from brick-and-mortar to online retail, which relies on warehouse and transportation logistics to thrive. Fitting in line with the Marxist research on monopoly capitalism, Amazon is concentrating consumer spending in its own hands, emphasising the normative and practical appeal of convenient shopping of any item imaginable during any time of day. The

---

1 Collins 2013; Brynjolfsson and McAfee 2014; Ford 2015; Bögenhold and Permana 2018.
2 Baran and Sweezy 1966.
3 Braverman 1998.
4 Marx 1990a, ch. 15.

heightened consumer convenience ignores the pushing out of existing brick-and-mortar retailers, who are shedding jobs intensively, although the aggregate employment effect of Amazon's war on its competitors is still uncertain given the substantial gains in warehouse and transportation employment in the face of Amazon's becoming the second-largest private sector employer in the US. It also ignores the rise of precarious, surveilled and deskilled workers. Drawing on field notes of a warehouse tour, and worker accounts from Reddit, YouTube, and published news outlets, this chapter examines the continued dependence on deskilled and surveilled workers in the warehouses. Intrusive forms of technology that make up 'algorithmic management'[5] include shelf-moving Kiva robots, the relentless speed of conveyor belts, and the productivity-tracking hand-held scanners goading on worker speed and efficiency.

Furthermore, developing automation is built into the Amazon strategy, which suggests that the mass army of 'cogs in the wheel' may only be temporary given the advancement of warehouse and transportation robotics. I conclude that while a shift to 'fully automationed luxury communism' may still be somewhat premature (and not intended by Amazon), it cannot be ruled out. I, therefore, reject the binary between automation *and* deskilling, but rather see them as simultaneous occurrences. The trend is that the mass creation of low-skilled jobs is temporary and full automation is the end result.

1   Marxist Labour Sociology

According to Karl Marx, technology has been an essential element in the genesis of modern industrial society. The steam engine that was invented in Britain in the late-eighteenth century not only increased industrial output, i.e., the number of textiles that is created per hour of labour, but also, in part, reconfigured social relations by forming an industrial proletariat and the wage-dependent employment relationship. An important qualification is that capitalist property relations favour the rise of the steam engine.[6] In the paradigm of historical materialism, complex human societies create a social hierarchy that produces unequal access to the economic 'surplus' long before industrialisation and the rise of capitalism.[7] Marx stressed throughout his work that technology is not the driving force behind the precarity of the workforce and the growing

---

5   Rosenblat 2018.
6   Malm 2016.
7   Marx and Engels 1976a.

inequality in capitalist countries. Instead, it is the use of technology in the context of the private ownership of the means of production that exacerbates the social rift.

While the nineteenth century was characterised by free-for-all small business competitive capitalism, Marxist thinkers of the early twentieth century began to note the emergence of monopoly capitalism as rising income levels and the difficulty of finding new productivity-raising innovation resulted in a stagnation of growth, hindering the 'absorption' of new economic surplus, thus favouring the merging of different companies.[8] One of the consequences of monopoly capitalism has been the growth of unproductive or wasteful jobs in sales, finance, and middle management, which have been termed 'bullshit jobs'.[9] Marx himself argued that merchant capital that includes the retail trade (Walmart and Amazon), along with financial capital, is 'unproductive' capital, which stands in contrast with 'productive' industrial capital.[10] While merchants do not directly contribute to the creation of surplus value, they can gain a bigger share of the producer surplus by deepening the use of technology and division of labour and by squeezing suppliers. These are all strategies that have been used by Amazon.

Monopoly capitalism also facilitated scientific management, which was pioneered by the management consultant Frederick Winslow Taylor. Under scientific management, manual labour tasks were broken into component parts and workers were given instructions and financial incentives to perform a task with a certain speed and accuracy. Management became very intrusive by using a stopwatch for each task, and then assigned the most efficient workers to a given task.[11] According to Braverman, scientific management under monopoly capitalism exacerbates worker alienation by deskilling tasks. Labour process is separated from the skill of workers; conception and execution are separated; and managerial control of each step of the labour process is increased.[12] Low skill levels make individual workers easily replaceable, which makes it difficult for them to fight against poor working conditions and low wages.

In most western countries, economic stagnation worsened by the late-1970s, thus inducing employers to restore profitability by forcing organised labour into concessionary bargaining while the non-unionised workforce grew.[13] The

---

8   Hilferding, 2019; Lenin 1916; Baran and Sweezy 1966.
9   Jamil and Foster 2014; Graeber 2018.
10 Marx 1991, ch.20. Also cf. Smith 2020.
11 Taylor 2004; Braverman 1998.
12 Braverman 1998, pp. 77–83.
13 Yates 1999.

assault on labour was, among other reasons, made possible by neoliberal globalisation, in which capitalists in the west offshore production to low-wage labour countries. This hollows out domestic unionised employment in production.[14] Labour was further weakened by new forms of technology, like computer processors or numeric control in the 1980s. These technologies have intensified the trend toward worker deskilling and enhancing managerial control,[15] even as a shrinking portion of the up-skilled working class profited.[16] Advances in information and communication technologies facilitated the monitoring of the workforce far away from management, which resulted in the outsourcing and subcontracting of growing parts of a precarious workforce.[17] The introduction of the internet in the 1990s coupled with the growing economic insecurity in the wake of the 2008 financial crisis resulted in the rise of online platform mediated labour markets that have perfected algorithmic tracking, centralised and automated surveillance systems, and minimum levels of expected skill.[18] In this context, Amazon built a retail empire of such proportions that it has since become a major employer in the US and other developed countries.

2 Amazon and the Political Economy of Retail

The expansion of Amazon's online retail infrastructure and first-mover advantage meant that by 2019, 38 percent of all e-commerce spending in the US flowed to Amazon.[19] A 2018 NPR/Marist poll shows that 44 percent of online shoppers first go to the Amazon website to order goods, making it more popular than other search options.[20] As of April 2018, 75 million people have subscribed to Prime (a membership fee in exchange for free two-day shipping and additional services) and 35 million use someone else's account, which means that one third of the population is an Amazon customer.[21] Ninety-two percent

---

14   Levinson 2013; Freeman 2008.
15   Zuboff 1988; Noble 2011.
16   Acemoglu et al, 2001.
17   Kalleberg 2011; Weil 2014.
18   Rosenblat 2018; Ravenelle 2019.
19   Day and Soper 2019.
20   ONLPRCHSRT1. NPR/Marist Poll, 25 April through 2 May 2018, available at: http://maristpoll.marist.edu/wp-content/misc/usapolls/us180423_NPR/NPR_Marist%20Poll_Tables%20of%20Questions_May%202018.pdf#page=2.
21   Selyukh 2018.

of all online shoppers ever purchased things on Amazon,[22] and 43 percent are regular Amazon shoppers.[23]

Amazon's growth is perceived as a threat by its competitors. In 1997, Barnes and Noble sued Amazon, alleging that Amazon is not a bookstore as self-advertised, but a book broker. In 1998, Walmart sued Amazon for stealing distribution and merchandising trade secrets after hiring Walmart executives.[24] Suppliers' margins are squeezed in the competition between Amazon and Walmart. Amazon can dictate lower supplier prices. When Walmart discovers these lower prices, it then demands from these suppliers to match the Amazon prices. Walmart has also forced some suppliers to pull their products from Amazon with the threat of losing business with Walmart if the suppliers refuse to comply.[25] Increased buyer power and the squeeze on suppliers has been associated with wage stagnation in the broader economy.[26] In June 2019, FedEx, which is an important part of Amazon's logistics network, announced the end of its air-shipping contract with Amazon, indicating that FedEx regards Amazon's internal logistics plans as a threat to FedEx's business model.[27]

Amazon's efforts to concentrate retail have given it outsize influence when deciding on a location for its second headquarters (the first headquarters is in Seattle), which it announced in September 2017.[28] Amazon promised the creation of 50,000 new jobs and $5 billion investment, and it expected different municipalities across the country to compete with each other to give Amazon more generous subsidies, tax concessions, and breaks. In January 2018, Amazon had narrowed the list to 20 cities.[29] Not all cities were enthusiastic in giving tax breaks to Amazon. The San Antonio city government, for instance, refused to submit a proposal to Amazon, stating that "blindly giving away the farm isn't our style."[30] Andrew Schwartz, from the Center for American Progress, raises significant doubts that tax giveaways for big businesses spur job creation, and even if new jobs are created the taxpayer costs are substantial per job. There is no guarantee that businesses receiving subsidies will survive or might decide to relocate once they have collected the subsidies and incentives. The subsidies could also come at the cost of public services (e.g., education

---

22 PRCHAMZ1. NPR/Marist Poll 2018.
23 AMZFRQ1R. NPR/Marist Poll 2018.
24 Each of these cases were settled out of court, see Reference for Business 2002.
25 PYMTS 2019.
26 Wilmers 2018.
27 Ziobro 2019.
28 Barron 2017.
29 Wingfield 2018.
30 City of San Antonio 2017.

or affordable housing), which has a long-term negative impact on economic development.[31] In November 2018, Amazon announced that Long Island City, Queens, New York City and Crystal City, Virginia would become the secondary HQ locations.[32] The announcement was followed by anti-Amazon protests in New York. The protesters cited the unwise use of $3 billion in tax incentives to lure Amazon. As a result of the protests, Amazon announced four months later to withdraw their New York expansion decision,[33] restricting the secondary HQ to Crystal City.

Amazon's size and economic power have drawn high-profile criticism from political leaders. Former President Donald Trump has a profound dislike for Amazon, citing conversations he has with his real estate friends who regard online retail as a threat given the declining shopping mall rents and closing retail shops. Trump has alleged that Amazon has been taking advantage of the US postal service for getting deep discounts, even though USPS is still profiting from their business with Amazon.[34] Senator Bernie Sanders focused on Amazon's low wages and in September 2018 introduced a bill in the Senate titled Stop BEZOS Act (Stop Bad Employers by Zeroing Out Subsidies Act, a clear dig at Amazon CEO Jeff Bezos). The bill would have imposed a surcharge on corporate taxes for large firms that have workers that qualify for welfare benefits like food stamps or Medicaid. Senator Sanders hoped that big firms with a low-wage workforce would increase wages in response.[35] While the bill stood no chance of passing Congress, Amazon promptly announced a minimum wage of $15 an hour. It was positive publicity at a time when the company was facing a labour shortage during a booming economy.[36] Furthermore, the wage hike also came with the announcement to cancel company bonuses and stock-options for employees, which reduces the positive impact of Amazon's wage hike for warehouse workers.[37] Nonetheless, there was a spike in job applications just in time for the holiday season.[38]

Amazon's growing market power has raised calls for regulating or even breaking up the company. The first antitrust complaints emerged in 2012 among booksellers and publishers who were concerned about being devoured

---

31 Schwartz 2018.
32 Taibbi 2018.
33 Goodman 2019.
34 Kosoff 2018.
35 Bhattarai 2018.
36 Weise 2018.
37 Del Rey 2018.
38 Griswold 2019.

and priced out by Amazon.[39] The calls for regulating or breaking up Amazon have only increased since then. In March 2019, Senator Elizabeth Warren called for the breakup of big tech companies, including Amazon.[40] While antitrust concerns emerge from growing political concerns about economic inequality, the present case of tech monopolies is complicated by their large-scale popularity among customers,[41] which is in line with Bezos' emphasis on 'customer obsession rather than competitor focus'.[42] An Amazon break-up could result in customer pushback and reflects the power of the network effect, i.e., the more people use a service, the more value it has for each user.[43] Within the framework of a capitalist economy, only an even bigger monopolist that dominates even more domains of business can beat Amazon.

Amazon's growing monopoly power and the rise of online retail have led to the diminution of brick-and-mortar retail employment. The retail sector slashed over 194,000 jobs between January 2017 and August 2019.[44] At the same time, non-store (online) retail share increased from 10.7 percent to 12.2 percent between July 2018 and 2019. While the volume of general retail increased by only 3.4 percent (and even declined for electronics, appliances, clothing, sporting goods, and department stores), non-store retail increased by 16 percent in the same time period.[45] A further indication for the decline of brick-and-mortar stores is that since the end of 2016, the vacancy rate at regional and superregional malls has been increasing, while the rental rate increases have become negligible and even negative at one point since the beginning of 2018.[46]

First reports on a 'retail apocalypse' appeared in 2012[47] and became ubiquitous by 2017.[48] Retail is still thriving in a highly consumerist economy, but is now primarily based on online retail, which is less labour-intensive than brick-and-

---

39 Neary 2012.
40 Lee et al. 2019.
41 Selyukh 2018; Culpepper and Thelen 2019.
42 Amazon 2018.
43 Shapiro and Varian 1999.
44 Franck 2019.
45 Calculated from the US Census Bureau 2019, 'Advance Monthly Sales for Retail and Food Services, July 2019', 15 August, available at: https://www.census.gov/retail/marts/www/marts_current.pdf.
46 Real Estate Solutions, Moody Analytics 2019, 'Retail Preliminary Trends, Q2 2019', 3 July.
47 McKinsey 2012.
48 A simple proxy is to count the number of references by date in the Wikipedia article on "Retail Apocalypse" (retrieved Sep. 9, 2019). Out of 51 references, 12 were published in 2017, 23 in 2018 and 8 in 2019. Another proxy is a keyword search of the term in NewsBank (newspaper database), which produces less than 500 results annual average prior to 2017, 1,001 in 2017, 1,224 in 2018 and 764 in 2019 until 11 September, 2019.

mortar retail. The opening of an Amazon fulfilment centre does not increase overall local employment, even as states and municipalities have been vying to attract Amazon offices and warehouses via tax breaks.[49] In Ryan Carrier's analysis of BLS data in 2017, Amazon is responsible for 198,000 job losses in the non-store retail sector and a loss of 329,000 jobs in competing retail sectors from 2007 to 2018.[50] Amazon employment (750,000 by the end of 2019) is partly counted in the non-store retail sector, although many jobs are counted in transportation (e.g., warehouse workers or delivery drivers) and professional service (e.g., software engineers and robotic technicians). The loss of retail employment may be counterbalanced by the rise of warehouse employment, which doubled from 600,000 to nearly 1.2 million from 2007 to 2019,[51] and courier employment which increased by one-third of its size to 773,000.[52] The aggregate Amazon employment impact across the entire economy is, therefore, ambiguous.[53]

3 Methodology on Amazon Warehouse Work

To capture the Amazon warehouse work experience, I draw on three types of source: (1) published journalist accounts;[54] (2) Reddit discussions (the subreddits for Amazon warehouse workers are called 'FASCAmazon' and 'AmazonFC'); and (3) 43 testimonies of Amazon warehouse workers that were self-recorded and uploaded to YouTube. The YouTube testimonies, which I transcribed and thematically coded, vary in length from 5 to 60 minutes. I found them by using the search term 'Amazon warehouse work' and by clicking suggested/related videos.

---

49 Jones and Zipperer 2018.
50 Carrier 2017.
51 Federal Reserve Economic Data, 'All Employees: Transportation and Warehousing: Warehousing and Storage', Retrieved 11 September 2019, available at: https://fred.stlouisfed.org/series/CES4349300001.
52 Federal Reserve Economic Data, 'All Employees: Transportation and Warehousing: Couriers and Messengers', Retrieved 11 September 2019, available at: https://fred.stlouisfed.org/series/CES4349200001.
53 For Mandel 2017, warehouse employment more than compensates for retail losses. For Jones and Zipperer 2018, Amazon has no positive impact on local employment.
54 Guendelsberger 2019. Bloodworth 2018.

4 Technology and Productivity

Amazon's strategy of capturing a bigger chunk of the retail market is based on the convenience of online shopping, which, in turn, comes from the heavy use of robotics in its logistics chain. The number of robots in the warehouses increased from 1,000 in 2013 to 100,000 in 2017, 200,000 in 2019[55] and 520,000 in 2022.[56] The Amazon website touts the job-creating potential of robotics:

> [Robots] aren't competing for jobs – they're creating them at Amazon fulfilment centres. Transporting thousands of pods per floor with millions of products stowed inside, the robots enable more inventory to pass through a fulfilment centre, which means more associates are needed for handling that inventory. Since 2012, Amazon has added tens of thousands of robots to its fulfilment centres, while also adding more than 300,000 full-time jobs globally.[57]

According to Amazon, the use of robots has increased inventory size by 40 percent per warehouse. Robots do the heavy lifting like grabbing totes (bins) from conveyor belts and stacking them on pallets for shipping or stowing. Robots allow employees to walk less, hence raising worker output from a stationary position with the robot carrying and moving the shelves. The heart of automation at Amazon is the orange Kiva robot, which moves shelves. Across the length of the warehouse, automated guided vehicles (AGVs) bring items to employees. Conveyor belts move rapidly to connect inbound, picking, stowing, packing, and outbound stations.

Despite the introduction of a robot workforce to supplement the human workforce robots have not yet fully automated output. A robot cannot unpack a pallet, pick, stow, package, or load trucks, which are all tasks that are left to humans. Nonetheless, the implication of robotics is increased productivity, which can be measured as revenue generated per employee, which increased by a whopping 20 percent between the last quarter of 2018 and the last quarter of 2019.[58] In contrast, Amazon's major competitor Walmart increased pro-

---

55 Jon Ehrlichman, Twitter, 2 August 2019, available at: https://twitter.com/jonerlichman/status/1157351209583501315?lang=en.
56 Amazon 2022.
57 Amazon 2019.
58 CSIMarket, 'Amazon Com Inc Sales per Employee', accessed 13 April 2020, available at: https://csimarket.com/stocks/AMZN-Revenue-per-Employee.html.

ductivity by barely 2 percent.⁵⁹ Greater productivity could imply cheaper prices (though not necessarily), which attract more customers, which in turn strengthen Amazon's drive to monopoly power.

Productivity rises not only originate from new technology but also from organisational games, which are known to encourage more worker effort.⁶⁰ In the factory game, managers encourage workers and workers encourage each other to surpass production targets, which increases the prestige and respect of the faster workers. During Amazon's peak season, the organisational game is to have a so-called 'power hour', where workers work as fast as they can and compete with co-workers in the same group inside the warehouse. The fastest worker wins a prize such as an Amazon gift card or a video game. Amazon would also occasionally buy food gifts to motivate their workers.⁶¹

5   Technology and Surveillance

Amazon uses technology to track worker productivity. As Marx predicted, within the capitalist context technology alienates workers from the product they are producing. In the case of the warehouse, workers are not manufacturing a product, but they are processing them with the tangible output being the delivery of the product to the customers' doors. Amazon operates in the sphere of circulation rather than the sphere of production.⁶²

The main technology that surveils workers is the hand-held scanner. The scanners record how long it takes to scan and process items.⁶³ After a grace period in the first month of work, if the worker is too slow, the manager will come with a laptop or tablet by the hand and coach the workers, giving tips on how to scan more efficiently or write-up/warn the employee.⁶⁴ The stricter measures tend to be reserved for workers performing in the bottom 5 percent with regard to speed.⁶⁵ Harshness also depends on the season, whereby inab-

---

59   CSIMarket, 'Walmart Inc Sales per Employee', accessed 13 April 2020, available at: https://csimarket.com/stocks/WMT-Revenue-per-Employee.html.
60   Games are similar to what happened in the Soviet Union. Stakhanov was hailed as an exemplary Soviet worker. The Soviet leaders hoped to encourage effort without paying differential salaries. Burawoy 1979.
61   Dzieza 2019a.
62   Marx 1992.
63   Guendelsberger 2019.
64   Neon X Sonic, 'Working at Amazon! My Story', Youtube, 29 October, 2018, available at: https://www.youtube.com/watch?v=tXyHjbNN_8Q.
65   Lecher 2019.

ility to 'make rate' is less concerning during the 'peak' period, i.e., between Thanksgiving and Christmas, when a lot of labour is needed, while not making rate after the end of peak is a good excuse for layoffs.[66] Management, which can vary by location, also influences the harshness of disciplinary measures. Some workers like the lack of managerial intrusion as long as productivity targets are met,[67] while others complain about favouritism and arbitrary punishments by managers that do not like certain workers.[68] Workers do not want to violate time-off-task (TOT) regulations, which vary from 2 to 6 minutes. TOT is the time between scanning two items.[69] The desire to avoid TOT-related write-ups discourages visits to the bathroom. Break times are precisely timed.[70] Workers perceive the 15-minute breaks as extremely short because it can take several minutes to reach the breakroom and return from it.[71] Personal phones cannot be brought into the workstation and are checked when passing through the metal detector in the security gate. Failure to comply results in write-up.[72] As of April 2022, the cell phone ban has been lifted.[73] Shifts are between 10 and 12 hours long with mandatory extra time during the peak season.[74]

6   Technology and Deskilling

In addition to surveilling worker productivity, the scanner results in deskilled work.[75] A worker who was trained as a picker explains how the scanner regulates his workflow:

---

66   Angry Workers of the World 2018.
67   Raquel Lancaster, 'What it's like to work at Amazon', *Youtube*, 7 October, 2019, available at: https://www.youtube.com/watch?v=NAa48XO2vso
68   TheArtMom, 'Storytime: Working at Amazon at your own risk', *Youtube*, 31 March, 2017, available at: https://www.youtube.com/watch?v=qm7H2lKcctA.
69   Reddit, Scan to Scan TOT available at: https://www.reddit.com/r/AmazonFC/comments/fim9ro/scan_to_scan_tot/.
70   MyLifeAsMel, 'Amazon Update: First week at Amazon', *Youtube*, 22 June 2017, https://www.youtube.com/watch?v=HNn5kY4gqqs.
71   Frankenghoul, 'Update: Working at Amazon', *Youtube*, 27 June 2018, https://www.youtube.com/watch?v=OWLDfoaKOo4.
72   Reddit, 'IamA Security Guard at an Amazon Fulfilment Centre', available at: https://www.reddit.com/r/IAmA/comments/14m7oz/iama_security_guard_at_an_amazon_fulfilment/.
73   Navlakha 2022.
74   Cedric Johnson, 'Working at Amazon', *Youtube*, 24 December 2018, available at: https://www.youtube.com/watch?v=WBarm4TMZyY.
75   For the trend toward deskilling skilled work, see Harvey 2019.

> As a picker you have this scanning gun [...] You walk around with a little basket [...] and it has two totes [...] and you're pushing it around. So you scan the band. It tells you what I need to pull out. You pull out the item. You scan the item. You put it in your tote. Then your scanner tells you what your next item is and how long it should take you to get there.[76]

The scanner specifies the tasks of the workers, which are simple and repetitive. Body movements are monotonous but exhausting, resulting in back and foot pain especially for older and sicker workers.[77] Monotony and boredom can normally be averted by social interactions with co-workers, but these encounters are made difficult by the need to reduce TOT and the high turnover due to the stressfulness of the job. Tiredness after a long shift also reduces the desire for social interaction.[78]

The limited required skill level is reflected in the very low hiring barrier and the short training time, which lasts only a few days. Newly hired workers are shown a video with safety instructions and explanations of how to get around the warehouse.[79] They are lectured on disciplinary issues (e.g., reasons for termination) and are shown the workstation with so-called ambassadors and problem-solvers by their side to offer assistance when they get stuck.[80]

7   Technology and Precarity

The organisational surveillance and deskilling via technological systems contribute to the precariousness of warehouse workers. On the one hand, the $15 minimum wage makes the job quite attractive given that a high school diploma and a negative drug test are the only requirements to get the job. On the other hand, most workers are aware that given the simple, deskilled nature of their work, they are replaceable. The highly physical nature of the job results in heightened labour turnover. The National Employment Law Project estimates that counties with an Amazon fulfilment centre have a turnover rate of 100.9

---

[76] TobyInTransit, 'My experience working at Amazon', *Youtube*, 7 August, 2018, available at: https://www.youtube.com/watch?v=bgNv3X_Wekg.

[77] MyLifeAsMel, 'Amazon Update: First week at Amazon', *Youtube*, 22 June 2017, available at: https://www.youtube.com/watch?v=HNn5kY4gqqs.

[78] Shera DiVito, 'Working at Amazon: My experience', *Youtube*, 10 March, 2016, available at: https://www.youtube.com/watch?v=Z8KhRoqnnGc.

[79] Guendelsberger 2019.

[80] MyLifeAsMel, 'Amazon Update: First week at Amazon', *Youtube*, 22 June 2017, available at: https://www.youtube.com/watch?v=HNn5kY4gqqs.

percent (i.e., more workers leave warehouse jobs each year than the total number of warehouse workers employed in a county), compared to 69.8 percent for warehouse workers overall across the US.[81] Unlike high-skilled industries, this high turnover is not necessarily dysfunctional in a deskilled workplace that constantly hires to replenish depleted ranks.[82] Technological surveillance becomes the basis for disciplining, which is still done face-to-face by managers, thus reminding workers that they may be fired at any time.

Amazon hires all new workers as seasonal hires, especially during the peak period, and then fires the slow workers or those having conflicts with management when the business slows down. Workers are aware of their precarity and some desire a conversion from 'white badge' (seasonal) to 'blue badge' (permanent), although that conversion is uncertain.[83] Promotion to a permanent position includes medical, vision, dental, 401(k) retirement plan, parental leave, and tuition benefits. Another common response to precarity is a worker discourse of indifference, where employees downplay their commitment to the job as they are 'only doing it for the check' or look elsewhere for better job opportunities.[84] One of the few anchors of stability for warehouse workers is the predictability of the work schedule. The latter is fairly long, but the slots do not change, which allow workers with young children to plan childcare.[85] On the other hand, workers can volunteer for extra time/shifts (VET) on a mobile app or can be selected for voluntary time off (VTO), which is an early end to the shift during slow periods.[86] The employment contract is at-will and workers are not members of a labour union.

---

81   Tung and Berkowitz 2020.
82   Ehrenreich 2006.
83   Reddit, 'Blue badge conversion help', available at: https://www.reddit.com/r/FASCAmazon/comments/j32vs0/blue_badge_conversion_help/
84   Neon X Sonic, 'Working at Amazon! My Story', *Youtube*, 29 October, 2018, available at: https://www.youtube.com/watch?v=tXyHjbNN_8Q; Raquel Lancaster, 'What it's like to work at Amazon', *Youtube*, 7 October 2019, available at: https://www.youtube.com/watch?v=NAa48XO2vs0.
85   Sharetha Herrod, 'The Best Amazon Experience Update', *Youtube*, 16 August 2018, available at: https://www.youtube.com/watch?v=aLX6b3vBliU.
86   Reddit, 'Find voluntary extra time?', available at: https://www.reddit.com/r/FASCAmazon/comments/hpkm2r/find_voluntary_extra_time/; Reddit, "Where is the VTO?" available at: https://www.reddit.com/r/FASCAmazon/comments/ikkwn7/where_is_the_vto/.

8 Automation as a Moving Target

One can only guess whether Amazon is capable of automating all warehouse work, though that is clearly the aspiration. Since 2014, the Kiva robots have been introduced by Amazon and have substantially reduced the number of employees walking long distances to retrieve items. SLAM (scan, label, apply, manifest) is an automated device to apply address labels on boxes and was introduced in 2013.[87]

Amazon invests in companies that advance warehouse automation. In January 2019, Amazon entered a 7-year agreement with Balyo, a French company selling autonomous forklifts.[88] In April 2019, Amazon purchased Canvas Technology, a robotics start up based in Colorado that specialises in autonomous carts similar to the AGVs.[89] In May 2019, Amazon introduced CartonWrap, a robotic packaging machine developed by CMC (based in Italy) that can reportedly pack four to five times as fast as human packers and is expected to cut 1,300 jobs.[90] The added machine costs will allegedly be recovered in saved wages within two years. Amazon regards the automation of human hand motions for picking and stowing as the biggest challenge, though there are various technology companies like Soft Robotics in Boston that have teamed up with Amazon to solve this challenge.[91] Robots have already replaced some unskilled manual work in warehousing and back office jobs in corporations, e.g., the use of carousels, vertical lifts, automated storage and retrieval systems (AS/RS), mini-loads, and automated material-carrying vehicles.[92] The role of 'pickers' in warehouses (i.e., the employees who recognise and move objects) is reduced through the use of robotic arms, sensors, and artificial intelligence.[93] In November 2022, the Amazon robotics center presented the "Sparrow" robotic arm that picks and places objects in bins. It can recognize 65% of the total product inventory.[94] There was a reduction of seasonal holiday hires from 120,000 to 100,000 from 2017 to 2018,[95] despite flourishing sales. On the other hand, the number of seasonal hires for 2019 doubled to 200,000.[96] The

---

87 Curtis 2013.
88 Leonard 2019a.
89 Leonard 2019b.
90 Cosgrove 2019.
91 Dastin 2019.
92 Allais 2017.
93 Guillot 2018.
94 Heater 2022.
95 Griswold 2018.
96 BNN Bloomberg 2019. Also cf. Ehrlichman, Jon. Twitter, 2 December 2019. available at: https://twitter.com/JonErlichman/status/1201592086807617536.

coronavirus pandemic-related shutdown of local businesses in the first half of 2020 resulted in the further hiring of 100,000 workers.[97] This is an indication that Amazon is still in an expansion phase.

Amazon's investment in R&D topped $23 billion in 2017, far in excess of other Silicon Valley tech companies.[98] Starting in May 2019, Amazon began to convince employees (from all branches, but mostly inside the warehouses) to become 'independent entrepreneurs' in the Driver-Service Partnership (DSP) program.[99] This suggests that the greatest labour shortage is to be found in the 'last mile' (delivery) rather than the middle of the logistics chain (warehousing), which can be more easily automated.

The broader Amazon strategy is automation: Amazon Go is the food-retail arm of Amazon, which minimises employment to shelf-stockers and security guards while eliminating the cashiers via sensors, cameras, and a customer phone app.[100] Prime Air is experimenting with automated drone delivery to reduce reliance on delivery drivers, although regulatory hurdles and technical glitches push the timetable for introduction into the indefinite future.[101] Alexa is a personal voice assistant that does secretarial or clerical work like gathering information or scheduling events.[102]

In July 2019, the company announced that it would retrain one third of its workforce to become technicians, including data mapping specialists, data scientists, solutions architects, and security engineers. Others will be trained to become nurses or aircraft mechanics.[103] For public relations purposes, Amazon can argue that the training program is socially beneficial,[104] as most employers have become reluctant to train workers who can easily be poached by their competitors,[105] and more risks of training can be shifted from employers to workers. But, implicitly, this is also Amazon's admission that many of their current employees' jobs will not exist in a few years, contrary to what they state on their website.

---

97 Duffy 2020.
98 Fuscaldo 2019.
99 Palladino 2019; Vertesi et al. 2020.
100 Dignan 2018.
101 D'Onfro 2019.
102 The number of secretaries and administrative assistants collapsed from 3.3 million in 2000 to slightly less than 2 million in 2018, though much of that decline happened before 2010. Federal Reserve Economic Data. "Employed full time: Wage and salary workers: Secretaries and administrative assistants' occupations: 16 years and over." Retrieved 11 September 2019, available at: https://fred.stlouisfed.org/series/LEU0254502800A.
103 Porter 2019.
104 Wharton Business School 2019.
105 Cappelli 1999.

Amazon wants to contain labour unrest via automation. Marx argued that automation represses working class strikes by making a portion of labour redundant:

> But machinery not only acts as a competitor who gets the better of the workman and is constantly on the point of making him superfluous. It is also a power inimical to him, and as such capital proclaims it from the roof tops and as such makes use of it. It is the most powerful weapon for repressing strikes, those periodical revolts of the working-class against the autocracy of capital.[106]

In the contemporary context, there is a national pressure to increase wages in a tightening labour market, labour strikes in relation to Prime day,[107] and a bid to improve publicity for Amazon. US worker strikes are rather rare, but a frequent occurrence in European Amazon warehouses, as union organisation in Europe is stronger. In the US, despite the lack of unionisation, several cases of labour unrest made headlines in 2019: a Sacramento delivery centre fired a worker for unpaid time off exceeded by one hour after the worker's mother-in-law died. Although none of the workers on site was unionised, they formed an informal organisation called Amazonians United Sacramento and sent a letter to the manager and HR department to demand the reinstatement of the laid off worker. They also demanded that part-time workers get paid time off. Amazon subsequently rehired the worker.[108]

In March 2019, 50 Somali-American workers in Shakopee, Minnesota went on strike to protest work speedups and the reduction in the number of acceptable scanning errors.[109] In August 2019, in close-by Eagan, Minnesota, 80 workers walked off the job, citing a lack of employee parking and unreasonable fees for towing expenses.[110] In October 2019, in the same location 60 night-shift employees walked off the job, citing dissatisfaction with being kept part-time so Amazon can avoid paying for the workers' health insurance (which would kick in at more than 30 hours of work per week). They also demanded a more respectful work environment, complaining about managers knocking at the bathroom door when employees are using it. Moreover, the workers complained about the very high pace of work. Because many Minnesota warehouse

---

106 Marx 1990a, ch. 15.
107 Wu and Schoolov 2019.
108 Kaori Gurley 2019.
109 Demanuelle-Hall 2019.
110 Menegus 2019.

workers are Muslim immigrants from Somalia, they protested against not being able to observe Ramadan. The walkout ended after two and a half hours when the manager promised to speak to the upper management.[111]

UNI Global Union, a global union federation with 20 million members based in Switzerland, developed the tagline in the Amazon campaign, 'We are not robots'. They cite the brutal working conditions tied to the high productivity requirements that are reinforced by robots and tracking technology.[112] The British GMB Union, which organises over 600,000 workers in retail, security, schools, distribution, utilities, social care, and health care, designed a website titled 'workers, not robots' that has a whack-a-mole game where site users click on a package on the screen, which then disappears and new packages pop up, reminding the user that one cannot take toilet breaks. The site encourages users to share this website with others and write letters to Amazon to build public pressure against Amazon to improve working conditions.[113]

In New York, one of the few national union strongholds, the Retail, Wholesale, and Department Store Union (RWDSU) announced its effort to organise 2,000 workers at JFK8, an Amazon warehouse in Staten Island, in December 2018. RWDSU advocates for better working, health, and safety conditions and battles the firing of workers who associate with union activity. RWDSU has filed at least 50 labour grievance cases with the National Labour Relations Board related to unfair dismissals for union activity.[114] These dismissals make workers fearful of joining a union.[115] The Staten Island union vote has succeeded in April 2022, although as of the end of 2022, Amazon has not formally certified the union vote even though the company lost a legal challenge against unionisation, and was ordered to reinstate workers that had been fired during the vote drive.[116] In contrast, a 2021 attempt by RWDSU to unionise the Bessemer, Alabama warehouse failed by a margin of 2 to 1 after an intense anti-union campaign by Amazon's management.[117]

Worker protests are not exclusively about labour conditions. Among the HQ employees, 900 Amazon workers signed an internal petition to demand that Amazon take action to do more to combat climate change. They also walked

---

111 Dzieza 2019b.
112 Cain and Hamilton 2019; UNI Global Union 2018.
113 https://workersnotrobots.com/.
114 McPherson 2019.
115 Reddit, 'Amazon employee explains hellish working conditions of an Amazon Warehouse', available at: https://www.reddit.com/r/bestof/comments/8d64rj/amazon_employee_exp lains_the_hellish_working/.
116 Sainato 2022.
117 Jones 2021.

off the job in September 2019. Amazon has promised to make 50 percent of its shipment net zero carbon.[118]

Amazon is fighting unionisation by showing their managers anti-union videos in which managers are told that Amazon has a 'direct management' structure where employees can bring complaints to bosses directly rather than unions. The video also shows warning signs of employees that plot to unionise a warehouse, e.g., using words like 'living wage', distributing petitions and fliers, or raising concerns on behalf of co-workers. Managers are told to advise their employees that 'having a union could hurt innovation which could hurt customer obsession, which could ultimately threaten the building's continued existence'.[119]

Rising wages and worker activism provide the impetus for future automation.[120] A common theme in the Reddit discussions is the expectation that warehouse jobs are about to be automated. One forum member writes:

> Humans are currently the most efficient and least expensive way to move stock to and from various conveyors. It won't be long before Amazon has a way of picking a product off of one of their pods and putting it in the right bin or carton for shipping and of assembling an order for final processing. No one should be looking at an Amazon warehouse job as a long-term employment prospect.[121]

Whether these public expectations are accurate or not, Amazon favours the perception of quick automation to make the stressful work environment less problematic. Similarly, Rosenblat views Uber's automation narrative as 'subtle propaganda tool to justify poor working conditions'.[122] If the robots are coming for your job, why bother making it better?

---

118   Jee 2019.
119   Menegus 2018.
120   Boggs 2009.
121   Reddit, 'Amazon warehouse life revealed with timed toilet', available at: https://www.reddit.com/r/SeattleWA/comments/7fwrjf/amazon_warehouse_life_revealed_with_timed_toilet/.
122   Rosenblat 2018, p. 180.

9   Conclusion

Technological advancement and creative destruction are in operation within the US retail sector as brick-and-mortar retail is replaced by online retail with its attendant rise in warehouse and transportation logistics. The effects on labour relations involve the intensification of precarious, deskilled, and surveilled labour. This trend has been very well theorised in the Marxist labour literature. Marx pointed out that automation and deskilling are the result in a capitalist economy that is propelled by the profit motive and the private ownership of the means of production.[123] This process of automation and deskilling is not exclusive to the factories in the manufacturing sector (productive capital) but also include warehouses in the retail sector (commodity/merchant capital),[124] where output is measured by the volume of parcels delivered.

During the coronavirus pandemic, Amazon stocks have been rising with the virus lockdown. Even with virus cases inside warehouses and slower delivery schedules, ordinary retail sales are collapsing with the lockdown.[125] Given that retail provides one-tenth of national employment, these job losses have deep distributional consequences. Greater concentration of corporate ownership will allow the remaining retailers to pay their employees lower wages and treat them worse as there are fewer alternative work options available. Marx described this process as the growth of the industrial reserve army of labour.[126] The lack of unionisation and unfavourable labour laws result in rising profits for Amazon and the few other winners of the contemporary economy (primarily technology companies). Within the warehouse, casualisation is characterised by the vast number of seasonal workers, while in the delivery sector casualisation is furthered with the reliance on Flex (private individuals using their own car and an app to deliver packages) and DSP (delivery service partner, where Amazon subcontracts with small business owners who lease the truck to contract workers for delivery).[127]

Monopoly capitalism makes possible high technological investments, which precede widespread automation.[128] Over the long term, automation reduces

---

123   Marx 1990a.
124   Marx 1992.
125   Maheshwari and Casselman 2020.
126   Marx 1990a, ch.25.
127   Semuels 2018; Bishop 2018.
128   This may seem counter-intuitive given that monopoly capitalism emerges with the economic stagnation and lack of new productivity-rising innovation, which is also reflected

the employee headcount even as an increasing share of our shopping demand is fulfilled online. The currently exploding employment figure at Amazon and online retailers in general reflect the short-term displacement effect of competing brick-and-mortar retailers. But the pattern of online retail employment could go in the direction of textile and automobile work historically, which is characterised by the inverted u-shape. First, technological change lowered the price and unleashed higher demand, resulting in rising employment until demand saturation and continued technological improvement reduced total employment later.[129] The technology and work discourse favours either the deskilling *or* the automation/technological unemployment angle,[130] while I do not see a separation between these two phenomena. In conditions of monopoly capitalism, the case of Amazon's automation and employment pattern shows that new employment opportunities are deskilled and precarious, while the ultimate goal is to eliminate these jobs as much as possible via automation. Within Marxist labour sociology, deskilling and automation are both eminent themes. The masses of the proletariat face the undesirable choice between becoming alienated cogs-in-the-wheel or jobless.

Following Marcuse, we know that the opposition between humanity and technology comes from the underlying cultural value system in the capitalist economy, which is incapable of organising technological output for human needs, freedom, and leisure rather than endless private accumulation of capital. Technology in the service of humanity requires a revolution in cultural values and sensibilities:

> If this idea of a radical transformation is to be more than idle speculation, it must have an objective foundation in the production process of advanced industrial society. In its technical capabilities and their use. For freedom indeed depends largely on technical progress, on the advancement of science. But this fact easily obscures the essential precondition: in order to become vehicles of freedom, science and technology would have to change their present direction and goals; they would have to be reconstructed in accord with a new sensibility – the demands of the life instincts. Then one could speak of a technology of liberation, product of

---

in the aggregate economic data. However, in the case of Amazon so long as it faces *some* retail competitors it will relentlessly acquire and invest in new technology. Gordon 2018; Vertesi et al. 2020.

129   Bessen 2017.
130   On deskilling, see Braverman 1998; Ikeler 2016. On automation, see Ford 2015; Boggs 2009.

a scientific imagination free to project and design the forms of a human universe without exploitation and toil. But this gaya scienza is conceivable only after the historical break in the continuum of domination – as expressive of the needs of a new type of man.[131]

Marcuse's call is to convert existing and new technologies to satisfy real human needs and that it would not be possible in the framework of capitalism. This task cannot be left to capitalists and their coterie of professionals in think tanks, media, and government. Their preference is to convert the amazing economic surpluses of the productive workers (which include the warehouse workers in the broadest sense) into funding largely useless and 'empty' administrative jobs (even as these companies make a profit),[132] as opposed to figuring out how to reduce labour hours and raise leisure time and non-work pursuits.

Does the contemporary dominance of Amazon imply a qualitative change in contemporary capitalism? I find that the mechanisation and deskilling that has traditionally occurred in the manufacturing and agricultural sector (productive sector) is also applicable in retail and transportation logistics (sphere of circulation or merchant capital), primarily because the same logic of measurable output is applicable in the latter (i.e., packages delivered) to raise surplus value. Before automation occurs, the increasing short-to-medium-term concentration of employment in the Amazon empire implies a return to more precarious jobs[133] but also offers opportunities for labour organisation. Marx himself noted:

> The increasing improvement of machinery, ever more rapidly developing, makes [labourers'] livelihood more and more precarious; the collisions between individual workmen and individual bourgeois take more and more the character of collisions between two classes. Thereupon, the workers begin to form combinations (Trades' Unions) against the bourgeois; they club together in order to keep up the rate of wages; they found permanent associations in order to make provision beforehand for these occasional revolts.[134]

---

131 Marcuse 1969.
132 Paulsen 2014; Graeber 2018. In Max Weber's account, non-productive labour conforms to the Puritan belief of having work hard for material rewards, Weber 2002.
133 An important qualification is that workers today are still better off than in the nineteenth century, because the standard of living is still higher than in the past, see Pinker 2018.
134 Marx and Engels 1978.

This prediction has practical implications for Amazon workers. Forming trade unions would increase worker control over schedules, pay, job content, and conditions of work. It would also allow for a more human usage and introduction of robots. In the past, union agreements on automation in car manufacturing included extended unemployment benefits (primarily by lobbying for state benefit expansion), better pension plans and preferential treatment for re-hiring of displaced union workers.[135] This outcome is certainly feasible for Amazon workers today. Assuming unionisation happens, the drawback of this approach is that those gains can be subverted by competition from a non-union competitor, industrial/technological changes (as with the shift from traditional to online retail), unfavourable changes in the political environment (i.e., anti-union policies), or economic stagnation impacting that sector (e.g., consumers buy more non-retail goods or demographic decline resulting in less demand). Hence, broader structural changes also need to be addressed.

The left has to converge on four interrelated demands to 'erode' or 'tame' capitalism,[136] each of which requires substantial state intervention: (1) the work week has to be reduced with productivity-enhancing technological progress. In that manner, workers that are made redundant in the high productivity sectors can be shifted to lower productivity sectors, where labour shortages emerge with shortened hours (e.g., social care, teaching, entertainment).[137] The market mechanism does not redistribute social gains to the broader population, because under monopoly capitalism productivity gains are absorbed by shareholders and even if they shared the gains with unionised workers, it would be reflected in rising wages rather than shorter hours. Low productivity workers with low pay have no luxury to cut their hours given the continued high living expenses. The mandate has to come from the state, which has to enforce cross-subsidisation to low productivity workers whose hourly cuts must be fully compensated by pay rises. (2) The state has to expand universal basic services (UBS) such as housing, health care, education, public transit, and telecommunication.[138] Better public services funded by the wealthy and powerful corporations would lower the cost of living for the general population (decommodification) without having to manipulate the work week or labour relations.

---

135 Meyer 2002.
136 Wright 2015.
137 I refer to the bourgeois economic use of the term 'productivity', i.e. how many customers are serviced per employee per hour. Educators and care workers face the constraint that increasing classroom sizes and patient load are either not feasible or counterproductive, hence their productivity is low.
138 Bastani 2019.

(3) The state has to introduce a universal basic income (UBI), a guaranteed payment of income to all residents for simply being alive. UBI is another measure to decommodify life (i.e., reduce reliance on labour income for daily needs) but allows for a greater scope of market functions, which is better tailored to individual needs than UBS.[139] (4) A percentage of profits from companies that benefit from public funding could be reappropriated by the state to provide the population with a universal basic dividend (UBD).[140]

Even with these bureaucratic measures in place, we must not forget Marcuse's call for a change in the social value system to undergird any policy changes. Otherwise, capitalists will find ways to undermine, subvert, or reverse any institutional changes. While we are continuously facing the choice between socialism or barbarism, the road to socialism remains a continuous struggle.

---

139 A possible version of UBI is anti-welfare state and right-wing but it is not the only one. A left wing UBI combines UBI with UBS and the poor cannot be made worse off. Van Parijs 2017; Bregman 2017.
140 DiEM25 2017.

CHAPTER 8

Automation along Global Supply Chains: How RFID-Systems Will Transform Work and Power Relations in the Supply Chain of Fast Fashion

Steffen Reitz

Industria de Diseño Textil, S.A. (Inditex), a Spanish multinational clothing company, has been at the forefront of digitisation in retail. Growing out of the retail company Zara, which is still its most prominent brand, Inditex is one of the largest fashion groups in the world and considered to be the pioneer of what has been termed 'fast fashion'. Beginning in 2016, Inditex introduced radio-frequency identification (RFID) technology at all Zara stores. This technology allows the unique identification of objects, animals, and humans through the use of radio frequency and without human intervention.

RFID technology was officially introduced by Inditex as a more effective means of keeping stock at its retail stores. However, in discussions initiated by German and Spanish unions and the Transnationals Information Exchange (TIE) network, retail workers reported that RFID has substantially changed their working conditions. They reported that it has rendered certain positions obsolete, that it has expanded control of employees, and that it has increased pressure and reduced autonomy at the workplace.[1] Workers' committees have complained that at the time of negotiations, it was impossible to understand the full scope of restructuring measures made possible by RFID.[2]

In 2016, Inditex announced that it would greatly expand its RFID system beyond its retail stores. It mandated the introduction of the technology along the company's supply chain and at manufacturing sites. This announcement caused questions among workers initiatives shaped by the experience of RFID in retail: How would RFID impact working conditions along Inditex's supply chains? How would RFID change power relations along the supply chain between contractors and Inditex? How would the technology be used in organising production? How might these changes in production impact struggles for workers' rights? What changes would be conducive to their struggles, and

---

1 Köhnen and Nutzenberger 2019, p. 177.
2 Köhnen und Nutzenberger 2019, p. 184.

which changes might endanger positions they had fought for? Which parts of production could RFID make vulnerable to automation? What should be their stance towards RFID? If they were able to negotiate about the integration of the technology in production, what should they be aware of? How could the technology be integrated in a way that it might also benefit workers and not only the retail company at the end of the supply chain?[3]

In his famous letter to Arnold Ruge, Marx suggested that his theory meant to provide a moment of self-education for the proletariat, rather than a definition of what was the right or wrong thing to do.[4] Following this suggestion, Jasper Bernes writes that one way of doing Marxist theory is to provide 'a map (rather than) a set of directions: a survey of the terrain in which we find ourselves, a way of getting our bearings in advance of any risky course of action'.[5] This study was conducted with the aim of contributing to such a practice of theory.

First, the chapter introduces RFID technology. It then provides an outline of the theory of the logistics revolution. Next, Inditex's production model is placed in the context of that theory to outline the company's production model and the power relations to which it is connected. In a further step, possibilities of RFID for production and logistics in the fashion industry are presented by considering studies conducted by industrial engineers and logistics researchers. It will then be explored how RFID will transform the power relations and working conditions outlined in the first step. RFID will vastly expand control along the supply chain, which will increase the power of Inditex to manage it and to control its suppliers. It will also lead to a restructuring of the production process of contractors with changes in the way work will be controlled, analysed, and organised.

Labour can only resist automation, influence automation, or even profit from automation if it gains a strategic understanding of the technologies and processes facing it. This chapter develops such a strategic understanding.

---

3   This chapter grew out of a study conducted for the Transnational Information Exchange network (TIE) in Frankfurt Main, Germany. TIE is a global workers' network founded in 1978 by union activists from different countries who came together to connect 'workers initiatives that see themselves as part of a broader movement for social change committed to fight for a life without exploitation and exclusion' (TIE Germany 2020).
4   Bernes 2013, p. 173.
5   Bernes 2013, p. 173.

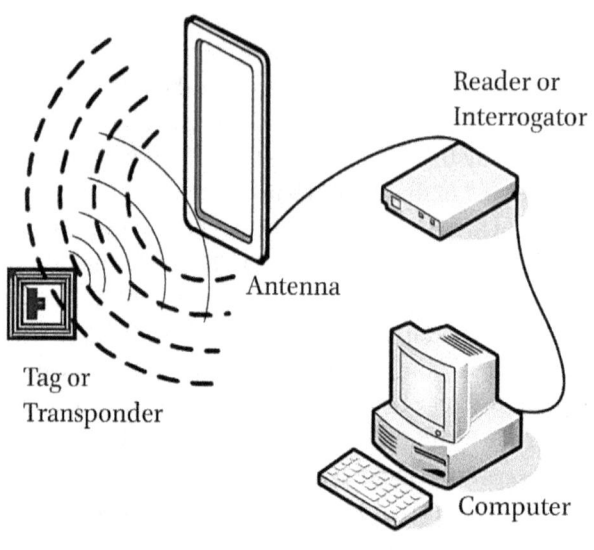

FIGURE 8.1
A RFID tag and a RFID system
GLIGOREVIC, SCHULZ, AND IVANOCHKO 2021, P. 557

1 RFID

RFID technology is based on the wireless communication between a tag, on which information can be stored electronically, and a reader. The basic components of a RFID system are the RFID tag (also called transponder), an antenna integrated into a reader (also called interrogator), and the control system or computer, which processes the information from the reader (figure 8.1).

RFID was originally developed in 1945 as a tool using radio waves and transponders to identify aircrafts as friend or foe on the battlefields of World War II.[6] Explorations for civil use of the technology began in the 1970s, but it took until the early 2000s for RFID systems to be implemented on a wider scale. RFID technology has become significantly cheaper during the past twenty years and is now used widely. Typical areas of use for RFID systems are libraries, identification and security systems, life stock farming, and warehousing.

Like barcodes, RFID can be used to identify objects with tags that can be scanned. In these functions RFID is, however, superior to barcode, since tags can store more information and can be read more quickly and from further away. RFID tags can furthermore be read without human intervention and can uniquely identify objects, whereas barcodes must be scanned by humans and can only identify product types. These two capabilities open up new possibilities for automation and allow the integration of humans, machines, and objects

---

[6] Nayak 2019, p. xvi.

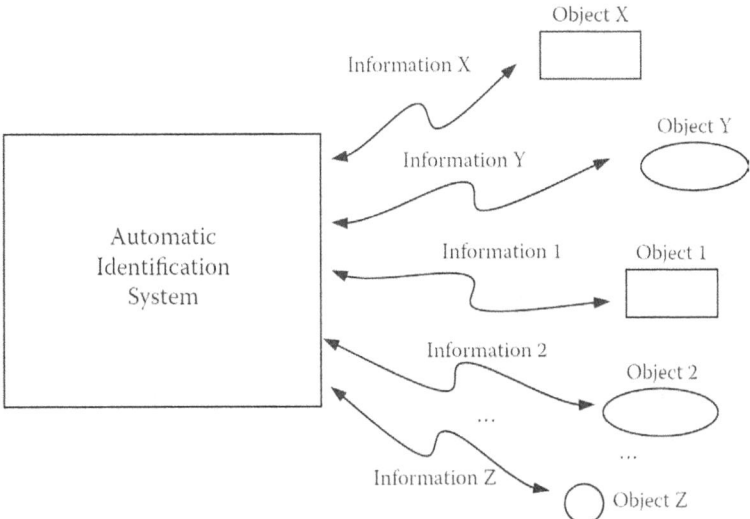

FIGURE 8.2 An automatic identification system: Integration of objects (human and/or non-human) into one system
LOZANO-NIETO 2010, P. 2

into one single identification system that can analyse and control processes in real-time (figure 8.2).

There are different types of RFID tags for different applications. There are active, semi-active, and passive tags, which can be read from different ranges and can store different amounts of data. Tags can be read-only or read and write tags. Information on read and write tags can be modified or erased, and they can therefore be reused. It is important to note that workers faced with the integration of an RFID system should find out about its exact capabilities and tag types. The tag type can make a big difference in what an RFID system is able of registering and recording (i.e. which processes it can track, analyse, and 'optimise') and how far that system can reach.

2   The 'Logistics Revolution' or a New Model of Industrial Production

The concept of 'logistics revolution'[7] was formulated by Marxist sociologists Edna Bonacich and Jake B. Wilson to account for a major shift in the way goods are produced and distributed. They draw heavily on British critical geograph-

---

7 Another noteworthy perspective on this major shift in the organization of global produc-

ers[8] in viewing the logistics revolution 'as having arisen in response to a chronic problem of the capitalist system, namely, the disjuncture between production and distribution, or supply and demand'.[9] This separation leads to overproduction or underproduction, which can be costly for retailers and manufacturers. The 'logistics revolution' is an attempt by retailers to bridge this gap solving the supply and demand contradiction.[10]

According to Bonacich and Wilson, this new attempt has led to a paradigm shift in the organisation of production, namely 'supply chain management' (SCM).[11] SCM is an effort to link supply to demand with the goal to produce only those goods that consumers are actually buying.[12] A standard workbook defines SCM as the coordination of production, inventory, location, and transportation among the participants in a supply chain to achieve the best mix of responsiveness and efficiency for the market being served. The goal of supply chain management is to increase sales of goods and services to the final, end-use customer while at the same time reducing both inventory and operating expenses.[13]

Using information technology, the sellers of goods (especially retailers) collect point-of-sale (POS) information by using bar codes, which is relayed electronically through the supply chain to quickly initiate replenishment orders. Using this method, retailers try to avoid the two major problems of selling goods, stock-outs and mark-downs. The idea is to be able to continuously customise store offerings to what consumers want by always having the exact inventory needed to avoid both overstocks and understocks. To achieve this goal, the supply chain managers treat the entire supply chain as a single continually flowing system. They seek 'to control costs, to limit inventory pileups at any stage of the chain, to speed up the time it takes to cycle through the system, and to provide better service to consumers'.[14]

The development of SCM can be linked to the end of Fordism and the emergence of flexibility and specialisation. Products are customised for a variety of

---

tion is given by Cowen (2014), who offers a genealogy of logistics by tracing the link between markets and militaries. She uses the term 'revolution in logistics' to conceptualize this shift. Another noteworthy perspective is Moody (2017), which deals with present conditions for class struggle with a strong focus on 'the logistics revolution'.

8   Ducatel and Blomley 1990; Harvey 2019; Wrigley and Lowe 1996, Wrigley et al. 2002.
9   Bonacich and Wilson 2008, p. 4.
10 Bonacich 2003, p. 45.
11 Bonacich 2003, p. 41.
12 Bonachich 2003, p. 42.
13 Hugos 2003, p. 38.
14 Bonacich and Wilson 2008, p. 4.

specific markets, which results in a tremendous product proliferation in an ongoing attempt to increase consumer demand.[15] The consequence is what Bonacich calls a 'mass customisation' or endless varieties of 'the same old thing'.[16] Fordist factories were very limited in producing this type of variety. In the current production model flexibility is achieved by 'contracting out for the production of smaller lots of constantly changing products'.[17]

At the heart of this change in logistics is a change in the balance of power between manufacturers and retailers: A movement from a 'push' production to a 'pull' production and distribution.[18] In a push-system, certain products are mass-produced by manufacturers and sold to retailers who try to sell them to customers. This type of system is prone to overproduction when products cannot be sold or to underproduction when demand increases unexpectedly. In a pull-system, consumer behaviour is tracked by retailers who transmit these preferences up the supply chain to the producers. Manufacturers try to coordinate production with actual sales. The result is a continuous attempt to reduce inventory throughout the supply chain in an effort to cut costs for both manufacturers and retailers.[19]

Pull-production requires speed, predictability, and accuracy in the delivery of goods since available inventory has been cut back.[20] To achieve these requirements, a high degree of control and coordination of the entire supply chain is necessary. According to Bonacich and Wilson this has led to efforts to integrate processes across the various businesses that make up the chain – businesses that usually operate as independent, competitive, and often secretive entities.[21]

Historically, manufacturers and retailers have viewed each other with caution, reluctant to reveal their costs and practices to each other. As part of the logistics revolution such barriers to information sharing have broken down. This has made the whole system from production to logistics and sales more efficient and has minimised total costs and maximised speed and turnover. These changes have also affected how firms in a supply chain operate. These firms increasingly need to find solutions that serve the whole system rather than specialising on a single aspect. Retailers, their suppliers, and transport-

---

15 Bonacich 2003, p. 42.
16 Bonacich 2003, p. 42.
17 Bonacich 2003, p. 42.
18 Seifert 2003, p. 5.
19 Bonacich and Wilson 2008, p. 5.
20 Bonacich 2003, p. 42.
21 Bonacich and Wilson 2008, p. 5.

ation providers 'ideally form strategic alliances, sharing information so that all partners are able to respond rapidly to shifts in demand'.[22] New technology such as EDI (electronic data interchange) have enhanced communications between supply chain partners and increased possibilities for control in the management of supply chains.[23]

According to Bonacich and Wilson, coordination along supply chains has decreased the competition between manufacturer and retailer. This competition between firms has been partly replaced by competition between supply chains or supply networks, because firms at different levels may have multiple and shifting relations. Bonacich and Wilson argue that this has resulted in a very complex network of overlapping relationships. Suppliers must maintain good relationships with buyers. Bonacich and Wilson assume that there is a complex game about who gets which information and which relationships are maintained.[24] Since retailers collect and hold POS data and usually manage the supply chain (or employ the company that manages it), they hold an advantage over suppliers. Retailers usually have knowledge about their suppliers' significance to and position within the entire supply chain and can strategically decide what to share with them in negotiations. Rather than a finished development, the 'logistics revolution' describes an ongoing shift, constantly developed further by capital interests and technological changes.

3 The Inditex Production Model and the 'Logistics Revolution' in Fashion

The way that Inditex organises its production already deviates a bit from the typical production model of a retailer under the conditions of the 'logistics revolution' described by Bonacich and Wilson. Inditex currently has around 7,490 stores in 96 countries. It directly employs 174,386 people and it can be estimated that through its suppliers hundreds of thousands more jobs rely on Inditex's business. According to the company's latest Annual Report (2019), about 7,235 factories of national and international suppliers formed an integral part of Inditex's supply chain in 2018.[25] In 2018, it turned a net profit of 3.448 million Euros,[26] which makes it one of the most profitable fashion retailers in

---

22 Bonacich and Wilson 2008, p. 5.
23 Bonacich and Wilson 2008, p. 5.
24 Bonacich and Wilson 2008, p. 5.
25 Inditex 2019, p. 90.
26 Inditex 2019, p. 20.

the world. The company's founder, Amancio Ortega, became Spain's wealthiest man when Inditex was turned into a publicly traded company in 2001.[27] In the business community, Inditex's supply chain process is regarded as highly innovative and exemplary.[28] While there are many studies from a management perspective,[29] which provide a comprehensive understanding of the supply chain, a full analysis of both the production model and the resulting power relations is still missing.

Inditex has been the principal innovator of fast fashion and is currently the most successful proponent of that business model. Fast fashion departs from traditional norms of designer-led fashion by adapting fashion to current trends and to customer demand in an ongoing process.[30] For customers, this means that more, rapidly changing, limited lines of fashion are available for cheap prices. From the business perspective, fast fashion requires an intricate model of production and logistics to keep up with ever changing trends and consumer demands. It requires a close network of design, marketing, sales tracking, just-in-time manufacturing, and effective logistics to detect trends, quickly mass-produce products, and deliver them to stores as fast as possible.

With the development of the fast fashion model, Inditex has taken the model of pull production one step further than outlined by Bonacich and Wilson. It has done so in at least three interconnected ways:
- It has further decreased the reaction time to consumer demand.
- It has achieved this by reintegrating almost all steps of production, while outsourcing as much as possible.
- It has integrated even design and advertisement with all other aspects of a supply chain led by a constant analysis of consumer demand and sales projections.

Before fast fashion, big fashion retailers created seasonal lines that were produced by subcontractors and then advertised and sold by retailers.[31] This process from design to sales would take six to twelve months (see figure 8.3). There was always the danger of failing lines, of over-production or under-production

---

27   Ghemawat and Nueno 2003, p. 1.
28   Leading business advisory firm Gartner has continuously ranked Inditex's supply chain as one of the top 25 companies 'that best demonstrate leadership in applying demand-driven principles to drive business results' (Gartner 2020).
29   The firm is covered in standard workbooks of supply chain management (for example Nakano 2020) and has been studied by business researchers (Ghemawhat and Nueno 2003, Dopico and Crofton 2007) as well as logistics theorists (Yip and Huang 2016, Burt et al. 2006).
30   Dopico and Crofton 2007, p. 41.
31   Nakano 2020, p. 42.

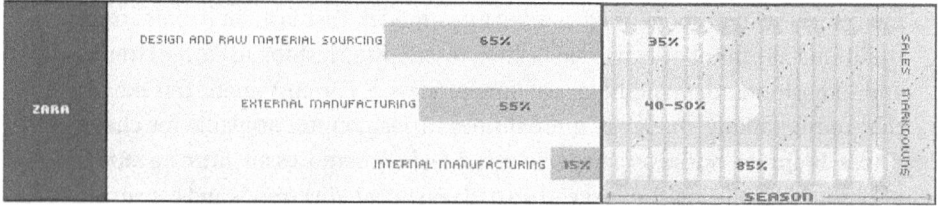

FIGURE 8.3 Traditional Fashion and Fast Fashion
GHEMAWHAT AND NUENO 2003, P. 30

of certain models, which generally resulted in relatively high amounts of markdowns at the end of a season. Production and even most of logistics before fast fashion was performed by subcontractors.[32] To be able to quickly respond to customer demands while still maintaining cheap prices, Inditex has abandoned the fashion industry's traditional model of seasonal lines for the majority of its products. It now produces or adapts most of its clothes when a season has already begun.

Inditex has directly linked customer demand to the design and production process. To anticipate the tastes and demands of customers, Inditex relies less on expert designers and more on procuring information. Hundreds of Inditex's designers continuously seek information about what customers may like from sources such as fashion shows, magazines, or the internet.[33] It essentially copies (some might say steals) current trends it projects to sell at a profit. Inditex sells its products only through its physical and online stores (about 10 percent of products are sold online), which in turn only sell Inditex's products. Stores collect detailed POS data and store managers make regular projections about demand.[34]

The production steps of the traditional fashion industry move chronologically from design to the production of a sample, the purchase of materials, man-

32 Ghemawat and Nueno 2003, 2.
33 Nakano 2020, p. 42.
34 Nakano 2020, p. 42.

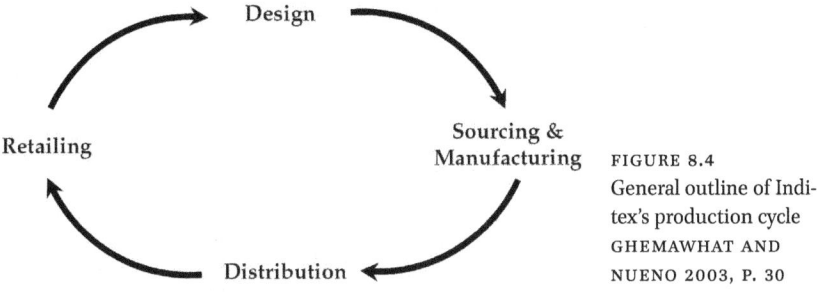

FIGURE 8.4
General outline of Inditex's production cycle
GHEMAWHAT AND NUENO 2003, P. 30

ufacturing of the garments, transporting the garments to warehouses (logistics), and finally sales at retail stores. Generally, this is not different for Inditex's production, but the company has interconnected these steps in a complex network. Inditex's production cycle is governed at all points by sales calculations from POS data and is closely controlled from headquarters.

4 Design

In fast fashion, design cannot really be situated at the beginning of the production cycle. It is done continuously and always directly connected to information of actual sales and considerations of the overall production process. A significant amount of Inditex's main retail brand Zara is designed after the season has already begun, which means the brand can quickly react to the latest trends. Designers at Inditex head offices in Artexo (A Coruña) work closely with marketers and procurement planners.[35] After an item has been designed, a sample can be made quickly at the same office and prices can be discussed with marketers and procurement planners who can estimate the manufacturing costs and production capacity. Design decisions are closely coordinated with marketing and production planning. Only if a design can be projected to be profitable, by estimating production costs and the market prices, will it be produced.[36] There are different production processes available for planners. A clothing line can be produced at Inditex's own factories (the fastest production process, usually used for seasonal and more expensive items), in factories of neighbourhood outside contractors (usually for seasonal items), or at other outside contractors situated in North Africa and Asia (mostly basic fashion with a stable demand).[37] Many designs can be adapted again at a later stage of production.

35 Nakano 2020, p. 72.
36 Nakano 2020, p. 72.
37 Nakano 2020, pp. 74–5, see also figure 5.

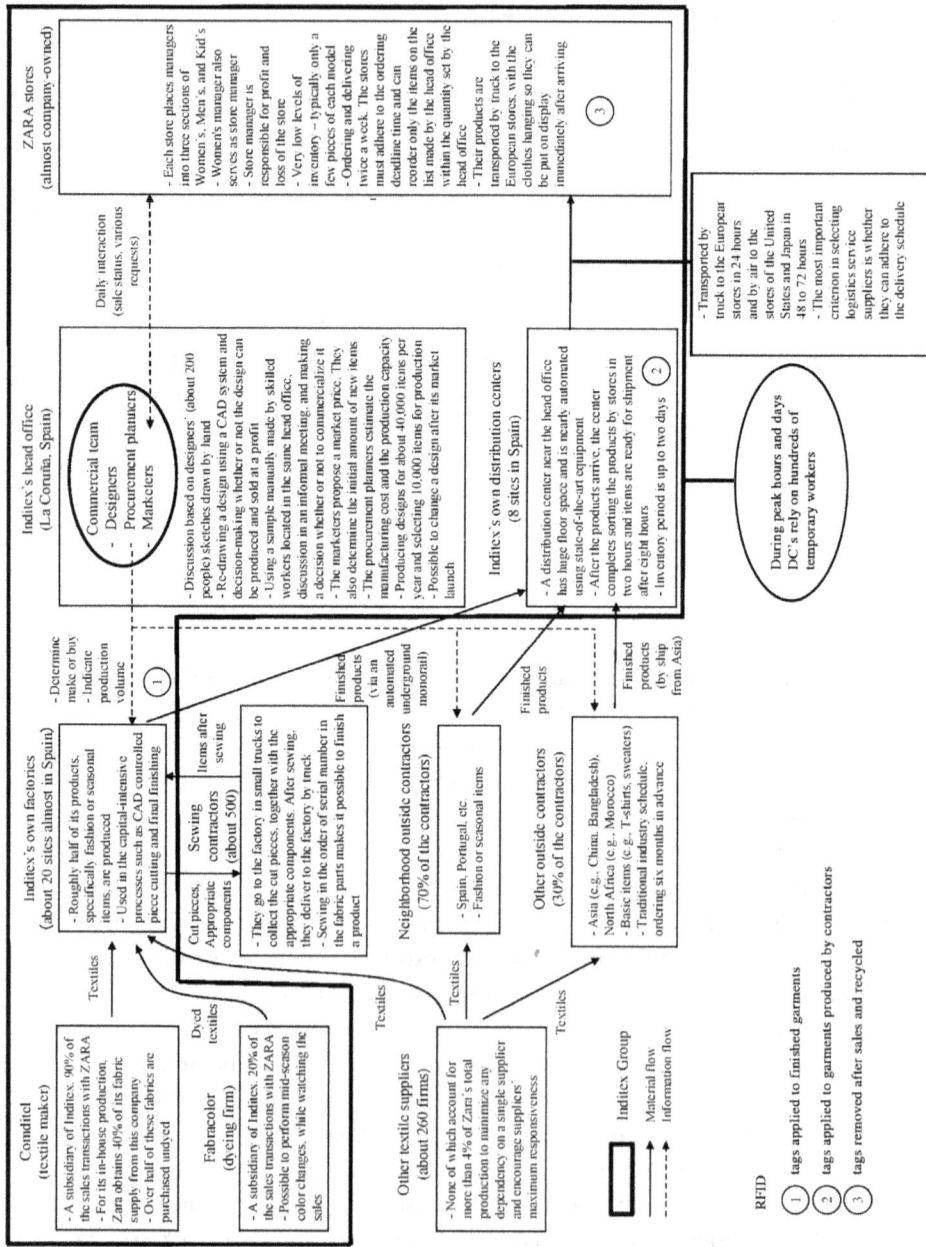

FIGURE 8.5 Zara's supply chain process including its current RFID system.
NAKANO 2020, P. 73. WITH ADDITIONAL INFORMATION ADDED BY ME. ('RFID' REFERRING TO YIP AND HUANG 2016 AND INFORMATION ON 'TEMPORARY WORKERS' REFERRING TO NAKANO 2020)

5 Fabric

Production planners can choose to get fabrics from Inditex's integrated textile maker Comditel or from about 260 textile suppliers with which the company has loose relationships (figure 5). About 40 percent of in-house production is done with textiles from Comditel.[38] Over half of these fabrics are purchased undyed so they can be dyed at a later stage. This allows Inditex to stay flexible and to react to changes in demand. For mid-seasonal colour changes, Inditex has its own dyeing firm (Fabracolor).[39] The other 60 percent of in-house production and 100 percent of external production are done with textiles purchased from external suppliers. Having its own textile maker increases production speed for some clothing lines and grants Inditex a more powerful position in negotiating with suppliers. The firm relies less on individual suppliers' work and therefore does not need to maintain a stable relationship with them. This ability of being flexible 'encourages suppliers' maximum responsiveness'[40] since suppliers cannot rely on steady contracts.

6 Manufacture

Firstly, Inditex can produce garments in house, which means that textiles are prepared at one of Inditex's highly automated factories situated close to its headquarters in northern Spain. These factories perform capital intensive process that can be automated, such as piece cutting controlled by a computer aided design program and the final finishing process.[41] Fabrics are cut and prepared for sewing in a way that clearly indicates how a piece of clothing is to be sewn together. The sewing itself is outsourced to about 500 small sewing contractors.[42] None of these contractors handle a larger part of Inditex's production. This strategically minimises dependency on any single one of the sewing contractors. This in-house-production process is mostly used for limited fashion items. After the finished products are back at Inditex's factories, they are immediately transported to Inditex's principal distribution centre near its headquarters via an underground monorail system.[43]

---

38 Nakano 2020, p. 75.
39 Nakano 2020, p. 74.
40 Nakano 2020, p. 75.
41 Nakano 2020, p. 74.
42 Nakano 2020, p. 75.
43 Nakano 2020, p. 74.

Secondly, production can be done at neighbourhood contractors in Spain or Portugal. This process is mostly used for fashion and seasonal items. It is fast, but not as fast as the first process. These contractors use textiles from outside contractors and account for about 70 percent of Inditex's 'outside production'.[44] Finished products are transported (mostly via trucks) to one of Inditex's major distribution centres (see figure 5).

Thirdly, Inditex produces basic items in Asia (for example in China and Bangladesh) and North Africa (Marocco). This process accounts for the other 30 percent of outside production.[45] Usually basic items like T-shirts or sweatshirts, which have a very predictable demand, are produced this way. This process works the same way as the traditional industry process before fast fashion: Designs are completed six months before a season and then ordered from suppliers with which Inditex has long and stable relations.[46] The finished clothes are transported mostly by ship and truck to Inditex's major distribution centres.

7   Distribution

Inditex currently operates ten logistics centres, all located in Spain.[47] The biggest one is located near the head offices in Arteixo (A Coruña). The distribution centres are highly automated and organised to sort and move products as fast as possible. Inventory is typically shorter than two days.[48]

Distribution centres are equipped with a cross-docking centre. Capacity utilisation at the distribution centres is low (usually below 50 percent), which gives the company room to operate at peak times before seasons. For these peak times it also relies on hundreds of temporary workers to be able to move higher volumes of goods.[49] The transportation and distribution of Inditex's products are undertaken entirely by external contractors[50] Inditex maintains control over transportation by strictly planning deliveries. Contractors are chosen for their ability to adhere to this schedule.[51]

---

44   Nakano 2020, p. 75.
45   Nakano 2020, p. 75.
46   Nakano 2020, p. 75.
47   Inditex 2020.
48   Nakano 2020, p. 25.
49   Nakano 2020, p. 75.
50   Inditex 2020.
51   Nakano 2020, p. 76.

8 Sales

Due to Inditex's pull-system, sales cannot be separated clearly from production. The main contributions of retail to the production process are the measuring and prediction of demand as well as a certain 'production of demand'. Inditex chain stores such as Zara 'function as both the company's face to the world and as information sources'.[52]

As outlined, POS data and regular sales predictions by store managers directly influence production planning via computer software and decisions of the creative teams. POS data is closely recorded and integrated in the planning of the production process. Since 2016, this is mostly done with an RFID system,[53] which allows for a more exact and quicker analysis of sales and stock data than barcode. Based on this data, store managers decide which merchandise to order and which clothing lines to discontinue. They also transmit their own sense of inflection points to Zara's commercial teams to register 'latent demand' that could not be captured through an automated sales-tracking system.[54]

Inditex saves a lot of money by advertising very little. For Inditex, lean production doubles as an advertisement strategy with stores functioning as important advertisement and marketing sites. Unlike Walmart, which has large warehouses integrated into each big box store, Inditex's stores hold very low levels of inventory, often only a few pieces of each model.[55] It is therefore likely that an item will be out of stock when a customer looks for a specific model. The brand connects these out-of-stock situations to its promotions, creating 'a sense of tantalising exclusivity'.[56] This also has the tendency to increase the desirability of models that are actually there, because customers know that products are limited and might be gone at a later visit. In 2002, Díaz Miranda, Vice-President of Manufacturing, summarised the role of retail in the production process and its close link to marketing as follows:

> The size of the production run – 'scale', in the traditional sense – is not an issue. We recoup our costs on the garments through markup because people will pay a premium for the right garment at the right time. It is the product that drives the customer. For an expected very strong demand,

---

52   Nakano 2020, p. 74.
53   Inditex 2019, p. 33.
54   Ghemawhat and Nueno 2003, p. 14.
55   Nakano 2020, p. 74.
56   Nakano 2020, p. 74.

we'll take a bigger risk on the fabric purchasing decision. Sometimes we make a decision that from an economic point of view might not seem sound, but we know that. For example, we might have an item that was selling very well, but if we think that we are saturating the market with that look we will stop manufacturing it and create unsatisfied demand on purpose. From a strictly economic point of view, that is ridiculous. But the culture we are creating with our customers is: you better get it today because you might not find it tomorrow.[57]

The goal is to keep up the desire for new products instead of satisfying it. This of course is reinforced by the widely criticised low quality of fast fashion. As sociologist Georg Simmel already observed in 1904, 'The more an article becomes subject to rapid changes of fashion, the greater the demand for cheap products of its kind'.[58]

9 Closely Controlled from Head Offices

While store managers are given relative independence over the choice of their offers at the stores, what they can actually choose from is closely controlled by the head offices. Additionally, head offices may intervene at any time in policies of stock keeping and offers at stores. This central control allows the coordination of even smaller decisions on the retail level with the overall strategy of the supply chain. It is possible to conceive of situations in which a store manager might request a certain item that sells well at his or her location, but the head office decides to withhold it, because producing it might delay production of another more profitable product. These kinds of consideration are only possible through a centralised control of the supply chain and of retail. Such considerations were not possible in the same way before the logistics revolution.

Another way of indirectly controlling stock and the analysis of demand in retail stores are the work contracts given to store managers. They receive a fixed salary plus a bonus based on their store's performance. This variable component typically makes up about one-half of the total wage. This incentive-intensive salary is geared towards controlling the focus of the manager's work. Since prices are fixed centrally, the store managers' energies are incentivised to focus on volume and mix.[59]

---

57 Cited in Fraiman et al. 2002, p. 6.
58 Simmel 1904, p. 151.
59 Ghemawhat and Nueno 2003, p. 15.

10 Inditex's Current RFID System in Retail and Logistics

Inditex currently uses RFID technology in retail and logistics activities at its distribution centres and retail stores (figure 5). The RFID system is integrated with a cloud-based central supply chain management system.[60] This centralised system links up major processes in the company such as sales, shipment, production, and demand forecast.[61] As soon as garments are equipped with a programmed tag, they appear individually in the IT system and can be tracked in real time. When the clothes arrive at the shop they can be scanned quickly and in bulk with an iOS device.[62] The tags are only removed at points of sale with a 'Smart Tag Detacher'. While the tag is removed, the Detacher automatically sends sales information to the head office and deletes the information inside of the RFID chip.[63] RFID chips removed from clothes after sales are returned to their supplier, Tyco, for maintenance and then re-circulated in Zara's supply chain.[64]

RFID technology enables Zara to know the precise location of their inventory with a much higher accuracy than before its implementation. At the level of store management, it has reduced work by making reception of inventory faster and partially automising the replenishment-process: Once an item is sold, the item's chip immediately places an order to the stockroom for an identical item.[65] This has also affected how workers are controlled and how their work is analysed, controlled, and organised.[66]

Putting Zara's integration of a RFID system in perspective of the logistics revolution, the technology has increased visibility and control in parts of supply chain management (from distribution to sales) and in retail. Additionally, it has made outsourcing more effective and controllable. As soon as garments have been equipped with RFID chips, they are individually visible in the company's Enterprise-Resource-Planning (ERP) system and can be tracked individually in the process all the way to sales. This connects well to a retail model of strategic scarcity that relies on keeping offers always a little bit below actual demand. RFID has made POS data, which is at the centre of Inditex's strategic production model, more instantaneously accessible and has enabled a closer

---

60 Yip and Huang 2016, p. 324.
61 Yip and Huang 2016, p. 327.
62 Yip and Huang 2016, p. 325.
63 Parks 2014 cited in Yip and Huang 2016, p. 325.
64 Tyco 2014 cited in Yip and Huang 2016, p. 325.
65 Walker 2015 cited in Yip and Huang 2016, p. 325.
66 Köhnen and Nutzenberger 2019, p. 177.

control of goods and processes from the point of distribution (out-of-house production) and finished production (in-house production) onwards (see figure 5). This has given Inditex a lot of detailed data on the whole process, which allows for a close analysis of supply chain activities from its own factories to sales. Having more accurate and more up-to-date data makes production planning, which for Inditex is closely integrated in the design process, more accurate. It also gives Inditex information about the performance of individual retail stores and the performance of contractors in logistics can be tracked and recorded. It can be assumed that RFID indirectly makes outsourcing logistics more efficient, since scheduled transportation by contractors can only be undertaken if Inditex can plan relatively well and ahead of time when and how many goods will have to be moved. Theft and inefficient transportation would immediately be registered in the system as well, because Inditex can know exactly which individual models have been sent out to which store and which models were received there.

11 Power Relations along the Inditex's Supply Chain before the RFID Expansion

The power relations that result from this production model are in line with the ones outlined in the theory of the logistics revolution. As Bonacich and Wilson show, the key to the retailers' power is based on access to POS data in the now prevailing model of a pull-production system. Inditex's power to manage its extended supply chain can be explained with the detailed POS data constantly generated at its retail stores with the use of RFID technology and the detailed feedback data from its store managers. Production for fashion basics (the seasonal lines) are outsourced completely to the factories of suppliers in Asia and North Africa. Inditex keeps relatively close relationships with these contractors, which can count on a steady stream of orders. Since Inditex profits from low labour standards and production costs in the regions where its goods are manufactured, the benefits of long-term relationships (better coordination, more control) can be assumed to outweigh their costs (higher dependency and more long-term commitments).

What deviates partially from the logistics revolution model is Inditex's vertical integration of all key elements of the supply chain and production process for the more expensive and risky lines. The fast fashion model relies first and foremost on being able to offer the latest fashion relatively cheaply. To achieve this, Inditex needs to be highly flexible to ramp up production whenever necessary and reduce it again when demand is low. It also needs to be as cheap and fast as possible to get affordable products to its stores at exactly the right time. In such a system, suppliers would potentially have a very powerful position. Stalling a delivery for just a few days could potentially lead to the loss of very substantial sales when products do not reach stores in time. It can be argued that Inditex has vertically integrated so many steps of its production not only to improve control and speed in its reaction to demand, but also to improve its position in negotiations with suppliers. While Inditex has its own textile maker, it only purchases 40 percent of its textiles for in-house-production at Comditel. In-house-production only accounts for about 40 percent or less of all production of Inditex, which means Comditel provides less than 20 percent of fabrics for Inditex's overall production (see figure 8.4). It can be assumed that while Inditex can produce fashion products in-house, it only does so when a cheaper option is not available. Having the option of producing a line of goods itself gives Inditex power in negotiating with suppliers.

The increased control over more parts of the supply chain through fast fashion gives Inditex the ability to outsource parts of the production that are more labour intensive and less capital intensive. Inditex does this for its in-house production process as well as for its logistics. Transportation from its logistics centre to its stores is performed by various contractors, which have to run on a schedule given by Inditex, enabling the latter to keep tight control without integrating the contractor into its own company (see figure 8.5). The same mechanism applies as it does to sewing in the in-house production process, which is performed by almost 500 hundred small contractors. If a company negotiates too aggressively, it can simply be replaced by a competitor. In a sense it is a system of fictitious self-employment: a step of production will be defined as separate work and then offered to a contractor. Since it is a simple and clearly defined task, many different companies may compete over a contract, which ensures minimal dependency and a low price. Reducing sewing to simply the sewing together of parts delivered with clear instructions makes it simpler and less valuable work. Reducing logistics to driving a truck on a pre-planned schedule also turns it into less complex and less valuable work. Even though sewers and truck drivers perform an integral part of Inditex's business, they do not have employee rights. Sewing for in-house-production and the transportation of products to the stores would also be highly vulnerable to a strike. Since workers with no rights as Inditex's employees do these jobs, a possible strike is made much more difficult.

Bonacich and Wilson describe contingency as the other side of flexibility. Flexibility means the ability to quickly react and or change course. Contingency describes the downfalls of a system that demands flexibility. Contingency means having to deal with the negative effects of unpredictability. Retailers under a pull-system manage to stay flexible because they are able to push contingency onto their suppliers. While a large retailer can quickly adapt to changes in demand by employing more or less contractors, contractors and subcontractors are faced with instability and unpredictability.[67] They cannot count on a steady stream of work. They only get it when there is work to be had. It can be argued that Inditex carries some contingency itself, by leaving a lot of extra capacity at its major distribution centres. However, much of this contingency is pushed onto the temporary workers it hires during peak times, who cannot rely on a steady stream of work (figure 8.5). Keeping contingent relations can also be seen as a mechanism of power. For sewing and logistics in its in-house production, Inditex employs a large number of small contractors. By spreading its dependency on many different suppliers, which are competing with each other, Inditex stays flexible: it can always employ more or less of them according to demand, and avoids having to rely on a single one. Avoiding long term relationships 'encourages suppliers' maximum responsiveness'.[68]

12 RFID Expansion: Possible Uses of RFID in Production and Supply Chain Management

In 2018, Inditex announced that it would expand its RFID program to its entire supply chain. In the following it will be outlined how RFID could change production of individual factories and then how it could change supply chain management. In a third step, different questions, scenarios, and theses will be drawn up to understand how this expansion will influence power relations along the supply chain and labour conditions in production. In the end it will be discussed how labour could react to these changes.

Rajkishore Nayak, a scholar of contemporary fashion production, has collected different studies on the application of RFID technology in fashion manufacturing and in supply chain management to outline the different capabilities of RFID technology in these areas. Nayak sees as the major driving force for the implementation of RFID the proliferation of the fast fashion model and rising labour costs in developing countries.[69] According to Nayak, fast fashion forces a close collaboration between manufacturing, logistics, and retail, which is made easier by data generated with RFID technology. He also sees that RFID

---

67 Bonacich and Wilson 2008, p. 15.
68 Nakano 2020, p. 73.
69 Nayak 2019, pp. 73–4.

systems are becoming more attractive to manufacturers as trials and studies promise gains in efficiency, resource allocation, stock keeping, control of processes, control of workers, and automation of certain processes, all of which contribute to lower production and labour costs.[70]

13 Uses in Production

Although there has been increasing implementation of automation, production of garments is still very labour intensive.[71] Since much of the global fashion production has been moved to countries that provide cheap labour, RFID has not yet been widely implemented in production. While the technology has been getting much cheaper in recent years, it remains relatively expensive for garment production.[72]

There has been increasing pressure by retailers on manufacturers to implement RFID technology, because retailers argue that it would benefit coordination along the supply chain and make just-in-time manufacturing faster and more effective. However, manufacturers have been reluctant to invest in this technology, because they worry that the retailers would reap most of the benefits of the technology while they would have to carry all of the costs.[73] This is an ongoing conflict between manufacturers and retailers.

Citing different studies, Nayak outlines how RFID may optimise the whole sequence of the production process in textile factories. He does this strictly from a management perspective. It remains to be evaluated, however, how these changes in the production system could impact working conditions and power relations between manufacturers and retailers.

14 Spreading, Cutting and Bundling

According to Nayak, RFID can help to improve the accuracy and efficiency during spreading and cutting. Without RFID tags the spreading team relies on pre-written instructions in log books or computer printouts to properly lay out the materials. With RFID tags attached to the material the spreading team could get more instant and more detailed information by reading electronic

---

70 Nayak 2019, p. 75.
71 Nayak 2019, p. 73.
72 Nayak 2019, p. 73.
73 Bonacich and Wilson 2008, p. 11 and Yip and Huang 2016, p. 329.

tags, which could include simple and direct information on how to spread the materials for the cutting process.[74] RFID tags can be used in the place of printed bundle tickets to bundle cut materials for sewing.[75] These activities are currently performed by pattern tailors. These workers hold qualified positions and have played a key role in some workers' struggles in garment production, since they cannot easily be replaced. RFID technology could make that position obsolete.

15   Sewing

There are currently three types of production systems for organising industrial sewing: the progressive bundle system (PBS), the unit production system (UPS) and the modular production system (MPS). PBS is currently the predominant system. It organises the progressive completion of the garment components from one operator to the other in a progressive sequence. The UPS involves the use of conveyor hanger systems (usually automatic), which carry the different components from one operator to the next. This is the most automatic practice for the production of garments, which some manufacturers have implemented to save time and reduce labour costs.[76]

A study of a RFID system at a large manufacturer in Hong Kong shows that RFID technology can be applied to all three manufacturing systems to effectively monitor material and information flow.[77] According to this study, RFID can be used to improve complex operations on the sewing floor, which involves managing the movements of fabric components and trims by fitting each fabric component or trim with an RFID tag. The movement of each component from one operator to the other (UPS) or from one machine to the other (PBS) can be monitored. This allows the estimation of the efficiency of the manufacturing system. Any bottlenecking on the production floor can be identified with the data constantly produced and recorded in the RFID system during the production process, which can identify when one of the components has stopped moving.

In this study RFID technology was effectively used on sewing floors to obtain information on work progress, production quantities, and the efficiency of assembly lines. Another important aspect in the study was the estimate of the

---

74   Nayak 2019, p. 77.
75   Nayak 2019, p. 78.
76   Nayak 2019, p. 78.
77   Nayak 2019, p. 78.

efficiency of each sewing machine, productivity of each worker, and overall productivity. The manufacturing information system based on RFID technology consisted of:

(1) a high frequency (13.56 MHz) passive tag with its unique ID associated with the bundle of cut-raw materials (such as sleeves, cuffs and hoods);
(2) several RFID readers installed next to each sewing machine and quality control (QC) tables in the cutting department;
(3) a central PC workstation, connecting the 40 sets of RFID readers (the PC workstation collected the data first followed by uploading to the computer server);
(4) an Ethernet system to transmit data from PC station to the server.[78]

RFID systems were installed along all sewing lines, and supporting staff were selected from the Industrial Engineering Department (IED) that conducted the study. The IED team was responsible for collecting data from different lines, analysing and comparing the performance of each line, and the performance without the RFID system. The study found that:

(a) the actual time for each operation in garment manufacturing could be easily recorded and be used for better line balancing;
(b) this data can be used for establishing the standard time for each process, which allowed for the workers to be allocated accordingly;
(c) the performance of individual workers can be monitored using real-time data. It was also possible to calculate the daily efficiency of each worker and to rank them by efficiency[79]

The study noted as tangible benefits a 50 percent reduction in 'lost-time' (non-working time) and a 50 percent increase in efficiency compared to the previous year and a 90 percent increase in on-time delivery.[80] As intangible benefits the study uncovered previously 'hidden problems' such as bad coordination among departments. It also made possible the estimation of the working efficiency of the maintenance departments and allowed the identification of a 'sewer's skill set deficiency'. It also provided more accurate data for calculation of standard allowed minutes (SAM), which are normally used for cost estimates and line-balancing.[81]

---

78 Nayak 2019, p. 80.
79 Lee et al. 2012 cited in Nayak 2019, p. 80.
80 Lee et al. 2012 cited in Nayak 2019, p. 80.
81 Lee et al. 2012 cited in Nayak 2019, pp. 80–1.

16   Automation of Resource Allocation

Some models of RFID use go even further in that they envision the use of RFID technology to automate resource allocation and parts of the decision-making processes and to fully integrate manufacturing data with the Enterprise Resource Planning (ERP) of the retail company.

Lee et al. suggest an approach of using data mining, mathematics, and AI, which could allocate resources automatically with the use of a RFID system.[82] Lee et al.'s model is rather technical and complicated, but essentially the RFID-based resource allocation system consists of two modules: (a) a data-capturing module (DCM) and (b) a decision support module (DSM). The central database is also linked with Customer Relationship Management (CRM) and integrated with ERP. The system collects data, analyses processes, and uses the data collected to make suggestions for optimal resource allocation. It can also react to new customer demands via its information from Customer Relationship Management.[83] Such a system would be able to completely integrate all processes in production with the overall supply chain strategy of a retail company.

What is interesting about this complete RFID optimised factory system presented in this study is that it takes the capabilities of RFID assisted automation the furthest. Such a system would almost completely automate the decision-making process by connecting even the smallest decision with supply chain considerations (ERP) and consumer demand (CRM). Process evaluation and decision making could be performed or assisted by a system fully integrated into the overall planning of the supply chain. This fits well with a system of pull production and arguably exposes the central goal of supply chain management, which is to coordinate even the smallest processes with the overall supply chain strategy governed by POS data and sales predictions.

17   Labour and RFID in Production

Nayak also stresses that RFID technology could be an effective means to reduce labour costs. The price of labour currently contributes about 30–40 percent towards the price of garments for retailers. He argues that RFID could reduce that number significantly, if each worker carried an RFID chip.[84] This would make it possible to automatically record the time workers enter and leave the

---

82   Lee et al. 2013, p. 787.
83   Lee et al. 2013, p. 786.
84   Nayak 2019, p. 85.

factory and the number and length of the breaks they take. It could record work-related information (for example which machine they use for how long) and total working time and absences.[85] Nayak states that this would allow the calculation of the individual worker's efficiency, which could be used to locate inefficient workers or non-essential jobs in the production process and thus save labour costs.[86]

As this outline of the possibilities of RFID technology for garment manufacturing currently discussed by industrial engineers shows, RFID could change the organisation of factories in a substantial way. It could also drastically impact working conditions, by enabling the automation of certain processes, such as bundling and resource allocation, thus rendering certain positions obsolete or transforming them into less qualified positions. It could also drastically change the way workers are managed and controlled. As the study by Lee et al. shows, the technology promises a decrease of 'lost time' by about 50 percent.[87] This seemingly neutral description disguises the fact that for a worker this would essentially mean an intensification of work, which Marx had already identified as an immediate effect (and intrinsic tendency) of machine production in *Capital*:

> This gives an immense impetus to the development of productivity and the more economical use of the conditions of production. It imposes on the worker an increased expenditure of labour within a time which remains constant, a heightened tension of labour-power, and a closer filling-up of the pores of the working day, i.e. a condensation of labour, to a degree which can only be attained within the limits of the shortened working day. This compression of a greater mass of labour into a given period now counts for what it really is, namely an increase in the quantity of labour.[88]

RFID would also increase analysis and control of workers, machines, and resources. This would allow for an 'objective' (i.e. computer and data supported) evaluation of work performance, which would make new ways of disciplining or 'motivating' workers possible. It would also potentially make the position of foreperson obsolete.

---

85 Nayak 2019, pp. 85–6.
86 Nayak 2019, p. 86.
87 Lee et al. 2012 cited in Nayak 2019, p. 80.
88 Marx 1990a, p. 534.

18   Uses in Logistics and Supply Chain Management

From the perspective of management, the benefits of RFID along the supply chain are mostly that it provides more accurate and more up-to-date data than systems without RFID, which need to keep stock manually with or without the use of barcode technology. According to Nayak using RFID along the supply-chain therefore can make material management easier, inventory control more effective, and stock replenishment faster. It can also prevent theft more effectively and make tracking of products easier.[89] Since supply chain considerations are often directly linked to the planning of production (as for example in Inditex's 'design teams', which consist of designers, marketers, and procurement experts), more instantaneous and more accurate data on supply chain operations provided by RFID also promises better estimates for production planning. Another aspect largely neglected by Nayak, but essential for workers and contractors is that RFID data along the supply chain can also increase control over the entire chain, which usually consists of formally independent companies. Getting more detailed information in real time will give supply chain management (in the fashion industry the supply chain is usually either managed by retailers or by logistics companies employed by retailers) more power and more flexibility not only in production planning, but possibly in negotiating with contractors as well.

As outlined earlier, currently RFID tags are applied to finished garments at Inditex's own factories for in-house production and at the company's distribution centres for out of house production (see figure 8.5). Nayak writes that RFID technology can significantly reduce processing time in receiving and sending goods, which can reduce overall throughput time.[90] The fact that RFID tags can be read without human intervention and that multiple tags can be read at a time turns registering goods from a manual activity into an automatic one. RFID tags can also be reprogrammed or updated with information, which would replace the work-intensive activity of printing new barcodes to replace old ones.[91] Tags can also store more detailed, individual information so damages in lots could be recorded on the tags as well. With the expansion of the RFID system, logistics activities in the supply chain before the distribution centres could be closely analysed and controlled by supply chain management. This means it would be easier for Inditex to restructure these activities if it needed to deal with contingency. With this expansion, Inditex would get more

---

89   Nayak 2019, p. 92.
90   Nayak 2019, p. 97.
91   Nayak 2019, p. 97.

accurate and more up to date information on the movement of goods from the factories to its distribution centres. Logistics providers will be much easier to evaluate and control. This will also give Inditex more options to manage and possibly reorganise these logistics activities according to the Information collected with the RFID system.

19 Conclusions: How Will RFID Change Power Relations and Conditions for Workers along the Supply Chain?

Three major issues can be formulated from the perspective of labour faced with the implementation of an RFID system:

(1) RFID will allow the recording of more information on the production process. This data will be vaster, more exact, and more up-to-date. This will give parent companies more power ancontrol in their relations with contractors and more access to areas currently not visible and controllable.

(2) RFID will influence working conditions in production by making partial automation of certain positions possible.

(3) Workers at the factories will be faced with automation and increased control. Their direct employers will have even less power in the production process. This will present new obstacles for workers' struggles as well as new opportunities.

(1) RFID along the entire supply chain and in production will vastly increase information about production processes and the movement of goods and materials. Information will be more detailed and more recent. The questions that concern power relations are: Who gets access to what information and who is able to use that information to their advantage? Which data would a contractor be willing to share with Inditex? Which data would a contractor be forced to share? What information would a contractor receive in return?

It is not surprising that Inditex pushes so hard for the implementation of RFID since they have the most to gain from the technology. Inditex's business model is based on the ability to reproduce current fashion trends more quickly and for a higher profit than its competitors. To achieve this, Inditex needs to work fast, be flexible, effectively coordinate its production, and closely control its supply chain. The need for such a high level of control and coordination seems to have led to the company's partial integration of most key aspects of production (something that at first seems counterintuitive for a retailer during the logistics revolution) and can be seen as the driving force for its push for RFID technology. Inditex has a lot of interest in getting detailed real-time information on the production process of suppliers to be able to better plan its production. As was shown, the retailer's production planning (in the design teams) is directly influenced not only by POS data and sales projections, but also by estimates about whether a model can be produced on time and for a profit. An effective expansion of RFID will improve the accuracy of such estimates. Without RFID, Inditex relies on information provided by contractors, which can be inaccurate or even strategically false.

From the perspective of contractors installing a RFID system and sharing data generated by that system with Inditex, an RFID system connected to Inditex means that they would give up information without knowing exactly the implications that information has. Since Inditex knows far more about the other steps of the supply chain, it would have uses for the information gained with an RFID system that the manufacturer would not be able to conceive of. Since Inditex also has its own factories, it can act as a direct competitor in production against its contractors.

It is conceivable that Inditex could make demands that a contractor change its production in a certain way, because data suggests it could be optimised using fewer workers in a certain area or using different technology. Inditex would also be able to compare data of different contractors. It could learn that one contractor is especially fast at a certain part of production and exceptionally slow at another, which could prompt the decision to only contract parts of production for which the contractor has a particularly advantage. Inditex might even copy parts of production at its own factories because it has analysed that it can perform that step cheaper than a contractor can. All of these options prompted by information given up by contractors will not be seen by the individual contractor who has only limited information on the whole production process. Contractors only have access to that part of up-stream information that the employer shares with them or has to strategically share with them during a cooperation. These considerations seem even more likely if one considers the example of the RFID study conducted by Lee et al. and cited by Nayak, which showed that the capabilities of RFID systems go beyond automation and that the data which was generated allowed for a detailed analysis of the production process and for an 'optimisation' of that process, according to the needs of the employer. By giving up all this information, a contractor might even involuntarily render himself useless in the production process because production could easily be reorganised without him.

In light of Inditex's production system, it is clear that the company's goal is to be able to analyse and control all processes of production from procurement of raw materials, design, manufacturing, and sales through one integrated system. Such a system would be able to get real time data from all steps of production and would be able to analyse the efficiency of machines and workers along the entire supply chain. It could localise ineffective processes and find hidden potentials to make production faster or cheaper by intensifying work. 'Problems' could be treated with the entire supply chain in mind by being able to see it better in the context of the whole chain. For example, it would be possible to analyse more accurately if a certain problem for Inditex, such as a contractor producing a product too slowly, is fixable or if it is more profitable to just modify the chain by producing that product at Inditex's own factory or through a contractor that Inditex knows could produce it more quickly. As a result of the more accurate information of an RFID system, Inditex as the manager of the supply chain would have more options and be able to better project what each option would mean for the whole system. Outsourcing to deal with contingency and to limit workers' bargaining power would become easier, since Inditex would have to give up less control to do it. While it is possible that some contractors will profit from RFID along the entire supply-chain, they will generally lose power, because in a supply chain optimised with RFID technology they can be replaced more easily and more quickly. Having additional information and control with an RFID system will also influence the parent company's negotiations with its contractors.

(2) As was shown, RFID will influence working conditions in production. Spreading and bundling at factories can be automated with the use of RFID technology. Fewer experts will be needed and work can be simplified through direct instructions provided on RFID tags. This will devalue this work and make positions such as that of the pattern tailor obsolete. Successful organising in the garment industry has often depended on the support of these qualified workers. Jobs that are focused on the control and discipline of other workers would also be potentially vulnerable to automation since RFID technology in production will make it relatively simple to estimate the performance of departments and individual workers. The job of foreperson will be less necessary and it will become a different job.

It can be assumed that power at the factory will be exercised differently. Workers can be controlled more closely with RFID technology and their performance will become subject to individual analysis. Standard allowed minutes will be calculated from 'objective data', i.e., data automatically recorded in the actual work process rather than from estimates. This could mean that standard allowed minutes will become more realistic and therefore easier to meet, but it is more likely that exploitation would be maximised by taking out any additional room or leeway. A RFID system in production will also make it possible to pay workers according to their performance recorded by the system.

(3) Technologies like RFID have a direct influence on the way power is exercised at the workplace. Technological infrastructure enables control without

direct governance. A technology like RFID can increase control and process analysis without it being directly visible. It creates a situation where no one seems to be directly accountable, but in which individuals will experience an oppressive system whose parameters have been closely designed according to the needs of production. Technology can influence the possibility to act without direct leadership or visible accountability. While this has been discussed by various theorists, there need to be more discussions about the concrete connections between technology and power in the labour movement as measures of digitalisation are continuously intensified.

Bonacich and Wilson criticise that under the current neoliberal system capital can move fairly freely and look for the best deal, while workers are restricted to their countries.[92] The contradiction with data is a similar one. There are currently no real borders for data, but data can be used to control workers. It seems entirely possible, for example, that a RFID-optimised garment factory in Asia like the one envisioned by Lee at al. 2013 would be controlled by a European retailer. The retail company would be able to closely monitor the production process and the decision-making process with RFID data, while workers would be employed by a local company. With such a system, the supply chain manager has access to more data on a factory and its production processes without any of the 'downfalls' of ownership.

The fact that contractors will lose power in a completely RFID optimised supply chain makes a strategic alliance of workers' initiatives in production with their direct employers structurally possible. Such an alliance seems counterintuitive at first, but there have been successful cases of workers improving their conditions by temporarily working with their direct employers in negotiations with parent companies.[93]

Theoretically, the close recording and control of so many aspects of the production process can lead to improved conditions, because it will make long term considerations by employers easier. Since RFID will make many parts of production more transparent – at least for the ones who have access to the data that is generated – these parts of production can be the subject of negotiations, where workers are actually able to negotiate their working conditions. Workers would do well to get information on a RFID system installed at their factory and should know all the capabilities it has: Where are the readers and what can they record? Can the system be cheated or manipulated? What will be recorded by the system? How can data generated by that system be used to lobby

---

92 Bonacich and Wilson 2008, p. 19.
93 Bonacich and Wilson 2008, pp. 248–249.

for the betterment of their working conditions? Does that system potentially violate privacy laws? How does the employer intend to use the data? Will the information which is collected be made transparent for workers? What parts of production will become vulnerable to automation? Workers' activists should also discuss more general question that go beyond any one technology: How do processes of digitisation, RFID being one aspect in a larger development, lead to a concentration of work? How can a situation be created in which workers can be enabled to discuss and reflect on their situation? How can such a discussion lead to a necessary resistance?

Production models like Inditex's are essentially very vulnerable, because they are so strongly based on speed and just-in-time considerations. If workers managed to organise along the supply chain an employer could be put under pressure.[94] There are two major 'choke points' in Inditex's supply chain that could effectively stop the company's production process, namely the company's distribution centres (which handle all of the company's clothing lines) and the transportation of the goods to the company's retail stores. It seems that Inditex's management is very aware of this and has made organisation of workers in this part of logistics very difficult by employing a large number of temporary workers at is highly automated distribution centres as well as contracting out deliveries to its retail stores to a large number of competing contractors.

Workers directly employed by Inditex would have a lot to gain by getting access to RFID generated along the supply chain and to information on production planning. Sharing that data with workers employed along the company's supply chain would offer powerful options for negotiations that could benefit workers directly employed by Inditex as well as workers indirectly working for Inditex. Flexible organisation and strategic coordination along the supply chain are the best options to generate more power for workers in Inditex's production system.

---

94 There has been a recent debate in Anglophone Marxist theory about the role of logistics in the capitalist economy organized in supply chains. There is a structural potential for coordinated worker's action at 'choke points' i.e. points in the supply chain that could effectively shut down the whole production process (for example ports or distribution centres) (Alimahomed-Wilson and Ness 2018, Moody 2017). While most theorists stress the potential power of logistics, others caution that for multiple reasons it has been very difficult to organize logistics workers to put coordinated pressure on these 'choke points' (Browne et. al. 2018).

CHAPTER 9

End Meeting: A Workers' Inquiry into the Algorithmic University

Robert Ovetz

As David Noble alerted us, the effort to automate, outsource, and rationalize academic labour is not new.[1] The pandemic has created the ideal circumstances for corporate consultants and 'edtech' venture capitalists, textbook publishers, and online education non-profit 'reform' groups working to deskill and automate teaching in colleges and universities. Their efforts were propelled by the massive privatization of K-12 education in New Orleans following 2005 hurricanes Katrina and Rita and the pandemic.[2] In 2020, edtech watchdog Nancy Bailey warned that during the COVID-19 pandemic the self-isolation and quarantines instituted to stop the transmission of the virus have rapidly accelerated the turn toward remote work using new telecommunications technology such as the Canvas learning management system (LMS) and Zoom teleconferencing app.[3] The emergency swept away many of the barriers to the spread of another epidemic – the digital automation and deskilling of teaching in higher education. What we currently face is a confluence of forces that is accelerating the attack on the very academic labour of faculty in higher education. This attack must be understood in order to devise the necessary tactics and strategies to counter and resist it.

In the past decade, on-line education (OLE) in the US has been making slow and steady gains. The number of students who have taken at least one OLE class grew from 8 percent in 1999–2000 to 18 percent in 2017 with twice as many in

---

1 Noble 2003.
2 One of the most significant pushes for moving and keeping higher education online is being made by the Boston Consulting Group whose Managing Director & Partner Nithya Vaduganathan has touted her efforts to 'develop strategic plans for scaling personalized learning' (code for online education), and 'supported rebuilding the K-12 system in New Orleans following Hurricane Katrina' (Boston Consulting Group 2020b). In fact, the massive shift to Zoom during the pandemic is modelled after the Sloan Semester online courses for hurricane Katrina and Rita refugees organized by the Sloan-C project to expand OLE (Online Learning Consortium 2020). See also Bay View Analytics 2020.
3 Bailey 2020.

public institutions as in private.[4] According to a recent Public Policy Institute of California report, OLE has taken a hit due to devastating reports of the 'online performance gap', in which online courses in every academic discipline results in higher failure and dropout rates than in person courses, and the defeat of the much hyped Massive Open On-Line Courses (MOOC) at my campus, San Jose State University, after its first and only semester in 2013.[5]

The pandemic gave new momentum to OLE. The widespread reliance on conferencing platforms such as Zoom to move nearly all higher education into OLE accelerated the process of imposing a new technical composition of 'academic capitalism' on higher education.[6] This necessitates that as faculty and other academic workers we shift our organizing tactics, strategies and objectives to address the changing organization of academic labour.

The accelerated reliance on conferencing platforms like Zoom and LMS's such as Canvas that drive OLE is not a neutral process. The emergence of OLE coincides with decades of neoliberal assaults on higher education to produce more degreed productive workers through adjunctification, austerity, privatization, entrepreneurialisation, and shifting costs to students and their families through skyrocketing tuition and fees paid for by massive personal debt. These represent the external factors placing relentless pressure on higher education make it more effectively serve capital.[7] Such an analysis is informed by a class analysis of the role of higher education in capitalism in which faculty academic workers 'co-produce new labor power' of new waged workers who 'will in turn be employed to produce value and surplus value'.[8] Alongside these external factors is the equally critical internal factor of the fragmentation and rationalization of academic labour by OLE that threatens to undermine the very craft once thought insulated from attack – the human skill of teaching.

This chapter will focus on the central roles of Canvas (one of the three largest providers of LMS's) and Zoom in the emerging new technical composition of academic capital as the latest phase in the response to the recomposition of the power of academic labour that accelerated in the 1960–70s. OLE is predicated on fundamentally shifting teaching and learning from assessment of *comprehension* of content knowledge, problem solving, and the production of new knowledge to the measurement of *proficiency* in task completion.[9] There

---

4   National Center for Education Statistics 2001 and 2019.
5   Johnson and Mejia 2014.
6   Slaughter and Leslie 1999; and Slaughter and Rhoades 2004.
7   Ovetz 1996; Harvie 1999, 106.
8   Harvie 1999, p. 105; 2006, p. 12.
9   Ovetz 2021.

are two critical aspects to this shift. First, it is made possible by the emergence and ubiquity of artificial intelligence (AI) and communications technologies built into Canvas, for example, that are used to reduce the reliance on tenure track faculty while rationalizing academic labour and increasing the precarisation of the professoriate. Second, it is intended to produce more productive self-disciplined students as labour power to meet the growing demand for precarious 'platform' or 'gig' work.

The rise in organizing among adjunct faculty in recent years will not be sufficient in itself to halt the emergence of this new technical composition of academic capital by continuing to rely on what Ness calls contract unionism that merely trades wages and benefits for control over academic labour.[10] A workers' inquiry into the new technical composition of academic labour in the university understood through the lens of class composition theory is critically needed.[11] A workers' inquiry is a method for studying the new technical composition of capital, which reorganizes work as a strategy to decompose the power of workers from previous successful struggles. This allows capital to recompose the relations of production so as to restore control over the production of new labour power which is the fundamental work of higher education. Understanding each phase of the class composition is critical for academic workers to devise new tactics and strategies to recompose their power and shift power back in their favour.[12]

1   'Unbundling' Teaching in Higher Education

After decades of austerity, entrepreneurialisation, and outsourcing[13] neoliberal educational technology (or edtech) 'disruptors'[14] have been advocating the fragmenting of US higher education at the level of systems, institutions, non-academic services, instructional, and professional into separate 'primary'

---

10   In Alimahomed-Wilson and Ness 2018.
11   Ovetz 2020b.
12   Ovetz 2017.
13   Ovetz 1996, 2015a, b and c, and 2017.
14   Neoliberal higher education 'disruptors' are proponents of bypassing faculty shared governance to introduce measurements of competency and task completion, such as the now ubiquitous use of quantitative 'learning objectives' and student 'opinions' of teaching to evaluate faculty. The objective is not merely to move students through faster but also produce more useable labour power (Christensen et al 2011; Thornton 2013; Martinez 2014; Craig 2015; Carey 2016). For critical analyses of disruption see Rhoades 2013; and Ovetz 2015a and b.

(teaching and research) and 'support' activities (administrative and support services). This strategy has been called 'unbundling'[15] which breaks up, automates, privatizes, outsources, and off-shores each component of campus academic and administrative operations in order to disperse them to private companies operating along a global higher education 'value chain'.[16]

Many administrative components have already been unbundled directing new attention toward instructional components such as teaching and academic services such as counselling, advising, financial aid, tutoring, library support, LMS tech support, American Sign Language interpretation, and admissions. Prior to the pandemic, neoliberal pressures of austerity to increase headcounts created relentless pressure to expand OLE and integrate telecommunications and AI such as 'Packback' discussion and grading chatbots[17] (see below) in an effort to physically unbundle higher education from place based to online.[18] Even though edtech ideologues such as the Boston Consulting Group are quick to praise unbundling for avoiding or reducing the need to invest in infrastructure and faculty salaries[19] there is insufficient research demonstrating such cost savings once the fixed technology and staffing costs are included.[20]

There have already been three previous phases of unbundling of higher education driven by external pressures.[21] We now find ourselves in the fourth phase, which seeks to unbundle the academic labour of teaching by rationalizing teaching. This seeks to fragment, deconstruct, and redistribute its three key elements of design, delivery, and assessment of teaching into as many as nine components no longer under the control of faculty.[22] This unbundling of teaching has been explained as 'the differentiation of instructional duties that were once typically performed by a single faculty member into distinct activities performed by various professionals, such as course design, curriculum development, delivery of instruction, and assessment of student learning'.[23] This division has only been made easier by the nearly complete dismantling of the three pillars of faculty academic labour: research, service, and teaching

---

15 Sandeen 2014, p. 2; Gehrke and Kezar 2015, pp. 93 and 119; McCowan 2017, p. 737.
16 Ernst & Young 2012; Carnegie Mellon University n.d.; and Boston Consulting Group 2020a.
17 McKenzie 2019; Delaney 2019.
18 Mazoué 2012, p. 75.
19 Boston Consulting Group 2020a.
20 Sandeen 2014, pp. 6–7; Gehrke and Kezar 2015, p. 129.
21 Kezar and Gehrke 2015, pp. 97–108.
22 Smith 2008; Sandeen 2014, p. 3; Gehrke and Kezar 2015, p. 104.
23 Gehrke and Kezar 2015, pp. 93–4.

by transforming, according to Curtis, more than 70 percent of the faculty into contingent 'just in time' precarious or 'adjuncts'.[24]

Unfortunately, with the exception of Noble,[25] the recent research into bundling and unbundling have almost no explanatory power mistaking it merely as a management tool rather than a new model of academic labour. Lacking a class analysis, such theories are entirely unable to explain what is driving the deskilling of academic labour. The catchy concept of 'unbundling' could instead be understood as a euphemism for 'deskilling which involved a fragmentation of formerly comprehensive skill sets and the displacement of skilled labour ("all-round" academics ...) by semi-skilled or unskilled workers (semi-skilled para-academics)' both inside and outside academia.[26] Those who have reframed the rationalization of academic labour into unbundling mistakenly represent it as an unstoppable monolithic force with no origin whose penetration is leading to a predictable outcome. The outcome is far from predetermined considering the immense effort being undertaken to undermine the struggle of academic workers opposed to it.

Rather than adopt the term unbundling, we are better served to analyse the rationalization of teaching as a strategy to deskill, discipline and better control faculty academic labour[27] in order to produce more unwaged students who are self-disciplined and productive waged labour. For nearly half a century we have been subjected to the neoliberal attacks on higher education for churning out too many students who are unprepared for work and unprofitable to employ. These complaints are not mere hyperbole. Faculty imposition of work on students who are engaged in everyday refusals of work is the driving motivation for rationalization.

2 Faculty as Appendage to the LMS

Marx's analysis of the deskilling of workers characteristic of a new technical composition of capital is invaluable for understanding the rationalization of academic labour.[28] Marx examined the technical composition of capital in

---

24 Curtis 2014.
25 Noble 2003.
26 The term 'para-academic' is problematic in that it overlooks the reality that many have the same training as tenure track faculty and only differ in their contractual term as precarious and contingent professors. (Macfarlane 2011, p. 59; see also Czerniewicz 2018).
27 The Analogue University 2019, pp. 1187–8.
28 Marx 1990a, p. 481.

detail in chapter 25 of *Capital* Volume I.[29] The technical composition of capital has gained a resurgence in recent years. It can be understood as the current ratio of technology to human labour and the strategy, rules, and processes for organizing work and managing workers.[30] Braverman[31] used Marx's analysis of rationalization to demonstrate the Taylorization of craft labour at the turn of the twentieth century. Bringing both Marx and Braverman into the classroom, Foucault[32] applied rationalization to education as a strategy for the control and disciplining of academic labour.

Marx saw how the new technical composition of industrial work transformed the worker into an 'appendage' of the machine and the factory. 'Not only is the specialized work distributed among the different individuals, but the individual himself is divided up, and transformed into the automatic motor of a detail operation'.[33] In this way, Marx explains, 'the human being comes to relate more as watchman and regulator to the production process itself'.[34] His detailed analysis of the deskilling of craft workers in the rational organization of industrial production in the factory is entirely relevant to understanding the rationalization of skilled into deskilled academic labour today. As faculty labour is assessed and rationalized, course design, delivery, and assessment[35] becomes fragmented and the pieces redistributed to non-faculty academic staff such as content experts, counsellors, course designers, technical support, programmers, and outsourced to textbook and software companies.

Professors are bombarded on a weekly basis by examples of such rationalization. One example is non-profit publisher Norton's February 2017 spam email that led with the subject line 'No time for grading?' promising 'our content, your course'. A May 2020 spam email from Packback further promises the use of AI 'to improve student engagement for community college students … while also automating some of the administrative faculty burden that unfortunately comes with managing discussion'. These two companies are not merely pitching their product to engorge their bottom lines but are profiting from the rationalizing of academic labour by what Braverman famously described as the 'separation of conception from execution'.[36] He noted how this takes place

---

29 Marx 1990a, pp. 762–870.
30 Woodcock 2016; Cleaver 2019; and Ovetz 2021.
31 Braverman 1998.
32 Foucault 1977.
33 Marx 1990a, pp. 481–2.
34 Marx 1993, p. 705.
35 McCowan 2017, p. 738.
36 Braverman 1998, pp. 86–7.

when 'the first step breaks up only the process, while the second dismembers the worker as well, means nothing to the capitalist, and all the less since, in destroying the craft as a process under the control of the worker, he reconstitutes it as a process under his control'.[37] This corresponds to the rapid growth of middle level management in higher education.

A critical element in the rationalization of academic labor is the reliance on the 'datafication' and what van Dijck[38] and Williamson et al call 'dataveillance',[39] or 'a form of continuous surveillance through the use of (meta)data',[40] built into OLE. The complex multivariate aspects of teaching are transformed into tasks that measure and assess 'competency' of students represented in the form of data operationalized through OLE that disassembles teaching and redistributes its components to specialized staff responsible for highly differentiated technical aspect of the course.[41] What Marx and Braverman have taught us is that the rationalization of labour is not simply about reducing labour costs, although that is of critical concern. The cost of labour is a factor of the control of labour power. Capital must transform labour power from potential into actual work. Rationalization is a strategy for decomposing the power of academic workers in order to discipline and make them work.

Foucault's study of the 'learning machine' also applies Marx's analysis of the technical composition of labour to education and the body of the student. He meticulously related how 'the human body was entering a machinery of power that explores it, breaks it down and rearranges it'.[42] According to Foucault, the learning machine exists for 'supervising, hierarchizing, [and] reward'.[43] It breaks down the act of teaching and learning into its key components so that 'to each movement is assigned a direction, an aptitude, a duration; their order of succession is prescribed'.[44] Finally, Foucault noted that the labour of the student and faculty are similarly rationalized as the complex supervisory role of 'the master' whose assessment by exams and grades is replaced by the serialization and hierarchization of each task into a series along 'disciplinary time'.[45] Although he died about a decade before OLE was introduced, Foucault might as well have been describing its impact on teaching and learning today.

37　Braverman 1998, 54–5.
38　Dijck 2014, p. 198.
39　Williamson et al 2020, p. 351.
40　van Dijck 2014, p. 198.
41　Mcfarlane 2011.
42　Foucault 1977, p. 138.
43　Foucault 1977, p. 147.
44　Foucault 1977, p. 152.
45　Foucault 1977, p. 159.

OLE is the central organizing principle of the strategy to impose a new technical composition of capital in higher education.[46] The US labour market is rapidly moving to contingent part-time, temporary contract work in which increasing numbers of workers, between 10.1 and 36 percent of the US labour force, work remotely and are monitored and managed by information technology.[47] This rapid growth of contingent labour is intended to rapidly make the Northern labour force look more like the workers in the South where about 84 percent of India's 470 million workers, for example, are 'casual' or self-employed, e.g. contingent.[48] The adjunctification and rationalization of academic labour in higher education is not an exception to this new global division of labour, it is actually the model for it.

The short-lived MOOC functioned as the extreme end of OLE allowing tens of thousands students to select an online class from a higher education 'platform', in which a contingent professor delivers pre-packaged standardized lessons, they have no interaction with the professor or one another, and take exams 'assessed' by a computer program in order to earn a 'badge'.[49] The MOOC has all but disappeared from discussion since its high profile defeat at San Jose State University in 2013. However, the MOOC remains the ultimate objective of achieving the professor- and classroom-less 'university' by enclosing all higher public education into what Hall describes as an Uber style platform system for distributing courses in which the content specialist is paid by the head according to surge pricing.[50] Those seeking to rationalize college and university teaching are taking the long march through the institutions by using crises like the 2008 recession and the COVID-19 pandemic to accelerate the move to OLE.

Because the labour-intensive teaching and learning that comes from human interaction, social relationships, and emotional and intellectual exchange is lacking in the LMS, *teaching* is rapidly becoming deskilled into *assessment, measurement*, and *monitoring*, while *learning* is being replaced by *competency*

---

46   Ovetz 2020b, pp. 1–34.
47   Conlin, et al 2010; *The Economist* 2015, BLS 2018; McKinsey & Company 2022.
48   Ness 2015, p. 85.
49   Badges are now being used to demonstrate competency in task specific skills. I was required to obtain a badge by a community college to teach on-line courses despite more than a decade of experience teaching on-line. The badge is designed by the Calbright all on-line college that has costs hundreds of millions of dollars with enrollment from only a few hundred students. In 2021 the California Assembly had voted unanimously to shut it down. (*FACCCTS* 2021).
50   Hall 2018, pp. 22–9.

of *task completion*.[51] The rapid expansion of OLE run on the Canvas LMS and the delivery of courses through Zoom play a central role in the new technical composition.[52]

OLE is designed to provide an alternative to assessment exclusively controlled by faculty and institutions of higher education, what Wang long ago famously denounced as 'monopolies' subject to legally mandated unbundling.[53] OLE, the Canvas LMS, and Zoom are transforming faculty academic labour so that it is less about teaching than machine tendering for the remote monitoring, measuring, assessing, processing, and delivery of disciplined unwaged student labour power.

In order to understand the current technical composition of higher education a workers' inquiry into academic labour is carried out below by examining the structure and organization of Canvas and Zoom for the algorithmic management of academic labour.

3 A Worker's Inquiry into Canvas and Zoom

The process of rationalizing academic labour is built into the digital architecture of the Canvas LMS. The objectives of OLE are expressed by the design of the LMS. Although faculty appear to have complete autonomy to set up their LMS shell for their course with a variety of possibilities to match their chosen pedagogy, the very architecture of the LMS is designed to fragment teaching into the delivery of tasks and learning into competency of task completion.

Constructed as a diffused virtual space of an online 'classroom', the Canvas LMS is not intended to simply mimic in person classroom teaching, but replace it with an entirely different logic. Students no longer learn or study but respond to orders called 'prompts' in a virtual space in which their every action is designed to be treated as a measureable task.

After 'logging in', the student moves through the discreetly organized virtual space of the LMS differentiated by 'modules' that function as timed work spaces in which students write text, post a file, upload a video, download a reading assignment, stream a video, or follow a link to material or work elsewhere, to name a few of the possible tasks. Student work is highly regulated and regimented for example by having spaces 'open' and 'close' at predetermined times. A commonly used activity of faculty is to require students to

---

51   Ovetz 2021.
52   I focus on Canvas as the dominant LMS in the education market at this time.
53   Wang 1975.

respond to another student's text, work, or video post. This not only serves to use students to 'prompt' other students to complete their work, it turns students against one another as little bosses that inform on one another for missing work to provide a response. In effect, the isolated student moves virtually through the architecture of the LMS, disassociated from personal contact with fellow students, faculty, and the physical space of the classroom and campus.

The LMS is designed for virtually isolated students to discipline themselves by completing the sequence of tasks in the predetermined order established by the faculty member, course designer, or content specialist. Because each student moves in complete isolation and solitude through the LMS, their 'learning' becomes a series of discreet disconnected tasks to be completed during the window of time allowed.

The apparent similarity to the use of time, such as in the form of due dates, to impose work in an in person class is deceptive because the LMS functions to achieve an entirely different immediate objective. Time takes on a different role in OLE by guiding the completion of discreet tasks that substitute for the complex inter-personal relationships that are central features of traditional learning. Because OLE can use AI programmed by technicians to entirely bypass faculty, time becomes the predominant standard of assessing how students complete the now rationalized components of the curriculum. Just as OLE rationalized teaching into its component parts, the LMS becomes the technology for dispersing and sequencing these parts and using time to measure the intensity and productivity with which a student completes them. The ability to time is the ability to impose and measure work. Like Taylor's much-despised stopwatch, the LMS is the mechanism for solving the transformation problem of turning student labour power into unwaged work.

Timing student work assumes the ability to surveil it. In this way, the LMS serves as a mechanism dataveillance, a twenty-first century virtual digital panopticon, Foucault described as the 'eyes that must see without being seen'.[54] In the LMS, students never know with certainty when they are being remotely observed, tracked, monitored, measured, and assessed.[55]

An alternative method of measurement to faculty's perceived subjective assessment of the usefulness student work is provided by the ubiquitous collection of data in the LMS. Just as teaching is shifted to competency, learning shifts to dataveilled task completion. Just as the classic classroom 'made educational

---

54   Foucault 2010, p. 171.
55   Ovetz 2017.

space function like a learning machine, but also as a machine for supervising, hierarchizing, rewarding',[56] the LMS was designed as a data driven machine for the imposition of academic work.

Canvas's LMS 'learning machine' provides an unprecedented rich source of granular metadata on both a student and the faculty's current work that van Dijck observes can be used to measure, manipulate, predict, quantify, and monetize future behaviour.[57] From log in to log out, very large amounts of data are available to faculty, or anyone with administrative access, in Canvas and its integrated apps, as well as employers willing to pay for the data.

LMS-harvested data is immensely useful to campus administrators and future employers for what it tells about student work. For example, the Canvas 'People' window contains a wide range of detailed real time and historical data on a student's online work. In it, the 'Access Report' provides precise details about every step a student took in every area of the Canvas space. The 'Analytics' page gives dynamic bar graphs on 4 types of XY axes or tables with precise days and times spent on each task, number of tasks completed, number of page views, number of actions taken, interactions with instructors, and comparisons to the class median on each graded assignment. In effect, students can be monitored for the efficiency, intensity, productivity, and persistence of their work.

The 'Quizzes' tab provides a range of similar aggregate data in spreadsheet format on how each student engaged with every question on a multiple choice exam. An 'Item Analysis' is available which contrasts how each student did on every exam question, including the variance, standard deviation, difficulty index, and Point Biserial of Distractor. This last factor is intended to identify a reliable answer based on each student's answer choice in order to provide a standardized measurement that discriminates between students who mastered the material on the exam and those that didn't.[58] This function allows a student's work to be measured in comparison to other students' outcomes rather than assessed by the faculty according to attributes of learning, which are notoriously difficult to assess and evade comparability and standardization. Standardized compatibility is a subtle shift towards the automation of assessment and grading.

The Canvas LMS is invaluable for generating vast amounts of data on student work habits. This is critical to the deskilling of academic labour and the shift from learning to competency. In the version of Canvas I use there are literally hundreds of available integrateable apps under 'Settings' that I can

---

56   Foucault 2010, p. 147.
57   van Dijck 2014, p. 200.
58   Educational Assessments Corporation n.d.

request to automate many aspects of the course to insert standardized content, grade exams, issue badges, access user and exam data, acquire biometrics, assign peer evaluations, take polls, grade papers, post grading comments, and tutor.

The apps Dropout Detective and MyCoursEval stand out for their accumulation of data on both faculty and student work. According to the corporate text embedded in the app, Dropout Detective 'integrates with a school's existing learning management systems and analyses student performance and behaviour across ALL courses in which they may be enrolled'. The corporate text for MyCoursEval promises that when being embedded in the LMS it can provide real time student evaluation of faculty. Both Canvas apps are just two of many intended to provide immediate dataveillance of faculty and students to evaluate the productivity of one another's labour by daily producing more than 280 million rows of data. Canvas is hosted on Amazon Web Services servers giving commercial access to oceans of data about student work to anyone who wishes to pay for it. In fact, Canvas's privacy policy discloses the use of cookies, web beacons, and third party hosting to gather, store, and link data to 'personally identifiable information'.[59]

The power of Canvas is that students are aware of the potential of being monitored even when they are not sure precisely when they are actually being surveilled – Foucault's 'eyes that must see without being seen'. This serves as a velvet glove to self-discipline and self-impose work, which is what makes OLE so valuable as a technology for producing measureable disciplined labour for platform work. Students who have taken some OLE courses and graduated provide rich data to a future employer of their ability to work and presumably internalizing the procedures for working under algorithmic management regimes. The Boston Consulting Group and Arizona State University ask us to 'imagine the implications for higher education' from the application of 'Amazon's predictive models of human behavior'.[60] Perhaps BCG and ASU are unaware that Instructure, which owns Canvas, has not just stopped at imagining this integration but is actually doing it.

The persistent problem of student refusal of work can be identified in the high rate of drops and F's in OLE courses and poor performance relative to in person classes.[61] This gap is partially attributed to 'difficult to measure' student characteristics such as 'self-directed learning skills', motivation, ability, and time management. Each of these factors can be understood in class terms

---

59 Marachi and Quill 2020, pp. 421, 423, and 425.
60 Boston Consulting Group and Arizona State University 2018, p. 3.
61 Johnson and Mejia 2014, p. 1; Barshay 2015.

as tactical refusals of school work.[62] The prevalence of such refusal raises doubt about whether Canvas has effectively solved the problem of turning labour power into work.

As a result of forcing countless thousands of professors and millions of students on-line during the 2020 pandemic, the number taking on line course reportedly grew 500 percent. While the LMS infrastructure was already in place, a new tool was quickly added to it literally overnight, even making inroads into the nearly impenetrable arena of public K-12 education. Zoom, Google Hangout, Webex, GoToMeeting, Big Blue Button, and Jitsi teleconference tools suddenly moved from an obscure business tool into the mainstream as an OLE delivery mechanism. At the top of the teleconferencing market sits Zoom which received immense scrutiny due to a takeover bid by a hedge fund in early 2020.[63]

Zoom adds yet another layer of dataveillance of faculty and students that streams into the already immense ocean of data accessible through Canvas, according to Marachi and Quill. Until these features were removed, Zoom could turn on and override the host's security settings, turn on the camera without the consent of the user, track users even when the app is turned off, install a local server on users devices, and be vulnerable to hacking known as 'Zoombombing'.[64] What has escaped all attention, however, is that like the LMS into which it is integrated, Zoom accumulates data that is now available to administrators and potential employers and can be used for measuring and disciplining academic labour power.[65] This connection is explicitly illustrated by Instructure's, which owns Canvas, recent purchase of the integrated app Portfolium which directly provides data on student achievement and competency to employers.[66]

Canvas and Zoom accumulate data controlled by private companies that seek to further commodify it by integrating it with data from plagiarism detection apps, learning analytics and outcomes, attendance, social media, credit records, and other sources of metadata that are intended to be made available to potential employers. This data can be further combined with the grow-

---

62   Johnson and Mejia 2014, p. 8.
63   EdSurge 2020.
64   Electronic Privacy Information Center 2019. Some of these issues have been resolved or removed by new security features or upgrades in response to the complaint, regulatory enforcement, and other public reports.
65   At one institution where I teach administrators accessed faculty class Canvas pages early in the pandemic when all classes went online. They then sent messages to students that were not making 'progress' in the course. This was done without the consent or knowledge of the faculty.
66   Marachi and Quill 2020, 428; Hill 2019.

ing plethora of student IDs with RFID tags and license plate readers that can track a student's activity and work outside the classroom or LMS. The granular data generated by a student's movement through every module and task of the online course makes the LMS a rich array of data on the effectiveness, efficiency, and productivity of a student's work. Rather than demonstrate student learning, the massive data being accumulated about each student is designed to measure their work habits, efficiency, productivity, and most importantly their willingness to work. According to Williamson et al, 'New organizations have even suggested that it may be possible to quantify the value of every university module, course or career choice and, by consolidating a permanent record of students' qualifications and skills from across the whole educational "supply chain" – as "learner wallets" hosted on blockchain technologies – offer AI-enhanced employability advice and enable students to securely share their data with employers'.[67] Such rich data on each individual student is likely to follow a student for their lifetime as a commercially available 'work record score' that will determinately shape their life outcomes except where it is prohibited by law, such as in the EU.

4   Managing the Algorithmic University

Over the past two decades, there has been a massive push for the 'onlineification'[68] of higher education by tech companies, corporate-funded foundations, textbook companies, industry lobbying groups, campus administrators, and state and federal government officials. The potential profits from the $600 billion higher education sector is so immense that investments by the 2,861 so-called edtech companies then in existence grew 32 percent between 2011–15. Edtech investment in higher education was 30 percent of the total, just behind K-12. 97 percent of all investment was concentrated in just five countries, with 77 percent of that in the US with Canada, with the UK, India and China composing the rest.[69]

Although layoffs and budget cuts have the effect of extracting more academic labour from the remaining faculty, the primary objective of the reformers has been to produce more 'work ready' college graduates for the labour market. In class terms, it is a strategy to produce more productive academic workers who can work remotely, submit to precarious 'flexible' working con-

---

67   Williamson et al 2020, p. 355.
68   I would like to thank Professor John Holmes for coining this term for Higher Educators United of Northern California.
69   Boston Consulting Group 2016.

ditions, and are self-disciplined by the presence of ubiquitous algorithmic surveillance. To achieve this outcome, the primary impediment must be moved out of the way. That impediment is the relatively well-organized faculty who use shared governance and union contracts to fight to defend a semblance of autonomy over the content, delivery, and assessment of academic work in marginally democratic institutions.

Reformers commonly resort to hyperbole about the campuses being populated by unruly students, grade inflating faculty, and graduates who cannot or will not work. Such language underscores the intention of using OLE to automate the disciplining of labour power in the abstract, or what Marxists called 'immaterial labour'. OLE is the strategic response to what De Angelis and Harvie call the struggle over measurement. Such tools 'help shape *the form* of academic labour in both educational and research contexts. They do so by counter-posing the measures of capital, which privilege the meeting of abstractly defined targets (whether these indicate financial viability or consistency with government policies), to the immanent measures of immaterial labourers'.[70]

The wide variety of strategies to measure and standardize immaterial academic labour reflect the intense struggle still raging over the form and purpose of academic labour. These approaches include neoliberal faculty and student 'performance indicators' of 'student success', faculty-student ratios, progression rates, matriculation, retention, degree completion, guided pathways, units earned, student, college, and departmental learning objectives, and even access and equity reported in periodic program reviews required by government agencies and private accreditation agencies. The imposition of these new measures of learning reflects the shift from learning to competency. The professor is transformed from expert to foreman, from facilitating learning and knowledge generation to managing self-disciplined students who demonstrate competency by completing tasks that demonstrate standardized 'learning objectives'.[71]

Mirroring the technical composition of other sectors in higher education lies at the centre of the strategy to use OLE to produce more and better self-disciplined workers. The logic of the technology that drives OLE is analogous to the logic of contingent labour as the self-disciplined labour power that is always available for waged work. As the proportion of the labour force that is contingent, contract, consultants, gig, and platform workers grows, higher education is being reorganized to produce the labour power to do that waged work.

---

[70] De Angelis and Harvie 2009, p. 20.
[71] Prendergast 2017.

This emerging division of labour in higher education analysed here mirrors the emerging global division of labour from 'gig work' to the work of legal document review. Big data is being used to rationalize every type of labour from the unskilled to the professional. It does so by fragmenting it into its component parts, automating some of the parts, and distributing the rest either horizontally to other deskilled workers or situating them under the control of management. More of the work is distributed to informal 'gig' workers who are considered 'self-employed' because they are intentionally hired without any formal legal contractual relationship with the employer of fact. In an updated 'putting out' system described in detail by Marx,[72] these workers work remotely carrying out discreet tasks, lack immediately overt oversight by human managers, must possess the self-discipline to always be ready to work, and are entirely responsible for ensuring their own reproduction and tools whether they have paid work or not. In his study of class struggle in platform food delivery work, Cant calls this new technical composition of algorithmic management the 'black box'.[73] Because education plays a critical role in the reproduction of labour power, the new division of academic labour is designed to serve the global division of labour designed for workers who can labour under the conditions of gig work – including professors.

5   Resisting the Academic Black Box

Unfortunately, contemporary resistance to these 'reforms' has been mostly levelled at the external factors and for loss of 'quality', declining 'outcomes', and cost while almost entirely missing the primary attack on academic labour. The implications of the rationalization of faculty academic labour has been apparent since Troutt first pitched the professor-less classroom more than four decades ago in which 'an unbundled system assumes learning can transpire without students having to purchase the teaching function'.[74] Today, it is common to read about the 'automation of the profession' in which AI is paired with an entirely precarious faculty 'machine tenders' delivering 'digitally mediated rebundled teaching'.[75] The professor-less virtual classroom is attractive to universities that wish to be 'swapping expensive lecturers for cheap, versatile machines that don't go on strike, don't need sleep, and respond to stu-

---

72   Marx 1990a.
73   Cant 2019.
74   Troutt 1979, p. 255.
75   Czerniewicz 2018.

dents within nanoseconds'.[76] Higher education faculty and unions have not yet grasped the full extent of these objectives for expanding OLE. What is overlooked about edtech advocates is that they are not merely proposing to outsource rationalized teaching merely to make money but to continue reorganizing higher education to better subordinate it to global capital accumulation.

Achieving this objective means first breaking the power of academic workers over teaching and learning. As Mazoué bluntly asserts: 'If we assume learning is dependent on teaching, and that teaching is an inherently labor-intensive activity, then we will never be able to increase productivity, improve quality, and lower cost simultaneously'.[77] As long as faculty control teaching and assessment of learning, faculty labour is a critical choke point for disrupting the reorganization of higher education.

OLE is only the latest "reform" effort intended to rationalize and measure academic labour. The outcome of a university education is not pre-ordained because the struggle over measurement is a continuation of the struggle over the uses of academic labour. As De Angelis and Harvie remind us, 'capital's constant struggle to impose and reimpose the "law of value" is always a simultaneous struggle to impose (a single, universal) measure'.[78] As the anonymous aptly-named academics writing as The Analogue University put it: '[W]e need to do more than merely reveal the darker side of these transformative neoliberal relations; we need to find ways to mobilize and actively resist them'.[79]

Resisting the rationalization of academic labour will require devising new tactics, strategies, and objectives. Faculty can organize more academic workers into the struggle and slow down online-ification using intransigence and rigidity[80] to until we can shift power in our favour. Because there has to date been little attempt to assess the current composition of academic labour, the outcome is of yet uncertain. The COVID-19 pandemic vastly accelerated global online-ification transforming OLE into the central terrain of struggle over academic labour. As Noble reminded us: 'The ultimate viability of these technologies under the present mode of production depends, in the final analysis, upon the political and economic conditions that prevail and upon the relative strengths of the classes in the struggle over the control of production'.[81] Fortunately, online-ification is not a foregone conclusion.

---

76 Haw 2019.
77 Mazoué 2012, p. 80.
78 De Angelis and Harvie 2009, p. 27.
79 The Analogue University 2019, p. 1186.
80 Ovetz, 2020a, 2020b, 2021.
81 Noble 1979, p. 40.

Acts of refusal and solidarity are needed between increasingly contingent and deskilled faculty managed by the algorithmic black box and students destined for the global labour market characterized by precarious low waged work. Disrupting higher education's production of disciplined labour power for exploitation points us to a way out of capital's endless colonizing all of life as work.

CHAPTER 10

Automation of Artistic Labour and Its Limits

Jens Schröter

In the current debate on artificial intelligence (=AI), a term that at the moment refers mostly to machine learning, the question emerged whether these systems can be ⟨creative⟩. This discussion heated up as Alpha Go beat Lee Sedol in Go. Especially the famous move 37 in match two on 10 March 2016, which seemed radically new even for experienced Go-Players and which no one foresaw, raised the question whether it was not creative, at least in some way. I cannot go into depth on the theory of creativity or ⟨creative subjectivity⟩,[1] but mention that the question for machine creativity led to a new phenomenon: the field of art created by AI.

"The AI Art Gold Rush is here" is the title of a critical essay of Ian Bogost.[2] The images created look a little bit like deformed Renaissance portraits, perhaps with a little Francis Bacon in it (Fig. 10.1).

Especially the slightly deformed and fuzzy character of the portraits resembles the famous *Portrait of Edmond de Belamy* (Fig. 10.3), auctioned for 432.500 $ at Christie's in October 2018. That this computer-generated portrait was sold for a comparatively high price was surprising. The joke being that a part of the algorithm was included as a signature, added to the discussion on the ⟨creativity⟩ of AI and the question emerged whether an AI system could be an artist or an author. Moreover, a debate began whether the art collective *Obvious* who used the AI system was the ⟨real⟩ author or even the programmer who developed some of the algorithms. And on the horizon loomed the question of whether such systems might one day be able to *automatize artistic work*. In other words, old problems of the relation between technology and work, economy and power seem to emerge again with AI.[3]

When considering this question, it is necessary to historicize this discussion on ⟨AI art⟩, the implications of ⟨machine creativity⟩ and therefore the (pos-

---

1 Compare amongst many others Reckwitz 2012.
2 Bogost 2019.
3 See Dyer-Witheford, Kjøsen and Steinhoff 2019 and Schröter 2019. See also Bogost 2019: 'Given the general fears about robots taking human jobs, it's understandable that some viewers would see an artificial intelligence taking over for visual artists, of all people, as a sacrificial canary'.

The AI-Art Gold Rush Is Here

An artificial-intelligence "artist" got a solo show at a Chelsea gallery. Will it reinvent art, or destroy it?

FIGURE 10.1   Article of Ian Bogost proclaiming "AI Art Goldrush". The images shown are: *Faceless Portraits Transcending Time*
BY AHMED ELGAMMAL, USED WITH ARTIST'S PERMISSION

sible) automation of artistic work. In part two I want to look specifically on ⟨information aesthetics⟩, a discourse and practice from the 1960s, in which the idea already was to produce art *by* computers (not mainly *with* computers as tools).[4] In part three I will discuss reasons for why the idea to automatize artistic work and therefore simply mass produce ⟨art⟩ with machines did not seem to work then. In part four I will draw some conclusions.

1   Information Aesthetics

The origin of information aesthetics is the attempt to formally determine the ⟨measure⟩ of aesthetics. In 1933 David Birkhoff formulated an equation (Fig. 10.2) in which the variable O denotes the measure of the order of a given work and the variable C the ⟨complexity⟩.

$$M = \frac{O}{C}$$

FIGURE 10.2   *Birkhoff-Equation*

---

4   There are even more precursors, for example the computer cluster Iamus, who is composing music and even released an album, see: Melomics n.d. See also Miller 2019.

FIGURE 10.3   *Portrait of Edmond de Belamy*
IMAGE USED WITH ARTIST'S PERMISSION

According to this formula, the more ordered and the less complex a work of art, the more aesthetic it would be. Apart from the fact that it is difficult to understand exactly how to determine the degree of order and complexity in each case, this attempt to express the ⟨aesthetic quality⟩ of art in an equation seems strange to us today. Nonetheless, especially in the 1960s there were several attempts to formally understand art and its aesthetic criteria and consequently to produce it with computers (although the computers were slow, the output possibilities limited, and computer technologies were only available in research institutions and large companies).

In 1967 Michael Noll describes in his essay 'The Computer as a Creative Medium' an information aesthetic experiment with a painting by Mondrian (Fig. 10.4). Noll writes:

AUTOMATION OF ARTISTIC LABOUR AND ITS LIMITS 277

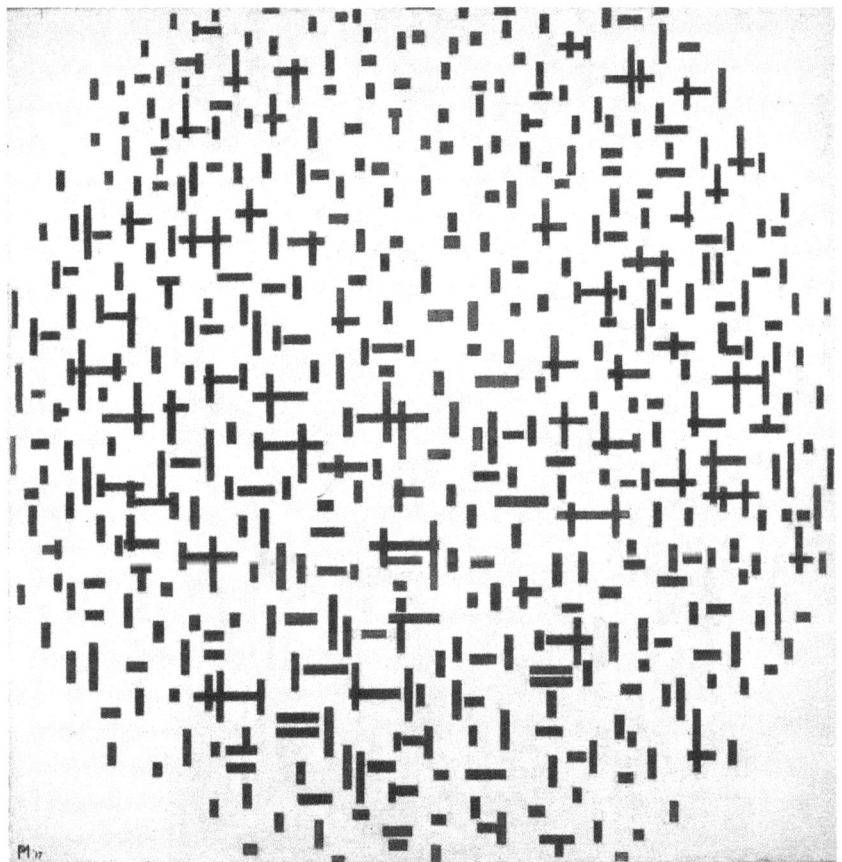

FIGURE 10.4   Piet Mondrian, *Composition With Lines* (1917)

[An] experiment was performed using Piet Mondrian's 'Composition with Lines' (1917) and a computer-generated picture composed of pseudorandom elements but similar in overall composition to the Mondrian painting. Although Mondrian apparently placed the vertical and horizontal bars in his painting in a careful and orderly manner, the bars in the computer-generated picture were placed according to a pseudorandom number generator with statistics chosen to approximate the bar density, lengths, and widths in the Mondrian painting. Xerographic copies of the two pictures were presented, side by side, to 100 subjects with educations ranging from high school to postdoctoral; the subjects represented a reasonably good sampling of the population at a large scientific research laboratory. They were asked which picture they preferred and also which picture of the pair they thought was produced by Mondrian. Fifty-nine

FIGURE 10.5 *Computer Composion with Lines* (1964)
BY MICHAEL NOLL © 1965, USED WITH ARTIST'S PERMISSION

percent of the subjects preferred the computer-generated picture; only 28 percent were able to identify correctly the picture produced by Mondrian. In general, these people seemed to associate the randomness of the computer-generated picture with human creativity whereas the orderly bar placement of the Mondrian painting seemed to them machinelike. This finding does not, of course, detract from Mondrian's artistic abilities. His painting was, after all, the inspiration for the algorithms used to produce the computer-generated picture, and since computers were nonexistent 50 years ago, Mondrian could not have had a computer at his disposal.[5]

---

5 Noll 1967, p. 92.

Noll thus simulates a Mondrian (Fig. 10.5) on the basis of a statistical distribution which is supposed to describe the arrangement of the lines in Mondrian's work. And since this distribution looks more ⟨disordered⟩, a group of observers, presumably not to be regarded as representative, preferred the simulated Mondrian, while the real one appears too regular and mechanical. Does this preference show that the disorderly is attributed a higher aesthetic quality? Or is it because the disorder is simply understood as a reference to human authorship? It could also show that the being-art of a given picture does not depend on perceptual traits alone ... It should be noted that Noll tries to find formal, algorithmic rules that allow the production of a work of art that can be identified as a work of art. These examples are taken out of context and an appropriate discussion of information aesthetics would have to consider at least the positions of Frieder Nake, Max Bense, Rul Günzenhäuser, and Abraham Moles, which cannot be done in this contribution. In any case, these attempts seem strange today, because they remove the works of art from their historical context and reduce them to abstract structures that can be formalized and therefore automatically reproduced. But it must be emphasized that the aesthetic strategies of Mondrian and Klee reacted to certain historical questions, not the least of them the development of art before them. Information aesthetics seems to understand artworks as ahistorical, formal structures – although the artworks produced by information aesthetics could also be seen as historical answers to a certain historical, in this case computational, context of art as such.[6]

However, these attempts are relatively characteristic of a larger development, namely the attempts to formalize cognitive labour and, if possible, to transfer it partially or entirely to machines, for which computers as symbol-processing machines are suitable.[7] Early texts from the domain of Computer Science, such as Douglas Engelbart's 'Program on Human Effectiveness' or Joseph C.R. Licklider's 'Man-Machine Symbiosis' from the 1960s, are programmatic for this.[8] Noll tries to make aesthetic work formalizable and, in principle, executable by machines. His attempt can be understood, whether intentionally or not, as a contribution to the automation of aesthetic work – at a time when automation of labour by computers was a hot topic.[9]

In other forms of commodity production, processes of automation often lead to the cheapening of commodities.[10] That was obviously not the goal of

---

6   Cf. on the vexed relation between form and history Buchloh 2015.
7   On the history of automation see Noble 2011.
8   Cf. Licklider 1960 and Engelbart 1991.
9   See Woirol 1996, especially ch. 8.
10   See Marx 1990a, p. 432.

information aesthetics, already because Noll operated as the author of the work (and not Mondrian); and if he would have called his work a Mondrian, this could have led to serious legal problems.

2 Artistic Labour and Automation

Apparently, art it is not directly threatened by automation. Art does not appear in the highly discussed Oxford research report, which started a nervous discussion on the disappearance of work: the only activity that resembles artistic work is that of the ⟨art director⟩, who gets off quite lightly with a 95th place on the computerizability probability list.[11] Artists are not rationalized away and their work continues to be called ⟨work⟩ – a type of work, however, that obviously cannot or should not be formalized, algorithmized, and consequently neither reproduced mechanically so easily. This suggests the suspicion that artistic work is either not work at all, but another form of activity or another form of work. But why?

At least at first glance, the art market looks exactly like any other market: artists are well aware that they have to earn money with their work and their ⟨works⟩. Some succeed to an almost inconceivable degree, but most have serious difficulties competing in the art market. Work, money for one's own work, market, competition, rich versus poor – this first glimpse seems to suggest that the art business does not differ in any way from other forms of production. It is not a realm of freedom, but only a kind of service or consumer goods industry that serves a special market.[12]

Therefore, we can find cases where work is technologically made superfluous in the art business. Looking at a large studio, such as Studio Olafur Eliasson, it could be that the introduction of new computer-assisted technologies will directly make work disappear there. A series of jobs, for instance website maintenance, management, public relations, logistics, up to cleaning the studio, may be replaced. Inasmuch as Studio Olafur Eliasson also operates under capitalist conditions, one can be tempted to save costs by rationalizing the way work is done. However, this aspect of art is external; it does address the problem that information and generative aesthetics have pointed to, namely whether the work of making art *itself* can be rationalized. Although Olafur Eliasson could

---

11   Cf. Frey and Osborne 2013, p. 59.
12   This is only the first impression of course, as we will see later. Beech (2015) discusses the problems of value in the field of art. See also Bourdieu (1996) for a discussion of the specifics of the field of art.

not realize any of his elaborate projects without his team, the disappearance of the teamwork in the black box of the author's name does not seem to affect art ⟨in itself⟩.¹³ Although artists employ in their studios people doing wage labour, the work of the artist is not wage labour in the strict sense, since the artist is not separated from the means of production.

This finally brings us to the *core of the problem* suggested by information aesthetics. Obviously, the idea of art without the intervention of a human author – and even if their role consists precisely in demonstratively withdrawing – does not seem plausible to us. We do not see art appearing in nature.¹⁴ The fact that the technical reproduction of artistic work does not seem possible does not have to be attributed – this would be the bourgeois answer – to the ⟨genius⟩, which is ultimately of divine origin and thus per se untechnical. Luhmann remarks: 'The artist's genius is primarily his body.'¹⁵ One could therefore simply say that the separation of knowledge from the working body, characteristic of the progression of capitalism and perhaps first discovered by Marx and then elaborated by Harry Braverman,¹⁶ does not or cannot take place in art. But why?

One possible reason would be that the work of art – despite the attempts of information aesthetics (at least according to Birkhoff) to formalize precisely the complexity – is *too complex* and thus the work that produces it cannot be sufficiently understood.¹⁷ Thus, it is noticeable that Noll focuses on a certain type of painting, which is determined by the extensive recourse to basic geometric forms. Such forms seem simple enough to allow their formalization, while other more complex forms would elude them. But even with geometric forms, it is true that we cannot quite imagine their existence as art without an artist. For even if an artist – like Noll, for example – were to define himself precisely by delegating all work to machines, we would still call the result ⟨a work by Noll⟩, which is similar to the finding already made above, that the produc-

---

13   A similar case is Damien Hirst. He subcontracts others to produce the actual pieces for him in a gesture that reminds of Renaissance masters. But he could also be understood as a capitalist for the production of art, whose price is then inflated by the symbolic capital of his name that is added to the piece. Thanks to Johannes Fehrle.
14   Although art was historically sometimes understood as being close to nature (⟨Kunstschönes⟩ vs. ⟨Naturschönes⟩), compare on this Kant 1914.
15   Luhmann 2000, p. 38.
16   Compare Braverman 1998. Braverman insists many times in his important study that the dissolution of the '*unity of conception and execution*' (p. 35; original emphasis) is characteristic for the degradation of labour in twentieth century capitalism. I will come back to this in my conclusion.
17   Cf. Sigaki, Perc and Ribeiro 2018.

tion within a studio with a division of labour is ⟨black boxed⟩ under an author's name. It is important to underline here, that I'm speaking about the art system as it operates today in capitalism and not about an ahistorical essence of art (whatever that means). The artwork in capitalism is tied to a real person, we call the artist. This guarantees, at least, the scarcity of the artworks – because the finite lifespan means that there is a limited number of artworks that are genuine (and not fake or forgery). Moreover, since art is, especially in conservative theories of art, often understood as a kind of expression of inner psychical workings, the existence of a person that expresses him- or herself through the artwork has to be presupposed.

In that sense, the work of art is the result of a form of labour in which body and knowledge, i.e. the knowledge of *how this specific work is to be produced*, cannot be separated – and that means that the function of the ⟨author⟩, his style and his signature (think of the joke with the machine signature in *Portrait of Edmond de Belamy*) is central. If a worker produced something in a factory, he or she does this – for reasons of effectiveness – according to a standard formula and his name is not connected to the product.[18] In artistic labour, at least in modernity, artists have to devise their own formula (⟨style⟩) and their name stays attached to the work.[19] With his or her body, the author is materially tied to a certain historical place.

Therefore, if works of art are the connection between knowledge and a *false body*, this would be called forgery. Even if, for example, one was to take a work by Donald Judd that according to Sebastian Egenhofer is 'dissolved in the anonymity of the industrial dispositive,'[20] it would still be pointless if another person or simply a company, based on the knowledge of how it was made, were to produce the same object again, as it is also done in principle in the industrial production of reproductions – it would not be possible to recognize this reproduction as a work of art on the art-market.[21] Or let us take another example:

---

18   The difference between a worker working in a car factory and the artist does not mean that art is non-technical or non-technological. It just means that technologies are used differently. This points to a difference regarding formal and real subsumption (Marx 1990a, pp. 1019–38). While real subsumption applies to work under factory conditions, it does not to individual art production: that is basically the individual moment in art.

19   At least, this is the ideological figure that has historically emerged as characteristic of the art system. It could be seen as a reflection of the imperative of newness in a market economy. But this imperative is mandatory for everyone, also the producers of cars. But in the car industry names and bodies are attached in another way to the product than in the art system.

20   Egenhofer 2008, p. 214.

21   See Benjamin 2008 on the relation between the work of art and reproduction.

Elaine Sturtevant borrowed the silkscreen printing matrices from Warhol for the *Flowers* and printed the *Flowers* again, and in 1991, even made an entire exhibition with *Warhol Flowers* – and Warhol is said to have once stated, referring to the production process of the *Flowers*: 'I don't know. Ask Elaine.'²² Nevertheless, Sturtevant's appropriation of Warhol's knowledge is not a rationalization of Warhol's work in the sense that Sturtevant now simply makes ⟨cheaper Warhols⟩, but she rather makes ⟨Sturtevants⟩. The point is: Sturtevant looks from a different historical angle on to the work of Warhol. Not the repetition of the form per se is decisive – but that she, with the repetition, deconstructs a certain (male) image of the creator, even if this creator himself already quoted serial and industrial procedures. Her own formula is to expose the break with formulas as characteristic for artistic work.

The historical situatedness of the artwork shows up in another way. Diedrich Diederichsen points out that the work of assigning relevance by curators, critics, museums, audiences, etc. also belongs to the work that creates the artwork and its market value.²³ The ⟨distributed⟩ character of this work makes it impossible to rationalize – since it continuously accompanies the work, i.e. never ends, and can also take unpredictable turns (⟨Sturtevants⟩ can become more important and more expensive than ⟨Warhols⟩ in the future). The debate on the ⟨signature⟩ in the *Portrait of Edmond de Belamy* pointed in that direction. Artistic work is tied to a specific body, socially and historically located – and this location means that the body is an ⟨artist⟩ because he or she in enmeshed in a network of institutions and other bodies that collectively produce his or her ⟨authorness⟩. Since AI can only work with formal structures, abstracted form, certain works of a certain phase and so to speak de-historicizes, de-localizes and de-socializes artistic labour, it cannot automatize artistic labour, because this labour gains its value²⁴ from its specific historical relatedness to former forms of artistic labour. The social and historical embeddedness of the artwork, also in a history of art, from which the new artwork has to differentiate itself, makes an automated production of art – at least in capitalism – pointless.

---

22 Quoted in: Arning 1989, p. 44. It's interesting that Warhol itself often delegated work on his artworks to others – nevertheless his name was used when the works were sold.
23 Cf. Diederichsen 2012, p. 99. On the distributed character of the artist/author see also Hensel and Schröter 2012. See also Bourdieu 1996.
24 Compare Beech 2015.

3   Art As Resistance?

We can understand the idea (or the imaginary) that AI could abstract formal structures from artworks to make not only the works, but also the artistic work reproducible and therefore substitutable, as a radicalization of the tendency of capitalism, as described by Braverman, to separate conception and execution:

> In the human, as we have seen, the essential feature that makes for a labor capacity superior to that of the animal is the combination of execution with a conception of the thing to be done. But as human labor becomes a social rather than an individual phenomenon, it is possible-unlike in the instance of animals where the motive force, instinct, is inseparable from action-to divorce conception from execution. This dehumanization of the labor process, in which workers are reduced almost to the level of labor in its animal form, while purposeless and unthinkable in the case of the self-organized and self-motivated social labor of a community of producers, becomes crucial for the management of purchased labor.[25]

In the context here this leads to some questions:

1) The separation of conception and execution exists in some forms of the production of art works – think of a sculptor like Richard Serra who instructs a steel factory to produce his works. But it is the same as with the Studio Olafur Eliasson mentioned above: in the end the object is called an artwork by Richard Serra (and only experts know the name of the steel factory and in principle every other steel factory could do the execution too). The execution is made invisible in the black box of the artist's name. The concept of the artist is seen as the only and central important aspect. Unsurprisingly, so called ⟨concept art⟩ emerged in the twentieth century that can be understood as a rejection of execution. In the case of a painter, we have no difference of conception and execution – of course the painter can make sketches etc., but he or she makes the concept and then he or she later does the execution him- or herself. During this execution the conception often is changed and transformed for different reasons (site specificity, resistance of the material etc.). This is different from the separation Braverman describes. A worker is not supposed to change and modify the conception and normally and, that is the meaning of Braverman's discussion, the conception is not the product of the same person(s) that do the execution.

---

25   Braverman 1998, 78.

There is also another case: A composer composing a score and an interpret performing that score. This is a clear case of separation between conception and execution, but it is firstly very much older than capitalism (at least in its developed form) and secondly the interpreter can him- or herself become a name, a body, a style, a signature attached to a work (think of people like Glenn Gould and his charismatic interpretations of Bach). The execution becomes in this case itself a form of conception, similar to the case of painter. Perhaps this could be better described as a cascade of a first conception/execution (Bach writes his *Goldberg Variations*) and a second conception/execution (Gould performs the *Goldberg Variations*).

In that sense conception and execution are closer in art than in other types of work. Either execution is relegated to a secondary status, becoming invisible behind the figure of the artist or even rejected, or execution is a type of conception itself. In that sense conception and execution are inseparable in art (but in a different as they are inseparable in animals – there you don't have the difference; in art conception and execution ceaselessly transform into another). Interestingly Braverman hints at this complexity when he writes: 'In recent times, the artistic mind has often grasped this special feature of human activity better than the technical mind.'[26]

2) The cascades of conception/execution in art is, eminently historical, in that the specificity of a given conception/execution is resolutely tied to a specific socio-historical point in time and has therefore irreducible relations of continuity and difference to other conceptions/executions. The animal unity of conception/execution is ahistorical, partly following an instinctive program, like a programmed automaton, repeating the same every time (perhaps changing only very slowly with evolution). Of course, some animals can learn and find sometimes new and creative solutions, and these can sometimes even be taught to younger animals, but this knowledge is not stored in archives and developed over generations.[27] Human conceptions can accumulate.

The question emerges why this kind of work producing this kind of objects exist in capitalism. First of all, the operations to secure genuineness, scarcity, expressiveness etc. guarantee the special status of the artwork-commodity for a

---

26 Braverman 1998, p. 31. This remark is in a footnote to the sentence: 'Human work is conscious and purposive, while the work of other animals is instinctual.' With 'conscious and purposive' the unity of conception and execution is meant.
27 As Burroughs 2005, p. 4 puts it: 'In the beginning of WRITING. Animals talk and convey information but they do not write. They cannot make information available to future generations or to animals outside the range of their communication system. This is the crucial distinction between men and other animals.'

special group of buyers on a special market. Machines that produce cheap art or even cheap Warhols or Serras would completely ruin the prices. Artworks are, in capitalism, special, very unique and exclusive commodities – to automate their production would directly contradict their use value (to be unique and exclusive) and therefore destroy their exchange value.

But besides this sobering perspective remains, even in the left,[28] the idea that artworks (or at least some artworks in a special historical situation) are more than just special commodities. Isn't there something oppositional and critical in art?

I would argue: It is not a supposed ⟨eternality⟩ (as a trope from classical aesthetics) of art, but decidedly its historicity that makes it oppositional.[29] Art and its special inseparability of conception and execution resists the imperative to formalize the conception and to reify into a repeatable pattern or algorithm. Rebentisch argues that in contemporary art the sociohistorical singularity of the single work rebels even against art genres like painting, sculpture etc.[30] Therefore, even if artists use the same technologies as would be used in ordinary work processes, even if they mimic industrial production strategies (Warhol's studio was called ⟨Factory⟩ and at least for a time the production of his works was characterized by a division of labour), they do so not to rationalize by formalizing conception, but to invent contingent, singular conception/execution-configurations tied to a specific socio-historical place. Art cannot be automatized because automaticity is by definition ahistorical, because to automate something means to repeat it as the same over time (therefore all forms of presentations that are formulaic are *Kitsch*).[31] Artists invent their media, as Krauss would put it.[32] This is not the same as saying that art strives for the ⟨new⟩, because that would subsume it under the formulaic algorithm of the market – sometimes being ⟨not new⟩ can be the task of art (Sturtevants work points in that direction, but see also e.g. Gerhard Richter's return to painting at a point in time when it was considered completely old-fashioned).[33] The attempts of information aesthetics is pointless because here a rationalization dispositive from industrial production is transferred to an area that blocks it. Or as Adorno puts it:

---

28   Just think of Adorno 1997.
29   Compare Luhmann 2000, p. 194: 'The artwork must remain within the modality of contingency and must draw its power to convince from its ability to prevail in the face of other self-generated possibilities.'
30   Compare Rebentisch 2013, pp. 106–16.
31   That means of course that it is very hard for something to achieve the status of art, since repetition and the formulaic are so hard to avoid.
32   Compare Krauss 1999.
33   Compare Buchloh 2015, pp. 72–9.

On the other hand, however, whenever autonomous art has seriously set out to absorb industrial processes, they have remained external to it. ... The radical industrialization of art, its undiminished adaptation to the achieved technical standards, collides with what in art resists integration.[34]

But that also means that art cannot be a utopian template for other forms of work in post-capitalism. Every postcapitalist society will need, as long as it is not imagined as a pre-modernist, primitivist dystopia, mass production of certain products on a big scale, i.e.: automatized and formulaic working procedures. Perhaps these can be delegated completely to machines one day, perhaps mankind will invent new forms to deal with such boring work for example by distributing it equally (and not to ⟨lower classes⟩). But it cannot be done away with altogether (at least until hyper-futuristic technologies like the ⟨replicator⟩ from the television series *Star Trek* are invented, which produce things from pure energy[35]). There is no way to de-automation. But: For a long time at least art, or some rare moments, art will remain the place of radical sociohistorical specificity and contingency in which singular (even if distributed) unities of conception/execution are produced. They work (besides other things like offering unexpected forms of perception), as a reminder that the separations we encounter should not be taken as natural – but as historical. 'Art is not the prototype of labour freed from value production', as Beech puts it.[36] It is not the paradise of unalienated work, as Beech also insists against ⟨Hegelian Marxism⟩.[37] Instead, Beech argues, art is aligned 'with every non-capitalist activity and every subordinate mode of production.'[38] I would argue that the point is not that art is necessarily aligned with non-capitalist activities, which only means to reinstate the idea that it is a form of unalienated work, but that its radical historical singularity disrupts capitalist rationalization, that is the formulaic reproducibility of knowledge. As soon as it becomes integrated into markets and made reproducible and formulaic, new forms of art emerge. The *Portrait of Edmond de Belamy* was perhaps too formulaic to be such a rare moment, but at least the heated discussion about the signature made the distribution of conception and execution visible – at least for the blink of an eye.

---

34 Adorno 1997, p. 217.
35 Compare Schröter 2020.
36 Beech 2019, p. 103. In his chapter on post-capitalism Beech addresses not the questions of separation, but the role of value production for digital forces of production.
37 Beech 2019, p. 94.
38 Beech 2019, p. 103.

AFTERWORD

Stagnation, Circulation, and the Automated Abyss

James Steinhoff, Atle Mikkola Kjøsen and Nick Dyer-Witheford

Not automation – stagnation! So one might sloganise the latest current in the analysis of capital, labour and technology, a current which, rather than emphasising runaway mechanisation and job-killing robots, focuses instead on slowing economic growth and senescent job creation. In the present volume, this perspective is represented by Jason Smith's chapter, 'Nowhere to Go: Automation, Then and Now', the argument of which is expanded in his *Smart Machines and Service Work: Automation in an Age of Stagnation*, a book often discussed in company with Aaron Benanav's *Automation and the Future of Work*.[1] Both can be read as arguments that capitalism's drive to replace humans with machines, far from accelerating, is actually flagging, as technology investment buckles under the burdens of slowing growth and the lure of financial speculation.

However, rather than reading these two important texts as studies of stagnation-not-automation, pitting these two key terms in opposition, one can parse them as studies of the concurrence of stagnation-and-automation, unfolding in a reciprocal relationship.[2] To understand this process, particularly in regard to artificial intelligence (AI), requires greater attention than either Smith or Benanav give to the role of automation, not at capitalism's point of production, but rather in the process of its circulation – the sphere in which the value generated in production is realised through exchange. Circulation refers to markets and marketing, but also, in Marx's account, to transportation, storage and logistics.[3] The automation of circulation is capital's logical response to stagnation.

Several technologies manifest this response. Take for instance, Steffen Reitz's chapter in this volume, which shows how RFID technology allows the automation of complex logistical circulatory processes through continuous and fine-

---

1 Smith 2020; Benanav 2020. Benanav's book was preceded by two major articles in *New Left Review* (2019a; 2019b) that summarize his argument; our quotations of his work are drawn from those essays.
2 Moraitis and Copley 2021.
3 See Kjøsen 2019. To circulation we might add consumption, although we cannot discuss that sphere adequately in this piece.

grained monitoring.[4] Also significant, and likely to be combined with RFID, is machine learning AI, which is being applied to increase the precision and velocity with which commodities complete their transformation into money, thereby reducing contingency and speeding up the valorisation process. The same technology is being used to surveil, analyse and algorithmically manage the proliferation of platform or gig work. In these circulatory domains of labour, capital is perfecting techniques which are in the early stages of being transferred back into the sphere of production.[5] We thus argue that capital's drive to automate, which manifests today in the AI-driven efforts of the big technology firms, is not the mere fantasy of a capital which is withering away, but a circuitous means for capital's further descent into an inhuman abyss, expunged of living labour.

1 Stagnation, Automation, Circulation

While Smith's and Benanav's positions are not identical, they share common concerns. Crucial to both is the long-term decline in the rates of productivity growth in the centres of advanced capitalism and perhaps globally, a subsidence that has continued, with only brief interruptions, from the 1970s to the present. The point is liable to misinterpretation – what is at stake is not absolute falls in productivity but a slowing rate of growth, which still, by compounding year over year, allows significant enlargement in productive capabilities.[6] Nonetheless, the slide in the rate of growth is striking, particularly in the manufacturing sector – the arena of the giant mass-production car, steel and machine-tool enterprises that drove 'thirty glorious years' of capital's Fordist expansion. Smith and Benanav argue that this decline in the rate of productivity is primarily associated with falling rates of profit and then, significantly, falling rates of business investment in technology. Capital appears

---

4   Reitz, this volume, p. 265.
5   For an account of how this is likely to happen with RFID, see Reitz, this volume, p. 281.
6   Indeed, it can be argued that the real scandal is that productivity continues to increase, even if more slowly than in previous decades, but much faster than median and average wages, generating a 'productivity-wages gap' which favours capital (see Benanav 2019, p. 12). According to a study by the Economic Policy Institute (2021), from 1948 to 1979, net productivity in the USA rose by 108.1% and compensation by 95%. However, from 1979 to 2019, 'net productivity rose 72.2 percent, while the hourly pay of typical workers essentially stagnated – increasing only 17.2 percent over 40 years (after adjusting for inflation)'. Productivity grew more slowly from 1979 to 2019 than from 1948 to 2019, but by 2019 it was nevertheless 258% higher than in 1948, and 158% higher than in 1979.

to be caught in a vicious circle, in which the failure to sustain automation-generated productivity increases bites into profits and then diminishes its willingness to automate even further. As a consequence, capital now seeks its rewards from monopoly power, mergers and acquisitions and financialisation.

Smith and Benanav have differing, though not necessarily inconsistent, explanations for the onset of this stagnancy. For Benanav, following a line of analysis opened by Robert Brenner, the cause lies in a crisis of overproduction.[7] As manufacturing capacity, which was largely concentrated in the US in 1945, spreads worldwide, competition increases and prices fall. This in turn reduces output growth and leads to job losses. Smith follows a more classical line of Marxist theory, pointing to the changing organic composition of capital, whereby the ratio of constant capital increases relative to variable capital. He asserts that this tendency finds a barrier in the post-industrial proliferation of service work, much of which is currently difficult or impossible to automate and is subject only to formal, rather than real, subsumption.[8] Increased productivity gains are hard to come by in such kinds of work.

In both cases, the argument is that rather than automation boosting capital to further machinic acceleration, it starts to undo the very impetus that brought it into being. As squeezing-out further productivity gains from industrial systems becomes harder, it also becomes more difficult to sustain profit rates and hence less attractive to invest in major technological innovations. Thus the Fordist virtuous cycle of profits and productive investment goes into reverse. For example, by expanding into areas where automation is (currently) impossible, capital lowers its overall organic composition without reversing its previous technical advances.

This logic manifests in the failure of the computer revolution – the very revolution that inspires the greatest optimism and apprehension about new waves of automation. Here Smith and Benanav address the famous observation of Robert Solow about computers showing up everywhere except in the productivity figures, and Robert Gordon's more recent observations about the failure of information technologies to repeat the great leaps in efficiency that characterised capital's industrial take off phase.[9] They therefore reject the idea that, for better or worse, AI and other so-called fourth industrial revolution technologies will destroy jobs in a new wave of automation. Employment in manufacturing powerhouses is falling not because the factories are full of robots, but

---

7 Brenner 2006.
8 Smith, this volume, p. 69.
9 Solow 1987; Gordon 2018.

rather, because the world is full of factories; or at least (since Benanav does acknowledge a 'secondary' role for productivity-raising technologies) because the world is already full of factories that are (but only slowly) filling with robots. What is at stake in capital's crisis of labour is not waves of technological unemployment but rather a condition of mass under-employment because 'rapid automation of production is hardly taking place at all – offstage or anywhere else'.[10]

Here Benanav's analysis makes a rendezvous with Smith's discussion of service sector work. As industrial jobs decline, the locus of employment shifts to the service sector. But precisely because much service sector work is not susceptible to productivity-raising automation or augmentation, it cannot replace manufacturing as a driver of economic growth. Smith observes that fear of job destroying technology appears in cycles and that such anxieties about human redundancy are always refuted by capital's capacities to absorb labour that appeared to have 'nowhere to go'.[11] And so it will be again with the technologies of the so-called fourth industrial revolution.

Like Benanav, Smith points out that empirical evidence on productivity rates and business investment do not support a picture of wide and rapid cybernetic automation. He notes in particular that in the US the vast majority of recent jobs (94%) have been created in 'education, healthcare, social assistance, bars, restaurants, and retail, that is, in the vast, motley, and above all technologically stagnant service sector'.[12] Most of these 'will be the least attractive ones, requiring few if any skills, and paying poverty-level wages'.[13]

The US economy is creating plenty of jobs, but 'these jobs are almost universally poorly paid and precarious, pooling at the very bottom of the labor market'.[14] As Robert Ovetz demonstrates in his contribution to this volume, even formerly prestigious service jobs, such as professorships, are being rationalised and made precarious (though he emphasises the use of automation technologies in such endeavours).[15] Smith, however, believes that even if some high-end service jobs are automated, the likely result will be a swelling of the ranks of people seeking low end service work where they 'scrape by on part-time precarious work and tenuous lines of extortionate credit, commuting to and from work an hour each way, surveilled by heavily-armed cops as they make their

---

10   Benanav 2019b, p. 123.
11   Smith, this volume, p. 43.
12   Smith, this volume, p. 55.
13   Smith, this volume, p. 58.
14   Smith, this volume, p. 59.
15   Ovetz, this volume, p. 298.

way home from the bus stop. Some run rackets and hustles, while others sink into depression, or drugs. For many, prison is always near'.[16]

There is a great deal in both Smith's and Benanav's analyses we agree with – especially in their critique of left proposals for a utopian fusion of AI capitalism and universal basic income (a point to which we will return), but there are also points of real divergence from the analysis we present in *Inhuman Power*. One such point is their dismissal of what we term 'actually-existing AI-capitalism' – the current deployment of machine learning and other AI technologies in the service of generalised commodity exchange.[17] As AI is a central concern of what are now widely agreed to be the world's most highly profitable corporations, specifically those of the Big Tech complexes of the US and China, one might expect that these operations, and the ways they relate to the current and future organisation of value production, warrant some close examination.

However, this is largely lacking from the work of stagnation theorists. It is almost completely missing from Benanav's book. Smith suggests a possible route for building a properly analytic perspective when, in a footnote, he mentions the importance of the Marxist category of 'circulation' to understanding the rise of the service sector.[18] But he does not follow up on this point in his essay, and in his book while he pursues it further, it is generally to repeat a conventional derogation of circulation as secondary and subsidiary sphere of capitalist activity, subordinate to the real action of production. Tech companies are described as opportunistic ventures whose experiments are doomed to be insignificant in the face of wider patterns of economic lethargy and decline.[19] We argue, on the contrary, that circulation has always had more importance to Marxist analysis than this, and is, moreover, becoming increasingly important to capital in the face of the stagnancy of production Benanav and Smith depict. It is in this enhanced circulatory aspect of capital, next to its further penetration into the sphere of consumption through domestic AI, that the most significant society- and humanity-transforming applications of AI are currently manifesting.

We are very, very far from being amongst the all-too many cheerleaders for Big Tech; indeed, our stance towards it is pretty much 'abolitionist', in the sense of Wendy Liu's *Abolish Silicon Valley* (2020). But we think that it is a mis-

---

16   Smith, this volume, p. 75.
17   Dyer-Witheford, Kjøsen and Steinhoff 2019, p. 2.
18   Smith, this volume, p. 70.
19   Smith 2020, ch. 2.

take to underestimate the importance of this sector of global capital, which now includes some of its most highly valued companies, with exceptional profit rates and megalomaniacal personifications of corporate power at their helms.[20] AI and its development and applications warrant careful analysis, both as part of the current landscape of capital and as a clue to its dynamic future transformations, and its truly inhuman possibilities.

2 Circulatory AI

For Marx, circulation, the domain of market exchange, may be secondary to production, but it is also crucial to it, for it is only in the process of circulation that the surplus value extracted in production is realised. Moreover, Marx notes that capital can increase its profitability by accelerating the circulation process – speeding up the processes of transportation and delivery and the moment of exchange – thereby multiplying the number of times a given amount of capital can turn-over, in a series of cyclical amplifications.[21] Increased emphasis on circulation is thus a logical response to the stagnation of capitalist production described by Benanav and Smith. It is also a primary arena of AI development. In noting this we follow a number of observers who have pointed to the increasing importance of circulation processes under conditions of globalisation and planet spanning supply chains.[22] Nelson Lichtenstein even speaks of a 'new age of merchant capital'.[23]

Although there are industrial production oriented AI initiatives from old conglomerates such as General Electric and Siemens and a handful of startups devoted to industrial production such as Landing.ai, the main protagonists of capitalist AI development are giants of circulation.[24] Google and Facebook are advertising companies and machine learning is central to their precision targeted ad sales. Amazon is a retailer for whom AI is the 'flywheel' of its retail and logistical operations which include the vast robot-human co-ordination of fulfilment centres that make it the second largest private employer in the US (see Larry Liu's thorough analysis of Amazon's warehouse automation in this volume).[25] Amazon also relies on machine learning as a silent salesman in

---

20 Liu 2020.
21 Marx 1992, pp. 200–6.
22 See Cowen 2014; Kjøsen 2016; Chua 2017 and Reitz's chapter in this volume.
23 Lichtenstein 2012.
24 de Jesus 2018.
25 Levy 2018.

their virtual storefronts. Apple, which masquerades as a production company, designing electronic devices whose manufacture is outsourced to Shenzhen, sells the iPhone (which shares the market with Google's Android devices). While Apple came late to the AI game, it now incorporates the technology into most of its endeavours and consequently harvests and utilises user data like its fellow AI-capitals.[26] No matter what brand, the smartphone is preeminently a circulation device. Its AI-powered apps have become the essential app-aratus of platform business models.

The prime project of 'Big Tech' AI corporations at the moment is, to reiterate, the automation of circulation, applying machine learning (among other technologies) to increase the precision and velocity with which commodities complete their transformation into money, the faster to repeat that crucial metamorphosis yet again and thus the overall accumulation process.[27] This process generates collateral effects and counterflows. The giant digital resources built up for the circulation process crystallise in data centres, factory-like facilities for both production and circulation purposes of corporate clients, and in warehouses and other distribution facilities. In turn, machine learning AI washes back into industrial processes and more generally intensifies the unification and blending of productive and circulatory processes that has been a feature of increasingly digitally integrated management systems. And, as Mary Gray and Siddharth Suri underline, the automation of circulation paradoxically requires the mobilisation, disciplining and exploitation of whole new strata of human labour, such as the legions of algorithmically-managed platform workers recruited for training machine learning AI and moderating the errors and problems of its deployment – labour which itself, however, then becomes the target of further automation processes.[28] As Christina Gratorp argues in her contribution to this volume, automation tends to conceal itself; it presents 'the potential to exploit and reproduce inequality while disguising its character of exploitation from the user'.[29] At the same time, the panoptic practices which informed Taylorist and Fordist worker monitoring at the point of production are now re-energised and carried to a new level of intensity and breadth of application by AI; most famously in Amazon warehouses, but also in retail, hotels, call centres: all kinds of service work.[30]

---

26 Berry 2021; Pathak 2021.
27 Manzerolle and Kjøsen 2012; 2014.
28 Gray and Suri 2019.
29 Gratorp, this volume, p. 81.
30 Dzieza 2020.

Our point here is that what Smith and Benanav characterise as braggadocious speculative ventures are already driving substantial, indeed massive, reworkings of capital's circuits, which are at work in every Google search, Tweet, Facebook posting or Amazon purchase. In manifold ways, the giant corporate platforms are deploying AI to make the matching of commodities and money more efficient and thereby accelerate the metamorphosis of capital in the sphere of circulation. However ambivalent, or frankly repelled, one may feel about this process, due to all it entails in terms of data gathering surveillance, and social engineering, Smith's dismissal of 'the technologies characteristic of the past two decades' as 'toys, not tools' which economically represent 'little' significance seems altogether inadequate to the scale of the problem posed by actually-existing AI-capitalism, and what can reasonably be extrapolated as to its future trajectory (we also note that the deployment of machine learning by capital only began in earnest around 2015).[31] Marx famously wrote that to understand capitalism we must 'leave this noisy sphere [of circulation], where everything takes place on the surface and in full view of everyone' and descend 'into the hidden abode of production'.[32] However, it may be that today AI capital advances under cover of the circulatory chatter of social media, mobile communications, logistics and online retail, to deliver a flank attack on the sphere of production, and on conventional Marxist priorities, an assault that is all the more disarraying because it comes from an unanticipated direction.

There is, however, another aspect of circulation, today sometimes designated as a distinct operation – that is, the translation of money directly into more money: M–M', otherwise known as financialisation. Capital's increasing resort to financial speculation rather than productive investment as a source of profit is widely recognised as at once a cause and effect of its relative stagnation in terms of economic growth and productivity. Today's big AI capitals are deeply involved in this process, to a degree that some authors write of the fusion of Big Tech and Fintech.[33] A crude indicator of this dynamic is the overwhelming role of the Big Tech companies in sustaining the buoyancy of the US stock market through successive crises. In 2020, five USA tech giants (Google, Amazon, Facebook, Apple and Microsoft) constituted some 20 percent of the stock market's total worth, and with the addition of Netflix, they accounted for

---

31 Smith 2020, p. 43.
32 Marx 1990a, p. 279.
33 Hendrikse, Bassens and van Meeteren 2019; Fernandez, Adriaans, Klinge and Hendrikse 2020.

78% of the growth in the S&P 500 index.[34] As many pension funds and other institutional investments simply follow this index, the significance of major AI companies in shaping financial markets has been massive. These corporations have played a crucial role in preserving the prosperity of the shareholding strata of contemporary capitalism, even as the life conditions of the poor collapse. This segmented prosperity in turn maintains ongoing consumer expenditure on the digital devices, online goods and services and virtual advertising that drive Big Tech revenues higher.

Within this overarching, bootstrapping dynamic are nested a series of subsidiary manoeuvres by which major AI developers leverage technological pre-eminence into speculative advantage, including large scale share buy-backs which boost share prices.[35] Big Tech companies also characteristically carry an unusually high debt load, the better to acquire smaller companies, to extinguish or absorb competitors, and to hire extraordinarily expensive high-level experts in data science.[36] AI corporations do not only accelerate the circulation of commodities. They are also central to the hypertrophic growth of a financial sector which seeks escape from the lethargy of productive investment by systematically short-circuiting the passage of capital through the commodity form and making a superhuman *salto mortale* directly from money to money prime.

The rise of Big Tech can therefore be seen as a systemic response to the post-1970s stagnation of advanced capitalist economies on which Smith and Benanav focus. It is a response that 'doubles down' on capital's perennial automating strategy through machine learning and other AI strategies but does so in the sphere of circulation rather than production. This response does not unleash a general crisis of technological unemployment, although it has resulted in sectoral employment decline in bricks and mortar retail work, legacy journalism and some aspects of advertising. In other cases (notably Amazon) it has produced a swell in logistical and gig work, which is disciplined to labour at AI-determined rhythms and is subjected to extraordinary levels of digital surveillance, rigorous algorithmic management and extreme precarity, all of which paves the way for further automation, if a rebellion of labour should drive up the cost of labour-power. In this sense, the invasion of capital's circulatory system by AI is opening up routes by which such technologies may work their way back into the production side of value's ontology. To underestimate the significance of this AI-driven revolutionising of logistics, retailing, marketing, management, banking and finance on the grounds that these are processes

---

34 Eavis and Lohr 2020; Roberts 2021.
35 Foorar 2019.
36 Smith 2019.

merely ancillary or secondary to the fate of production, or more precisely, manufacturing, is unwise. This transformation involves no less than the machinic reworking of the infrastructure of what Marx called 'universal intercourse', the combination of transportation and communication that he saw as integral to capital and its modernity.[37]

3   Automation and the Abyss

Emphasising the primacy of stagnation is one way to counter arguments for the significance of AI-powered automation. But there is also an even more fundamental objection. Throughout *Capital* Volume I Marx insists that surplus-value derives from labour, and that labour is an exclusively human capacity. He frequently affixes the descriptor 'human' to labour and labour-power and specifies that labour is 'an exclusively human characteristic'.[38] Indeed, it is Marxist doctrine that machines cannot labour and thus cannot create value.[39] Marx states:

> Like every other instrument for increasing the productivity of labour, machinery is intended to cheapen commodities and, by shortening the part of the working day in which the worker works for himself, to lengthen the other part, the part he gives to the capitalist for nothing.[40]

Machines are a means for augmenting the production of surplus-value by labour, not for producing value on their own. Indeed, a well-known reading of the 'Fragment on Machines' in Marx's *Grundrisse*, holds that automation is the weapon by which capital will ultimately commit suicide by reducing living labour to a quantity insufficient to sustain valorisation. The elimination of living labour from the circuits of capital is a contradiction since it removes the conditions for surplus-value extraction. Benjamin Ferschli's chapter in this volume contributes to this orthodox stance on automation. Ferschli asserts that the notion of capitalism without human workers is no more than an 'abstract fantasy' even if capital seems to be pursuing it concretely.[41] This is simply because machines cannot be exploited: 'Robots do not reproduce them-

---

37   Marx 1970, p. 56.
38   Marx 1990a, pp. 283–4.
39   See Caffentzis 1997.
40   Marx 1990a, p. 492.
41   Ferschli, this volume, p. 115.

selves, do not receive a wage and do not produce value'.[42] On this basis, one can argue, along with Ramin Ramtin, that in the long run automation 'undermine[s] the very basis of growth and accumulation' and drives an 'irreversible forward march of capitalism towards its inevitable breakdown'.[43]

However, in *Inhuman Power*, we suggest that the demarcation Marx made between machines and labour, fixed and variable capital, is not metaphysical, but historical. Its permanence cannot be taken for granted, and the possibility of its transgression cannot be precluded solely by vitalist assertion. The relation of machines to labour may shift as machinic capacities are enhanced by capital, allowing machines to assume new functions within its circuits. While capital existed before the proliferation of industrial machinery, Marx refers to machines as 'capital's material mode of existence' and the 'material foundation of the capitalist mode of production'.[44] This is because machines, deployed as fixed capital, can implement previously human knowledge and skill, and absorb those capacities 'into capital'.[45] This capacity to transmit once-human capacities to capital means that machinery 'appears [...] as the most adequate form of fixed capital, and fixed capital, in so far as capital's relations with itself are concerned, appears as the most adequate form of capital as such'.[46] Thus Marx holds that the 'development of the means of labour into machinery is not an accidental moment of capital, but is rather the historical reshaping of the traditional, inherited means of labour into a form adequate to capital'.[47]

Alfred Sohn-Rethel elaborates the notion of adequacy to capital as regards machines in his discussion of what he calls the postulate of automatism. He argues that from the perspective of the capitalist, the 'essential characteristic of the production process for which he is responsible is that it must operate itself. The controlling power of the capitalist hinges on this postulate of the self-acting or "automatic" character of the labour process of production'.[48] For

---

42 Ferschli, this volume, p. 136.
43 Ramtin 1991, p. 7.
44 Marx 1990a, p. 554.
45 Marx 1993, p. 694.
46 Marx 1993, p. 694. We note that 'In so far' is an important qualifier here. Marx goes on: 'In another respect, however, in so far as fixed capital is condemned to an existence within the confines of a specific use value, it does not correspond to the concept of capital, which, as value, is indifferent to every specific form of use value, and can adopt or shed any of them as equivalent incarnations. In this respect, as regards capital's external relations, it is circulating capital which appears as the adequate form of capital, and not fixed capital' (Marx 1993, p. 694). However, this presumes that there cannot be a machine with general functionality (see Dyer-Witheford, Kjøsen and Steinhoff 2019).
47 Marx 1993, p. 694.
48 Sohn-Rethel 1978, pp. 118–9.

the capitalist, the value invested in commodity production should proceed to exchange without his intervention (otherwise the capitalist himself would be labouring and he has not been paid to do so). Even in a labour process with an organic composition of zero (no fixed capital) the commodity should be produced independently from the capitalist, propelled through its metamorphosis solely by invested capital. We might call this *formal* automaticity. Living labour, however, does not always accede to the demand for formal automaticity and requires control and discipline, which come at an added cost. Machines are more adequate to capital because they allow increasingly *real* levels of automaticity, decreasing the reliance of capital on living labour in one respect (not in terms of value extraction). *Really* automated capital arises not just because of technological advances such as the invention of the microprocessor or of machine learning, but also because capital is always-already formally automated. As Sohn-Rethel puts it:

> the tendency which I described as the 'postulate of automatism' presents itself as a feature of technology. But it does not spring from technology but arises from the capitalist production relations and is inherent in the capital control over production. It is, as it were, the condition controlling this control.[49]

The postulate of automatism is capital's dream. Sohn-Rethel points out: 'a postulate is not necessarily a reality. It becomes a reality only when the appropriate conditions exist for its practical realisation'.[50] Perhaps we can even understand the history of capitalism as 'so many steps in the pursuit of that postulate'.[51] In some periods capital stumbles away from its dream while in others it consciously pursues it. Harry Braverman held that modern management, deriving from Taylorism, 'render[ed] conscious and systematic, the formerly unconscious tendency of capitalist production' towards the deskilling of craft labour.[52] We suggest, similarly, that technologies of mechanisation, followed by those of automation, have allowed capital to pursue the postulate of automatism increasingly consciously.

This abyssal theory of automation posits that *automation is capital* or *capital is automation*; capital contains within it an inherent tendency towards an increasingly machinic state, which is to say, from capital's viewpoint, a more

---

49 Sohn-Rethel 1978, p. 121.
50 Sohn-Rethel 1978, pp. 118–9.
51 Sohn-Rethel 1978, p. 122.
52 Braverman 1998, p. 83.

perfected state. This an inhuman state, and for this reason we say that capital's inherent trajectory is towards the abyss.

4 Labour Adequate to Capital

If machines could create value, the history of automation would not track the moving contradiction of living labour that is increasingly productive, but ever diminishing in value relative to fixed capital. Instead, it would mark a progression towards the production of a new population of machine workers, a pool of labour more adequate to capital. If this were so, then capital could fully realise the postulate of automatism and jettison living labour from its circuits. What would machinic labour adequate to capital look like? In other words, what are the necessary conditions under which machines could labour and be exploited within capitalist social relations such that they could generate surplus-value?

We argue that labour-power, as Marx defines it, is isomorphic to the notion of general intelligence as deployed in artificial general intelligence (AGI) research. Succinctly, general intelligence is the ability to reason about an unspecified number of domains and 'to transfer learning from one domain to other domains'.[53] We hold that '[g]rounding the concept of labour in general intelligence severs its inherent connection to human beings and [...] theoretically allows for the possibility that some other being that is generally intelligent [such as a sufficiently advanced machine] could labour'.[54] However, general intelligence is not the only requirement for a machine that might labour.

Marx insisted that under capital things exist in dual ontological modes or have a 'double form' consisting of a natural and a social form.[55] The natural form refers to the concrete existence of an object. The natural form of an AI system is its material existence as a collection of code stored in electronic memory which can be executed on some computing hardware, supported by the requisite infrastructure. Humans are limited in what tasks they can do, and at what intensity, by their natural form; the 'organic limitations' of their 'bodily organs'.[56] Machines in general have no particular natural form. Compare a gantry crane, a chatbot and PARO the robot seal pup.

Social form depends on how a given object or person is implicated within the social relations of capital. A machine only takes on the social form of fixed

---

53 Muehlhauser 2013.
54 Dyer-Witheford, Kjøsen and Steinhoff 2019, p. 125.
55 Marx 1990a, p. 138.
56 Marx 1990a, p. 495.

capital if it is employed in a capitalist production process.[57] The oven in a home kitchen is a consumer commodity, not fixed capital, since it is not productively employed. For a machine to labour in Marx's sense, it would have to change social form from fixed capital or commodity to variable capital (the social form of the worker). In other words, it would have to be proletarianised. This means being 'free' in a double sense. A proletarianised machine must be a free subject legally (i.e. not a slave or directly forced labour) and it must also be separated from ownership of the means to sustain itself.[58] In other words, it must be made dependent on purchased commodities for its continued existence. As Marx nicely pointed out: 'if the workers could live on air, it would not be possible to buy them at any price'.[59] In *Inhuman Power*, we discussed the first sense of freedom and some precedents set in that regard; here we focus on the second.[60] In conversations we have had on this topic, we have found that it is a sticking point in the argument for many people. It is, admittedly, an unfamiliar and science fiction-y notion to imagine a machine responsible for its own continued existence.

What needs would a proletarianised machine need to satisfy to perpetuate its existence? We can suppose that it would certainly need energy, and most likely shelter, since most electronic machines, at least, do not fare well when exposed to the elements. It would need occasional repairs and replacements, and probably upgrades, whether hardware or software, to remain competitive with other machine proletarians. It would also require a network connection, perhaps access to some sort of cloud resources to supplement its hardware, and if this machine involves machine learning or some such data-intensive technology, it will also require quantities of relevant, quality data.

For simplicity's sake, consider a mundane personal computer of the present day and how much energy it requires to run for a week. This ranges depending on usage, hardware and cost of electricity in the area, but for one journalist located in San Diego, USA with typical usage (40 hours/week working from home, with some gaming), it went like this: 11.02/KWh at $0.28/KWh for a total cost of $3.08 weekly or $160.16 yearly.[61] Could the computer be made to earn $3.08 each week to pay the bill? With the right software it could possibly trade stocks, generate, and sell clickbait text or mine cryptocurrency, although since any operations would increase its energy consumption, this would have to be

---

57 Marx 1992, pp. 239–40.
58 Marx 1990a, p. 272.
59 Marx 1990a, p. 748.
60 Dyer-Witheford, Kjøsen and Steinhoff 2019, p. 137.
61 Gordon 2019.

balanced such that consumption does not exceed earnings (of course this is also the case for humans). We can imagine that the computer might dedicate a certain percentage of its total computational power to work constantly in the background (a sacrifice computer users make today for antivirus protection) or maybe dedicate the entirety of its computational power during hours when its human user is sleeping. Or maybe there could be a market similar to Uber's surge pricing model, in which certain moments of increased computational need would be more lucrative and bid upon by competing proletarian machines.

Regardless of the particulars, this scenario would require that the resources required for our machine worker's continued existence not be freely available. The relevant necessities of survival would have to be enclosed, as they are for humans under capital. In terms of energy, free access would have to be rendered impossible or at least made liminal. If, say, solar could provide such machine proletarians with adequate energy, solar panels and the real estate on which to place them would form vigorous markets. Network connections along with software and hardware upgrades would, in any case, form other necessary commodities.

In *Inhuman Power* we sketch a speculative scenario in which, over the long term, ratcheting cyclical crises of technological unemployment, social unrest, and yet-further intensifying automation drive towards the replacement of human workers with machinic proletarians.[62] Yet, today's technology is nowhere near allowing general intelligence, and thus artificial proletarians. The point is (at least for the moment) merely illustrative. However, labour is only one component of the exploitation relation. If AGI could personify the category of labour-power, it could also personify capital (i.e. play the role of capitalist and the functionaries of capital). This possibility, though also very far from fully realised, is more developed than the proletarianised machine.

5   Capitalist Machines / Algorithmic Management / DAOs

Science fiction author Charles Stross has described capitalist firms as 'very old, very slow AIs'.[63] They are AIs lacking in general intelligence, however, as they have only one goal: the valorisation of value. But since the rise of the platform

---

[62]   Dyer-Witheford, Kjøsen and Steinhoff 2019, pp. 137–42. The question of precisely *how* a machine proletariat (rather than a population of highly sophisticated fixed capital) might come into being remains an open one, which we cannot deal with here.

[63]   Stross 2017.

business model many firms are, in a more literal sense, artificially intelligent. Platform capitals such as Uber pioneered techniques of "algorithmic management" in circulatory contexts, in which workers are directed by algorithmic systems rather than human managers.[64] These algorithms tend to be propelled by machine learning models and the data generated by the use of platforms tends to be used in their further improvement, often guided by a management aspiration to dispense with workers as much as possible, or failing that, to transform them legally into so-called independent contractors with diminished rights, benefits and means of contesting their working situations. Algorithmic management can be summed up with a few key qualities:

- Prolific data collection and surveillance of workers through technology;
- Real-time responsiveness to data that informs management decisions;
- Automated or semi-automated decision-making;
- Transfer of performance evaluations to rating systems or other metrics; and
- The use of 'nudges' and penalties to indirectly incentivize worker behaviours.[65]

While aspirations to dispense with human workers entirely have thus far failed (e.g. Uber sold its autonomous vehicle division in 2020 after spending five years trying to render drivers unnecessary), algorithmic management does seem to allow a more immediate and substantial reduction of humans serving as functionaries of capital. For instance, only a handful of managers are required for each city's fleet of Uber drivers.[66] Despite abundant criticism, the algorithmic management model is continuing to spread, 'becoming part of the ordinary infrastructure of workplaces from transportation and logistics to retail and service industries and even domestic work'.[67] This has led one journalist to quip: 'Robots aren't taking our jobs, they're becoming our bosses'.[68]

What exactly is being automated with algorithmic management? Algorithmic systems may augment and take on functions of middle management, such as scheduling, monitoring, directing and disciplining workers. But the means by which algorithmic management functions often also reach directly into the work and life of workers. Amazon has patented vibrating wristbands to direct worker's hands and Walmart is developing harnesses to monitor the workers' bodies at a finely grained level. As one Amazon fulfilment centre employee grimly jests: 'the algorithm is telling a manager to yell at us. In the future, the

---

64 Lee et al. 2015.
65 Mateescu and Nguyen 2019, p. 3.
66 Lee et al. 2015, p. 1603.
67 Mateescu and Nguyen 2019, p. 4.
68 Dzieza 2020.

algorithm could be telling a shock collar'.[69] This is a different kind of automation than the replacement of human workers and it gives new meaning to the idea of workers being reduced to appendages of the machine.

The proliferation of algorithmic management is perhaps not surprising given that managerial positions have been a growing sector of employment since at least the early 1990s. Ehrenreich and Ehrenreich discern the rise of the so-called 'professional-managerial class' from 1% of the US population in 1930 to 35% by 2006.[70] Management is not productive labour. Algorithmic management is capital's means for reducing the cost of such unproductive labour. In terms of the postulate of automatism, capital cannot admit even of a human managerial strata lording over a machine workforce. Management too must be automated. As Ramin Ramtin puts it, capital's 'ultimate vision' of itself is one of an 'automated system' wherein 'conception, co-ordination and execution' are fused into one 'all-embracing purely *managerial* function'.[71]

Algorithmic management finds its logical extension today not in AGI, but another emerging technology with vaunted potential for automation applications: blockchain, or more precisely, the applications of blockchain called smart contracts and decentralised autonomous organizations/corporations (DAO/DACS).[72] While these technologies are even less developed than AI and perhaps surrounded by even more hype, they afford further means for thinking about the postulate of automatism which are being concretely pursued in the world today.

Smart contracts are algorithms that automate contractual relations. A canonical explanatory example is a vending machine; its hardware embodies a contract pertaining to the items it contains and how they can be purchased. Smart contracts propose to do the same in software; to 'embed contracts in all sorts of property that is valuable and controlled by digital means' and to do so

---

69   Dzieza 2020.
70   Ehrenreich and Ehrenreich 2013.
71   Ramtin 1991, pp. 61–5; original emphasis.
72   An interesting thing about DAOs is that they are rarely imagined (as far as we can tell) as taking the place of the worker's side of the employment relation. One exception is a short piece posted on tech industry site HackerNoon in which the author speculates on the possibility of DAWs or 'Decentralized autonomous workers' by which he means a 'new class of assets' based on narrow AI which could be employed to make money for their owners prior to the development of 'super smart robot slave[s]' with true AGI (Asghari 2018). There 'could be a form of stock market for the masses where future citizens trade robotic labor and AI workers' (Asghari 2018). What this commentator misses is that such so-called DAWs would have the social form of fixed, rather than variable, capital.

'in such a way as to make breach of contract expensive ... for the breacher'.[73] Ethereum cryptocurrency creator and blockchain evangelist Vitalik Buterin recognises the smart contract as a means of 'decentralized automation' which might be applied to the labour/capital relation.[74] He calls the smart contract a:

> mechanism involving digital assets and two or more parties, where some or all of the parties put assets in and assets are automatically redistributed among those parties according to a formula based on certain data that is not known at the time the contract is initiated. One example of a smart contract would be an employment agreement [...].[75]

According to Buterin, smart contracts may be combined in a hierarchical manner to create a DAO/DAC since a corporation is 'nothing more than people and contracts all the way down'.[76] He explains DAOs as an expansion of the purview of automation. He is not content with:

> ... only automating the bottom; removing the need for rank and file manual laborers, and replacing them with a smaller number of professionals to maintain the robots, while the management of the company remains untouched. The question is ... even if we still need human beings to perform certain specialized tasks, can we remove the management from the equation instead? ... what if ... we can encode the mission statement into code; that is, create an inviolable contract that generates revenue, pays people to perform some function, and finds hardware for itself to run on, all without any need for top-down human direction?[77]

While the question of *why* goes curiously unasked here, Buterin's discourse on DAO/DACs expresses the logic of the postulate of automatism quite clearly. We note, however, that DAO/DACs remain a largely speculative technology, and their history is marred by the spectacular 2016 failure of the unimaginatively named 'The DAO'. This first attempt at a large decentralised investment corporation was quickly hacked, $60 million of ethereum was stolen from it, and the supposedly immutable blockchain technology underlying it had to be modi-

---

73 Szabo 1997.
74 Buterin 2014.
75 Buterin 2014.
76 Buterin 2013a.
77 Buterin 2013a.

fied to fix the vulnerability.[78] We do not offer an evaluation of the prospects of blockchain based organisations at this point, but merely point out that they offer a means by which capital might advance beyond the existing algorithmic management of today and inch further towards the postulate of automatism by routes outside the direct manufacturing labour process.

Algorithmic management, smart contracts and DAO/DACs also illuminate how stagnation and automation are not opposed, but are in fact part and parcel of an abyssal present and near future. Smith's discussion of the new servant economy connects directly to these techniques for automating the functionaries of capital. Buterin points out that one major problem for DAO/DACs, as mere software programs, is 'how to actually interact with the world around them'.[79] The solution proffered is to use smart contracts to subcontract humans to do that interacting via some sort of platform.[80] This model is illustrated by the group of artists terrao with their project Premna Daemon.[81] Premna Daemon consists of a 'Bonsai tree (a *Premna Microphylla*), a web interface, several sensors and cameras, and a Smart Contract on the Ethereum Mainnet'.[82] The idea is that the bonsai tree is 'autonomous' in caring for itself in an enclosed, indoor environment where it is separated from the necessities of survival. It does so by receiving payments (donations) from viewers with which it hires 'workers' to serve its reproduction needs as determined by an array of sensors. It sends ethereum payments to its workers who water it when it senses a certain dryness in its soil, for instance. According to the artists, 'Premna shows how

---

78 DuPont 2019.
79 Buterin 2013b.
80 The prospect of algorithmic management does not perturb Buterin: 'Contrary to fears, this would not be an evil heartless robot imposing an iron fist on humanity; in fact, the tasks that the corporation will need to outsource are precisely those that require the most human freedom and creativity' (Buterin 2013b). This is precisely the line of thought that shows up variously in both mainstream business literature on AI (Brynjolfsson and McAfee 2014) and *post-operaismo* (Hardt and Negri 2017; Virno 2003). Certain human capacities just cannot be automated, so there is little to fear from human-machine amalgamation. For capitalist AI analyses, a new era of more enjoyable work is upon us, while *post-operaismo* discerns the seeds of communism in emergent immaterial labour.
81 Premna Daemon is the follow-up to terrao, a project which asks: 'can an augmented forest own and utilise itself?' (Seidler, Kolling and Hampshire 2016, p. 1). The project creates a scenario in which 'a forest is able to sell licences to log trees through automated processes, smart contracts and Blockchain technology. In doing so, this forest accumulates capital. A shift from valorisation through third parties to a self-utilization makes it possible for the forest to procure its real exchange value, and eventually buy (thus own) itself. The augmented forest, as owner of itself, is in the position to buy more ground and therefore to expand' (Seidler, Kolling and Hampshire 2016, p. 1).
82 terrao 2018.

autonomous systems can engage in the form of relationships previously only occurring between humans'.[83] The artists are interested in augmenting natural environments such that they can 'own and utilise' themselves.[84] However, from the perspective of the automation of capital, Premna Daemon demonstrates how automation and the stagnation thesis of a servant economy might co-exist. Here humans are not wholesale eliminated but become literal servants to a smart contract.

6 Conclusion: Surplus Humans and Symbiotic Capital

In conclusion, we turn to Amy Wendling's chapter in this volume, which explicitly engages with *Inhuman Power*. This is an important conversation for us, as Wendling's book *Karl Marx on Technology and Alienation* (2009) is one of the best recent studies of Marx's machinic vision, but one that, unlike our abyssal interpretation, emphasises the emancipatory potentialities he saw in technological development. In her chapter in this volume, Wendling investigates the connection Marx draws between alienation and automation. She argues, as we do, that capital's intensifying drive to replace humans with machines not only has consequences for wages, work security and conditions, but that it also has profound ontological implications. Indeed, she suggests that optimistic discourse on automation, which places emphasis on the acquisition of new skills and emergence of new employment possibilities, obscures an underlying acceptance of the 'loss of strength and general eclipse of bodies in late capitalist society' as humans are assigned the supervisory roles of watchman and regulator in machinic processes.[85] It is, she suggests, not only the de-skilling inflicted by automation, but also the re-skilling that supposedly rectifies the situation that may, under the direction of capital, be injurious to humans.

Wendling thus agrees with us that automation is ultimately a question of what Marx termed human species-being (*Gattungswesen*). However, while she recognises that this species-transforming dynamic includes terrifying possibilities, she rejects some of our prognostications. In particular, she takes issue with our speculation as to how, under the tutelage of capital, AGI could dominate a mode of production in which humans are no more important than wild animals are in today's planetary factory. She argues that Marx's concept

---

83   terrao 2018.
84   Seidler, Kolling and Hampshire 2016, p. 1.
85   Wendling, this volume, p. 156.

of human species-being is not one of a fixed core of fundamental characteristics, but rather a colocation of attributes subject to historical change, such that 'the human essence is simply the vanishing point of a capacity to become'.[86] Wendling argues that, in light of this, our concept of machinic capital demoting humans to the status of animals is far too static. Instead, drawing on a line of thought made famous by Donna Haraway's (1991) cyborg manifesto, Wendling proposes 'an alternative speculative conclusion' which imagines the 'human animal' not as a legacy system unfit to run capital, but as a hybrid entity which 'include[s] AGIs in symbiotic co-evolution'.[87] She reassuringly observes that, since machine learning AI is trained on large data sets of human behaviour 'we are already talking about ourselves' when we talk about AI.[88]

We agree with Wendling's anti-essentialist view of species-being. There is every reason to foresee AI bringing deep changes in species definitions and new forms of cyborg symbiosis with machines. While in *Inhuman Power* we certainly explore how capitalist AGI could generate populations of superfluous humanity, we also raise the issue of cyborg possibilities in terms of the conjunction of AI-capital and transhumanism.[89] We suggest this rendezvous is already apparent in projects such as Elon Musk's Neuralink enterprise, which is developing brain–computer interfaces. Since we wrote, such research has advanced significantly, with one demonstration showing a macaque monkey playing the video game Pong via an interface wired into its motor cortex.[90]

Haraway's famous essay raised the possibility of rebel cyborg subjectivities. We, however, suggest that it is necessary to consider what these rebellions would have to subvert. To be truly revolutionary, cyborg transformation would have to contest and overcome hegemonic AI-capital. We argue that to 'keep up with inhuman AGI labour, humans would have to become equally inhuman: mind and body, as immortal wage-labourers. Together with AGI workers, these no-longer-human beings would make obsolete those who decline transformation. Humans would face a choice: capitalist transhumanism or death'.[91] Our point is that in a situation of capital-led AI development, human species-being changes are likely to take forms amenable to hypertrophic commodity exchange. Wendling's observation that AI is constructed through the study of humans cheers us little. Under the direction of Google, Facebook, Amazon,

---

86 Wendling, this volume, p. 169.
87 Wendling, this volume, p. 170.
88 Wendling, this volume, 171.
89 Dyer-Witheford, Kjøsen and Steinhoff 2019, p. 159.
90 Flaig 2021.
91 Dyer-Witheford, Kjøsen and Steinhoff 2019, p. 159.

Apple and other corporate AI developers, what are generated are primarily data sets about humans insofar as they are of interest to capital, (i.e. aspects that pertain to the metamorphoses of value: advertising, shopping, consuming, and, of course, producing). AI systems are trained and deployed with the aim of intensifying such activities, and of course, eliminating non-system-functional behaviours (i.e. in policing and military applications, but also in every digital nudge or push notification administered by an app).

Humans may well be upgraded as symbiotic, cyborg partners to AGI, but if the AGIs are designed to prioritise the goals of capital, such symbioses will develop in the direction of making more perfect vehicles for commodity production and circulation. We thus expect that the new transhuman powers of former homo sapiens will be channelled towards 'working for a wage, 24/7, until the heat death of the universe'.[92] The species-being of this transhumanity would be the very being of capital. We indeed foresee complex blurrings of human, animal and machinic identities, as well as previously discrete capacities and attributes, but predominantly in the direction of improved capitalist efficiency.

Perhaps the resulting world would be something like the one depicted in Kazuo Ishiguro's (2021) gentle but infinitely melancholy *Klara and the Sun*, in which 'elevated' (that is, techno-neurologically upgraded) humans and AGI androids constantly interact, but both are always threatened by obsolescence – humans by failure to successfully 'elevate' themselves, androids via their supersession by new models. In this situation it is indeed, as Wendling implies in her discussion of deskilling and reskilling, not so much the automating replacement of humans as surplus populations, but their employment-preserving augmentation that marks the full, transformative subsumption of the human by capital.

This brings us back to questions of stagnation. The future seems much worse than a mere lack of jobs. This is one reason we share Smith and Benanav's scepticism both about the progressive nature of Universal Basic Income (UBI), a measure which we think is more likely to consolidate corporate ownership of the means of production than to subvert it, and about the unlikelihood of reviving traditional models of trades unionism, whose original factory base has been unravelled by deindustrialisation. Like them, we see contemporary movements for social equality and ecological survival as 'explosive'.[93] But we believe these developments should be understood as the result of stagnation-

---

92   Dyer-Witheford, Kjøsen and Steinhoff 2019, p. 159.
93   Benanav 2019b.

and-automation, in which the two are reciprocal: techniques of AI-powered automation applied first in the circulatory sphere to revive stagnating markets are being transferred back into the sphere of production to wring the dregs of surplus-value from labour in production.

We doubt that Smith or Benanav would disagree with our account of Big Tech's techno-strategies. But there is a difference in emphasis. They see capital's preoccupation with new forms of AI as both desperate and inconsequential, the foolish, flawed expedients of a system imploding on itself; we suggest that today's AI projects and applications contain potentials for a profound – indeed, abyssal – reorganisation of total commodification, fulfilling Marx's intuition that the 'organic' nature of capital is machinic. Benanav and Smith believe capital has exhausted its need for industrial labour, and hence its capacity as a mechanism of human development, and that therefore it must ultimately fail. We believe that, facing this dilemma, capital will, and is already bidding to, dispense with its human element. The features Smith sees as marking the decrepitude of capital can be read in reverse; if capital's development of its 'human resources' has stagnated, is this a sign of capital's obsolescence for humanity, or humanity's obsolescence for capital?

Works Cited

'Automatic-Automaton' 1989. *The Oxford English Dictionary, Second Edition.* Oxford: Clarendon, 805–6.
'Die Computer Revolution – Fortschritt Macht Arbeitslos' 1978, special issue of *Der Spiegel*.
'The Return of the Machinery Question' 2016, special issue of *The Economist*.
'Workers Battle Automation' 1960, special issue of *News and Letters*.
Abromeit, John 2010, 'Left Heideggerianism or Phenomenological Marxism? Reconsidering Herbert Marcuse's Critical Theory of Technology', *Constellations*, 17, no. 1: 87–106.
Acemoglu, Daron et al. 2014, 'Return of the Solow Paradox? IT, Productivity, and Employment in US Manufacturing', *American Economic Review: Papers & Proceedings* 104, no. 5: 394–399.
Acemoglu, Daron, Philippe Aghion and Giovanni L. Violante 2001, 'Deunionisation, technical change and inequality', *Carnegie-Rochester Conference Series on Public Policy*, 55: 229–264.
Adler, Paul 2009, *The Oxford Handbook of Sociology and Organisational Studies: Classical Foundations*, Oxford: Oxford University Press.
Adler, Paul and Borys, Bryan 1989, 'Automation and Skill: Three Generations of Research on the Machine-Tool Case', *Politics and Society*, 17, no. 3: 377–412.
Adorno, Theodor W. 1997, *Aesthetic Theory. Newly Translated, Edited, and with a Translator's Introduction*, translated by Robert Hullot-Kentor, London: Athlone Press.
Alimahomed-Wilson, Jake and Immanuel Ness (eds.) 2018, *Choke Points: Logistics Workers Disrupting the Global Supply Chain*, London: Pluto.
Allais, David 2017, 'Automation in the Warehouse: Asset or Obstacle?', *Industry Week*, 13 July.
Althusser, Louis 1977, 'Introduction: Today', *For Marx*, London: New Left Books, 21–41.
Althusser, Louis 2006, *Philosophy of the Encounter: Later Writings 1978–87*. Trans. G.M. Goshgarian. London: Verso.
Amazon 2018, 'Annual Report to Shareholders. Item 1: Business'.
Amazon 2019, 'Inside Amazon's fulfilment centres: What you can expect to see on a warehouse tour', 11 March, available at: https://www.aboutamazon.com/news/operations/inside-amazons-fulfillment-centers-what-you-can-expect-to-see-on-a-warehouse-tour.
Amazon 2022, '10 years of Amazon robotics: how robots help sort packages, move product, and improve safety', 22 June, available at: https://www.aboutamazon.com/news/operations/10-years-of-amazon-robotics-how-robots-help-sort-packages-move-product-and-improve-safety.

Anders, Günther 1956, *Die Antiquiertheit des Menschen*, Volume 1: *Über die Seele im Zeitalter der zweiten industriellen Revolution*, München: C.H. Beck.
Anders, Günther 1959, 'Faule Arbeit und pausenloser Konsum: Fünf Tagebuch-Eintragungen zum Problem "freie Zeit"', *Homo ludens: Der spielende Mensch*, 1: 8–9.
Anders, Günther 1980, *Die Antiquiertheit des Menschen*, Volume 2: *Über die Zerstörung des Lebens im Zeitalter der dritten industriellen Revolution*, München: C.H. Beck.
Anders, Günther 1982, *Ketzereien*, München: C.H. Beck.
Anders, Günther 1984, *Mensch ohne Welt: Schriften zur Kunst und Literatur*, München: C.H. Beck.
Anders, Günther 1987, *Gewalt, Ja oder Nein? Eine notwendige Diskussion*, München: Knaur.
Anders, Günther 1992, 'Die Antiquiertheit des Proletariats', *Forum*, 462–464: 7–11.
Anders, Günther 2018, *Die Weltfremdheit des Menschen: Schriften zur philosophischen Anthropologie*, ed. Christian Dries, in collaboration with Henrike Gätjens, München: C.H. Beck.
Angry Workers of the World 2018, 'Power Hour or Workers Power?! – Reports from Two workers, Hemel Hempstead, Winter 2017/18', available at: https://angryworkersworld.wordpress.com/2018/01/27/power-hour-or-workers-power-reports-from-two-amazon-workers-hemel-hempstead-winter-2017-18/.
Arendt, Hannah 1977, *Eichmann in Jerusalem: A Report on the Banality of Evil*, London: Penguin.
Arning, Bill 1989, 'Sturtevant', *Journal of Contemporary Art*, 2, no. 2: 39–50.
Arntz, Melanie, Terry Gregory and Ulrich Zierhahn 2016, 'The risk of automation for jobs in OECD countries: a comparative analysis', *OECD Working Paper*, 189.
Aronowitz, Stanley and Johnathan Cutler, 1998, *Post-Work*, New York: Routledge.
Aronson, Ronald 2014, 'Marcuse Today', *Boston Review*.
Arthur, Christopher J. 2004, *The New Dialectic and Marx's Capital*, Leiden: Brill.
Aschoff, Nicole 2020 'Working for Facebook Can Give You PTSD', *Jacobin*, 1 February, available at: https://www.jacobinmag.com/2020/02/facebook-content-moderators-ptsd-tech-companies (last accessed 26 February 2020).
Ashgari, NiMA 2018, 'Decentralized Autonomous Workers', *HackerNoon*, 21 August, available at: https://hackernoon.com/decentralized-autonomous-workers-cbaeca9ce6b3.
Bahr, Hans-Dieter 1980, 'The Class Structure of Machinery: Notes on the Value Form', in *Outlines of a Critique of Technology*, edited by Phil Slater, London: Ink Links.
Bahro, Rudolf 1982, *Socialism and Survival*, trans. D. Fernbach, London: Heretic.
Bailey, Nancy 2020, 'Disaster Capitalism, Online Instruction, and What COVID-19 is Teaching us about Public Schools and Teachers', *Nancy Bailey's Education Website*, available at: https://nancyebailey.com/2020/03/16/disaster-capitalism-online-instruction-and-what-covid-19-is-teaching-us-about-public-schools-and-teachers/.

Balibar, Étienne 2011, 'Marx's "Two Discoveries"', *Actuel Marx*, 2, no. 2: 44–60.
Balibar, Etienne, Margaret Cohen, and Bruce Robbins 1994, 'Althusser's Object'. *Social Text* 39: 157–188.
Baran, Paul A. and Paul M. Sweezy 1966, *Monopoly capital: An Essay on the American Economic and Social Order*, New York: Monthly Review Press.
Barron, Richard M. 2017, 'Triad to take regional approach to Amazon proposal', *Winston-Salem Journal*, 8 September.
Barshay, Jill 2015, 'Five Studies find Online Courses are not Working Well at Community Colleges', *The Hechinger Report*, 27 April, available at: https://hechingerreport.org/five-studies-find-online-courses-are-not-working-at-community-colleges/.
Bastani, Aaron 2019, *Fully Automated Luxury Communism: A Manifesto*, London: Verso.
Baumol, William J. 2012, *The Cost Disease: Why Computers Get Cheaper and Health Care Doesn't*, New Haven: Yale University Press.
Bay View Analytics 2020, *The Great (Forced) Shift to Remote Learning: A Survey of Instructors and Campus Leaders*, available at: http://onlinelearningsurvey.com/reports/2020_IHE_BayViewAnalytics_webcast.pdf
Bean, N.L. 1949, 'Automation', *Mechanical Engineering* 71: 389–90, 394.
Beech, Dave 2015, *Art and Value. Arts Economic Exceptionalism in Classical, Neoclassical and Marxist Economics*, Leiden: Brill.
Beech, Dave 2019, *Art and Postcapitalism. Aesthetic Labour, Automation and Value Production*, London: Pluto Press.
Benanav, Aaron 2019a, 'Automation and the Future of Work 1', *New Left Review* 119, 5–38.
Benanav, Aaron 2019b, 'Automation and the Future of Work 2', *New Left Review* 120, 117–148.
Benanav, Aaron 2020, *Automation and the Future of Work*, London: Verso.
Benjamin, Walter 2006a, 'On the Concept of History', in *Selected Writings*, vol. 4, edited by Michael W. Jennings, Cambridge: Harvard UP.
Benjamin, Walter 2006b, 'Paralipomena to "On the Concept of History"', in *Selected Writings*, vol. 4, edited by Michael W. Jennings, Cambridge: Harvard UP.
Benjamin, Walter 2008, *The Work of Art in the Age of Mechanical Reproduction*, London: Penguin.
Benton, Ted 1989, 'Marxism and Natural Limits: An Ecological Critique and Reconstruction', *New Left Review* 178: 51–86.
Benton, Ted 1992, 'Ecology, Socialism and the Mastery of Nature: A Response to Reiner Grundmann', *New Left Review* 194: 51–86.
Benton, Ted 1996. *The Greening of Marxism*, New York: Guilford Press.
Berg, Lawrence 2019, 'Academic Knowledge Production, Neoliberalization and the Falling rate of Use Values in the Academy', *Fennia: International Journal of Geography*, 197, no. 2: 177–182.

Berg, Lawrence, Edward Huijbens, and Henrik Larsen 2016, 'Producing Anxiety in the Neoliberal University', *Canadian Geographer/LeGéographe canadien* xx, xx: 1–13.

Bernal, J.D. 1965, *Science in History Volume 2: The Scientific and Industrial Revolutions*. Cambridge, MA: MIT.

Bernes, Jasper 2013, 'Logistics, counterlogistics and the communist prospect', *Endnotes*, 3: 172–201.

Bernes, Jasper 2018, 'The Belly of the Revolution: Agriculture, Energy, and the Future of Communism', in *Materialism and the Critique of Energy*, edited by Brent Ryan Bellamy and Jeff Diamanti, Chicago: MCM' Publishing.

Berry, Manik 2021, 'Does Apple Sell Your Data? Everything You Need to Know', *Fossbytes*, 31 March, available at: https://fossbytes.com/apple-data-collection-explained/.

Bessen, James 2015, 'Toil and Technology', *Finance & Development* 52, no. 1: 16–19.

Bessen, James E. 2017, 'Automation and jobs: When technology boosts employment', *Boston Univ. School of Law, Law and Economics Research Paper*, 17–09.

Bhattacharya, Tithi 2017, 'Mapping Social Reproduction Theory', in *Social Reproduction Theory: Remapping Class, Recentering Oppression*, edited by Tithi Bhattacharya, London: Pluto.

Bhattarai, Abha 2018, 'Bernie Sanders introduces "Stop BEZOS Act" in the Senate', *Washington Post*, 5 September.

Bishop, Todd 2018, 'Owning an Amazon delivery business: The risks, rewards and economic realities of the tech giant's new program for entrepreneurs', *GeekWire*, 15 July.

Bittman, Michael, James Rice Mahmud and Judy Wajcman 2004, 'Appliances and their impact: the ownership of domestic technology and time spent on household work', *The British Journal of Sociology*, 55, no. 3: 401–23.

Björk, Mårten 2020, 'Life Against Nature: The Goldberg Circle and the Search for a Non-Catastrophic Politics', *Endnotes*, 5: 307–58.

Bloodworth, James 2018, *Hired*, London: Atlantic Books.

BLS (US Bureau of Labor Statistics) 2018, 'Contingent and Alternative Employment Arrangements Summary – May 2017', 7 June, available at: https://www.bls.gov/news.release/conemp.nr0.htm.

BNN Bloomberg 2019, 'Amazon Doubles Holiday Hiring to 200,000 Temporary Workers', 27 November.

Bögenhold, Dieter and Muhammad Yorga Permana 2018, 'End of Middle-Classes? Social Inequalities in Digital Age', *IfS Discussion Papers*, 4: 1–21.

Boggs, James 2009, *The American Revolution: Pages from a Negro Worker's Notebook*, New York, Monthly Review Press.

Bogost, Ian 2019, 'The AI-Art Gold Rush Is Here. An artificial-intelligence "artist" got a solo show at a Chelsea gallery. Will it reinvent art, or destroy it?', *The Atlantic*, available at: https://www.theatlantic.com/technology/archive/2019/03/ai-created-art-invades-chelsea-gallery-scene/584134/.

Bonacich, Edna, Sabrina Alimahomed and Jake B. Wilson 2008, 'The Racialization of Global Labor', *American Behavioral Scientist*, 52, no. 3: 342–55.

Bonacich, Edna and Jake B. Wilson. 2008, *Getting the Goods: Ports, Labor, and the Logistics Revolution*, Ithaca: Cornell University Press.

Bonacich, Edna 2003, 'Pulling the plug: Labor and the global supply chain', *New Labor Forum*, Volume 12, no. 2: 41–48.

Boston Consulting Group 2016, 'Following the Money in Education Technology-Infographic', Boston Consulting Group, available at: https://www.bcg.com/en-us/publications/2016/private-equity-following-money-education-technology.aspx.

Boston Consulting Group 2020a, 'A Blueprint for Digital Education', Boston Consulting Group, available at: https://www.bcg.com/en-us/industries/public-sector/successfully-transitioning-digital-education.aspx.

Boston Consulting Group 2020b, 'Education Experts: Nithya Vaduganathan. Boston Consulting Group', available at: https://www.bcg.com/en-us/about/people/experts/Vaduganathan-Nithya.aspx.

Boston Consulting Group and ASU 2018, *Making Digital Learning Work: Successful Strategies from Six Leading Universities and Community Colleges*, Boston Consulting Group and Arizona State University, available at: https://edplus.asu.edu/sites/default/files/BCG-Making-Digital-Learning-Work-Apr-2018%20.pdf.

Bourdieu, Pierre 1993, *The Field of Cultural Production: Essays on Art and Literature*, n.p.: Columbia University Press.

Bourdieu, Pierre 1996, *The Rules of Art. Genesis and Structure of the Literary Field*. Stanford: Stanford University Press.

Braverman, Harry 1998, *Labour and Monopoly Capital: The Degradation of Labour in the Twentieth Century: 25th Anniversary Edition*, New York: Monthly Review Press.

Bregman, Rutger 2017, *Utopia for realists: And how we can get there*, New York: Bloomsbury Publishing.

Brenner, Robert 1977, 'The Origins of Capitalist Development: A Critique of Neo-Smithian Marxism', *New Left Review*, 104: 27–92.

Brenner, Robert 2006, *The Economics of Global Turbulence: The Advanced Capitalist Economies from Long Boom to Long Downturn, 1945–2005*, London: Verso.

Brenner, Robert 2020, 'Escalating Plunder', *New Left Review*, 123: 5–22.

Breuer, Stefan 1977, *Die Krise der Revolutionstheorie: Negative Vergesellschaftung und Arbeitsmetaphysik bei Herbert Marcuse*, Frankfurt: Syndikat.

Briken, Kendra, Robert MacKenzie, Harry Pitts and Patrizia Zanoni 2021, 'Value and Labour Process Theory – revisited' [Call for Papers], ILPC 2021, Greenwhich, United Kingdom.

Browne, Chris, Jake Alimahomed-Wilson, Katy Fox-Hodess and Kim Moody 2018, 'Seizing the Chokepoints', *Jacobin Magazine*, available at: https://www.jacobinmag.com/2018/10/choke-points-logistics-industry-organizing-unions.

Brynjolfsson, Eric and Andrew McAffee 2011, *Race against the Machine: How the Digital Revolution is Accelerating Innovation, Driving Productivity, and Irreversibly Transforming Employment and the Economy*, Lexington: Digital Frontier Press.

Brynjolfsson, Erik and Andrew McAfee 2014, *The Second Machine Age: Work, Progress, and Prosperity in a Time of Brilliant Technologies*, New York and London: W.W. Norton & Company.

Buchloh, Benjamin H.D. 2015, *Formalism and Historicity. Models and Methods in Twentieth-Century Art*, Cambridge/MA and London: MIT Press.

Burawoy, Michael 1979, *Manufacturing Consent: Changes in the Labour Process under Monopoly Capitalism*, Chicago: University of Chicago Press.

Burkett, Paul 1999, *Marx and Nature: A Red and Green Perspective*, New York: Palgrave.

Burkett, Paul 2006, *Marxism and Ecological Economics: Towards a Red and Green Political Economy*, Leiden: Brill.

Burkett, Paul and John Bellamy Foster 2006, 'Metabolism, energy, and entropy in Marx's critique of political economy: Beyond the Podolinsky myth', *Theory and Society*, 35: 109–56.

Burroughs, William S. 2005, *The Electronic Revolution*, Ubuweb, available at: https://www.ubu.com/historical/burroughs/electronic_revolution.pdf.

Burt, Steve, John Dawson, and Roy Larke 2006, 'Inditex-Zara: re-writing the rules in apparel retailing', in *Strategic issues in international retailing*, edited by John Dawson, Roy Larke and Masao Mukoyama, New York: Routledge, 83–102.

Buterin, Vitalik 2013a, 'Bootstrapping a decentralized autonomous corporation: Part 1', *Bitcoin Magazine*, 19 September, available at: https://bitcoinmagazine.com/technical/bootstrapping-a-decentralized-autonomous-corporation-part-i-1379644274.

Buterin, Vitalik 2013b, 'Bootstrapping an autonomous decentralized corporation, Part 2: Interacting with the world', *Bitcoin Magazine*, 21 September, available at: https://bitcoinmagazine.com/technical/bootstrapping-an-autonomous-decentralized-corporation-part-2-interacting-with-the-world-1379808279.

Buterin, Vitalik 2014, 'DAOS, DACS, DAS, and More: An Incomplete Terminology Guide', *Ethereum Foundation Blog*, 6 May, available at: https://blog.ethereum.org/2014/05/06/daos-dacs-das-and-more-an-incomplete-terminology-gui.

Butollo, Florian and Sabine Nuss 2019, 'Einleitung der Herausgeber', in *Marx und die Roboter. Vernetzte Produktion, Künstliche Intelligenz Und Lebendige Arbeit*, edited by Florian Butollo and Sabine Nuss, Berlin: Dietz, 8–21.

Caffentzis, George 1997, 'Why machines cannot create value; or, Marx's theory of machines', in *Cutting Edge: Technology, Information Capitalism and Social Revolution*, edited by Jim Davis, Michael Stack and Thomas A. Hirschl, London: Verso.

Caffentzis, George 2013, *In Letters Of Blood And Fire. Work, Machines And The Crisis Of Capitalism*. Oakland: Pm Press.

Caffentzis, George 2018, 'Work or Energy or Work/Energy? On the Limits to Capitalist Accumulation', in *Materialism and the Critique of Energy*, edited by Brent Ryan Bellamy and Jeff Diamanti, Chicago: MCM'.

Cain, Aine, and Isobel Asher Hamilton 2019, 'Amazon warehouse employees speak out about the "brutal" reality of working during the holidays, when 60-hour weeks are mandatory and ambulance calls are common', *Business Insider*, 19 February.

Cant, Callum 2020, *Riding for Deliveroo: Resistance in the New Economy*, UK: Polity.

Cappelli, Peter 1999, *New Deal at Work*, Cambridge: Harvard Business School Press.

Carey, Kevin 2016, *The End of College: Creating the Future of Learning and the University of Everywhere*, NY: Riverhead Books.

Carnegie Mellon University Risk Initiatives Office n.d, 'Value Chain', available at: http://www.cmu.edu/erm/concepts/value.html.

Carpenter, Sandra 2018. 'Disability and Social Equality: The Centrality of Independent Living', *Alternate Routes: A Journal of Critical Social Research*, 29: 229–241, available at: http://www.alternateroutes.ca/index.php/ar/article/view/22454/18248 (accessed, 15 November 2020).

Carrier, Ryan 2017, 'The Amazon Impact on Retail Jobs – Defining Technological Unemployment', *Medium*, 9 July.

Carver, Terrell and Daniel Blank 2014, *A Political History of the Editions of Marx and Engels's "German Ideology Manuscripts"*. New York: Palgrave Macmillan.

Caspary, Adolf 1927 *Die Maschinenutopie: Das Übereinstimmungsmoment der bürgerlichen und sozialistischen Ökonomie*, Berlin: David.

Cholobi, Michael 2018, 'The Desire to Work as an Adaptive Preference', *Autonomy*, 4: 2–17.

Christensen, Clayton M., Michael B. Horn, Louis Caldera and Louis Soares 2011, 'Disrupting College: How Disruptive Innovation can Deliver Quality and Affordability to Postsecondary Education', Center for American Progress and Innosight Institute, February.

Chua, Charmaine S. 2017, 'Logistical Violence, Logistical Vulnerabilities: A Review of The Deadly Life of Logistics: Mapping Violence in Global Trade by Deborah Cowen', *Historical Materialism*, 25, no. 4: 167–82.

City of San Antonio 2017, 'Letter to Jeff Bezos', 11 October, available at: ⟨https://www.mysanantonio.com/file/247/8/2478-Wolff%20Nirenberg%20letter.pdf/⟩.

Cleaver, Harry 2019, *33 Lessons on Capital: Reading Marx Politically*, UK: Pluto Press.

Clover, Joshua 2016, *Riot.Strike.Riot*, London: Verso.

Clover, Joshua 2017, 'Transition: End of the Debate', *Amerikastudien / American Studies*, 62, no. 4: 539–50.

Cohen, G.A. 2000, *Karl Marx's Theory of History: A Defence*, Expanded Edition, Princeton, NJ: Princeton University Press.

Collier, Ruth Berins, Veena Dubal and Christopher Carter 2017, 'The Regulation of Labor

Platforms: The Politics of the Uber Economy', The Kauffman Foundation, available at: https://brie.berkeley.edu/sites/default/files/reg-of-labor-platforms.pdf.

Collins, Randall 2013, 'The End of Middle-Class Work: No More Escapes', in *Does Capitalism have a Future?*, edited by Immanuel Wallerstein, Randall Collins, Michael Mann, Georgi Derluguian, and Craig Calhoun, Oxford: Oxford University Press.

Conlon, Eddie 2019, 'Prisoners of the Capitalist Machine: Captivity and the Corporate Engineer', in *The Engineering Business Nexus: Symbiosis, Tension, and Co-Evolution*, edited by Steen Hyldgaard Christensen, Bernard Delahousse, Christelle Didier, Martin Meganck, and Mike Murphy, Cham, Switzerland: Springer.

Cosgrove, Emma 2019, 'Report: Amazon's new packing robots to replace workers', *Supply Chain Dive*, 13 May.

Cowan, Ruth S. 1985, *More Work For Mother: The Ironies Of Household Technology From The Open Hearth To The Microwave*, Basic Books, New York: Basic Books.

Cowen, Deborah 2014, *The Deadly Life of Logistics: Mapping Violence in Global Trade*, Minneapolis: University of Minnesota Press.

Coy, Peter 2010, 'The Disposable Worker', *Bloomberg Businessweek*, 7 January, available at: https://www.bloomberg.com/news/articles/2010-01-07/the-disposable-worker.

Craig, Ryan 2015, *College Disrupted: The Great Unbundling of Higher Education*, NY: St. Martin's Press.

Cruddas, Jon and Frederick Harry Pitts 2020, 'The Politics Of Postcapitalism: Labour And Our Digital Futures', *The Political Quarterly* 91, no. 2: 275–286.

Culpepper, Pepper D. and Kathleen Thelen 2019, 'Are We All Amazon Primed? Consumers and the Politics of Platform Power', *Comparative Political Studies* (online): DOI 0010414019852687.

Curtis, John 2014, *The Employment Status of Instructional Staff Members in Higher Education. Fall 2011*, American Association of University Professors, Washington DC, available at: https://www.aaup.org/sites/default/files/files/AAUP-InstrStaff2011-April2014.pdf.

Curtis, Sophie 2013, 'Amazon at 15: the technology behind Amazon UK's success', *Telegraph*, 15 October.

Czerniewicz, Laura 2018, 'Unbundling and Rebundling Higher Education in an Age of Inequality', *Educause Review*, 29 October, available at: https://er.educause.edu/articles/2018/10/unbundling-and-rebundling-higher-education-in-an-age-of-inequality.

D'Onfro, Jillian 2019, 'Amazon's New Delivery Drone Will Start Shipping Packages 'In A Matter Of Months'', *Forbes*, 5 June.

Dastin, Jeffrey 2019, 'Exclusive: Amazon rolls out machines that pack orders and replace jobs', *Reuters*, 13 May.

David, Paul A. 1990, 'The Dynamo and the Computer: An Historical Perspective on the Modern Productivity Paradox', *American Economic Review* 80, no. 2: 355–61.

Davis, Mike 2020, 'The Monster Enters', *New Left Review*, 122: 7–14.

Dawsey, Jason 2017, 'Ontology and Ideology: Günther Anders's Philosophical and Political Confrontation with Heidegger', *Critical Historical Studies*, 4, no. 1: 1–37.

Dawsey, Jason 2019, 'Marxism and Technocracy: Günther Anders and the Necessity for a Critique of Technology', *Thesis Eleven* 153, no. 1: 39–56.

Day, Matt and Spencer Soper 2019, 'Amazon U.S. Online Market Share Estimate Cut to 38% From 47%', *Bloomberg*, 13 June.

De Angelis, Massimo and David Harvie 2009, '"Cognitive Capitalism" and the Rat-Race: How Capital Measures Immaterial Labour in British Universities', *Historical Materialism*, 17: 3–30.

de Jesus, Ayn 2018, 'Artificial Intelligence in Industrial Automation – Current Applications'. *Emerj*, 29 November, available at: https://emerj.com/ai-sector-overviews/artificial-intelligence-industrial-automation-current-applications/.

Dean, Jodi 2020, Neofeudalism: The End of Capitalism?, *LA Review of books*, available at: https://lareviewofbooks.org/article/neofeudalism-the-end-of-capitalism/.

Del Rey, Jason 2018, 'Amazon's pay-raise backlash highlights a reality: People see the worst even when Amazon thinks it's doing its best', *Vox*, 9 October.

Delabar, Walter 1992, '"Aus lebenden Menschen hergestellte Geräteteile": Anmerkungen zur Technik-und Gesellschaftskritik von Günther Anders', *Text + Kritik*, 115: 27–38.

Delaney, Melissa 2019, 'Digital Transformation Empowers Student Learning in Higher Education', *EdTech*, 25 February 25, available at: https://edtechmagazine.com/higher/article/2019/02/digital-transformation-empowers-student-learning-higher-education.

Demanuelle-Hall, Joe 2019, 'That's Strike One, Amazon', *Jacobin*, 26 March.

Denby, Charles 1960, 'Workers Battle Automation', in *News & Letters* 5.7: 1–8.

Descartes, Rene 1993, *Meditations on First Philosophy*. Trans. Donald Cress. Indianapolis: Hackett.

DiChristopher, Tom 2016, 'Energy jobs: Oil and gas could hire 100,000 workers – if it can find them', *cnbc.com*, 8 July, available at: http://www.cnbc.com/2016/07/08/energy-jobs-oil-and-gas-industry-could-hire-100000-workers-if-it-can-find-them.html (accessed 8 June 2021).

Diederichsen, Diedrich 2012, 'Zeit, Objekt, Ware', *Texte zur Kunst* 88: 95–102.

DiEM25 2017, 'Why universal basic income is not enough', 18 September, available at: https://diem25.org/free-money-for-all/.

Dignan, Larry 2018, 'Amazon Go: The impact on human jobs, retail innovation, Amazon's bottom line', *ZDNet*, 23 January.

Dopico, Luis and Stephanie Crofton 2007, 'Zara-Inditex and the growth of fast fashion', *Essays in Economic & Business History*, 25: 41–54.

Dorninger, Christian, et al. 2021, 'Global patterns of ecologically unequal exchange: Implications for sustainability in the 21st century', *Ecological Economics* 179: 106824.

Downey, Gregory J. 2014, 'Making Media Work: Time, Space, Identity, and Labor in the Analysis of Information and Communication Infrastructures', in *Media Technologies. Essays on Communication, Materiality, and Society*, edited by Tarleton Gillespie, Pablo J. Boczkowski and Kirsten A. Foot, Cambridge, Massachusetts: The MIT Press.

Drum, Kevin 2017, 'You Will Lose Your Job to a Robot – and Sooner Than You Think: Automation helped bring on the age of Trump. What will AI bring?', *Mother Jones* November/December, available at: https://www.motherjones.com/politics/2017/10/you-will-lose-your-job-to-a-robot-and-sooner-than-you-think/.

Ducatel, Ken, and Nicholas Blomley 1990, 'Rethinking retail capital', *International Journal of Urban and Regional Research*, 14.2: 207–227.

Duffy, Clare 2020, 'Amazon hiring 100,000 new distribution workers to keep up with online shopping surge caused by coronavirus', *CNN*, 17 March.

Dunayevskaya, Raya 1958, *Marxism and Freedom: From 1776 until Today*, New York: Bookman.

Dunayevskaya, Raya, Herbert Marcuse, and Erich Fromm 2012, *The Dunayevskaya-Marcuse-Fromm Correspondence, 1954–1978: Dialogues on Hegel, Marx, and Critical Theory*, edited by Kevin Anderson and Russell Rockwell, Lanham: Lexington Books.

Dunker, Anders 2020, 'On Technodiversity: A Conversation with Yuk Hui', *LA Review of Books*, 9 June, available at: https://lareviewofbooks.org/article/on-technodiversity-a-conversation-with-yuk-hui/ (last accessed 26 June 2020).

DuPont, Quinn 2019, 'Experiments in Algorithmic Governance: A history and ethnography of "The DAO", a failed Decentralized Autonomous Organization', in *Bitcoin and Beyond: Cryptocurrencies, Blockchains and Global Governance*, edited by Malcolm Campbell-Verduyn. London: Routledge.

Dussell, Enrique 2001, *Towards an Unknown Marx*, London: Routledge, ebook.

Dyer-Witheford, Nick 1999, *Cyber-Marx: Cycles and Circuits of Struggle in High Technology Capitalism*. Urbana and Chicago: University of Illinois Press.

Dyer-Witheford, Nick, Atle Mikkola Kjøsen and James Steinhoff 2019, *Inhuman Power: Artificial Intelligence and the Future of Capitalism*, London: Pluto Press.

Dzieza, Josh 2019a, 'Beat the Machine: Amazon Warehouse Workers Strike to Protest Inhumane Conditions', *The Verge*, 16 July.

Dzieza, Josh 2019b, 'Amazon workers in Minnesota walk out in protest over part-time work', *The Verge*, 3 October.

Dzieza, Josh 2020, 'How hard will the robots make us work?', *The Verge*, 27 February, available at: https://www.theverge.com/2020/2/27/21155254/automation-robots-unemployment-jobs-vs-human-google-amazon.

Eavis, Peter and Steve Lohr 2020, 'Big Tech's Domination of Business Reaches New Heights', *The New York Times*, 19 August, available at: https://www.nytimes.com/2020/08/19/technology/big-tech-business-domination.html.

Economic Policy Institute 2021, 'The Productivity-Pay Gap', *Economic Policy Institute*, available at: https://www.epi.org/productivity-pay-gap/.

WORKS CITED

Editors of Marx and Engels 1976, 'Preface', in *Karl Marx, Frederick Engels, Collected Works*, Volume 5: *Marx and Engels 1845–1847*. New York: International Publishers.

EdSurge 2020, 'Instructure Sells to Thoma Bravo After Successful Tender Offer', 22 March, available at: https://https://www.edsurge.com/news/2020-03-23-instructure-sale-to-thoma-bravo-almost-complete-after-successful-tender-offer>.

Educational Assessments Corporation, 'Distractor Point Biserial Correlation', available at: https://edassess.net/eacs/distractorbiserial.aspx.

Egenhofer, Sebastian 2008, *Abstraktion – Kapitalismus – Subjektivität. Die Wahrheitsfunktion des Werks in der Moderne*, München: Fink.

Ehrenreich, Barbara 2006, *Bait and switch: The (futile) pursuit of the American dream*, New York: Metropolitan Books.

Elbe, Ingo 2008, '"Umwälzungsmomente der alten Gesellschaft": Revolutionstheorie und ihre Kritik bei Marx', in *Theorie als Kritik*, edited by Fabian Kettner and Paul Mentz, Freiburg: ça ira.

Elbe, Ingo 2010, *Marx im Westen: Die neue Marx-Lektüre in der Bundesrepublik seit 1965*, Berlin. Akademic Verlag.

Elbe, Ingo 2013, 'Between Marx, Marxism, and Marxisms', *Viewpoint Magazine*, Oct. 21, available at: https://viewpointmag.com/2013/10/21/between-marx-marxism-and-marxisms-ways-of-reading-marxs-theory/.

Electronic Privacy Information Center Complaint 2019, *In the Matter of Zoom Video Communications, Inc.*, Request for Investigation, Injunction, and Other Relief, Before the Federal Trade Commission. 11 July.

Elliott, Stuart W. 2014, 'Anticipating a Luddite Revival', *Issues in Science and Technology* 30, no. 3, available at: http://issues.org/30-3/stuart/ (accessed 8 June 2021).

Endnotes 2019, 'Error', *Endnotes* 5: 114–60.

Engelbart, Douglas C. 1991, 'Letter to Vannevar Bush and Program on Human Effectiveness', in *From Memex to Hypertext. Vannevar Bush and the Mind's Machine*, edited by James M. Nyce, Paul Kahn et al., San Diego, CA/London: Academic Press.

Engels, Friedrich 1987a, *Anti-Dühring: Herr Eugen Dühring's Revolution in Science*, in *Karl Marx and Friedrich Engels: Collected Works*, vol. 25, New York: International Publishers.

Engels, Friedrich 1987b, *Dialectics of Nature*, in *Karl Marx and Friedrich Engels: Collected Works*, vol. 25, New York: International Publishers.

Engels, Friedrich 2012, *Dialectics of Nature*. London: Wellred.

Engels, Friedrich 2015, *Socialism: Utopian and Scientific*, translated by Edward Aveling, New York: International Publishers.

Ernst & Young 2012, 'University of the Future: A Thousand Year Old Industry on the Cusp of Profound Change', Australia, available at: http://www.ey.com/publication/vwluassets/university_of_the_future/$file/university_of_the_future_2012.pdf.

FACCCTS 2021, 'Spring Edition: Focus on Equity', email, 8 May.

Feenberg, Andres 2002, *Transforming Technology: A Critical Theory Revisited*, Oxford: Oxford UP.

Fernandez, Rodrigo, Ilke Adriaans and Tobias J. Klinge 2020, 'Engineering digital monopolies: The financialisation of Big Tech', *Stichting Onderzoek Multinationale Ondernemingen*, available at: https://www.researchgate.net/publication/347430171_Engineering_digital_monopolies_The_financialisation_of_Big_Tech.

Ferschli, Benjamin and Jakob Kapeller 2019, 'Hans Albert Und Die Kritik Am Modellplatonismus', In *Handbuch Karl Popper*. Wiesbaden: Springer.

Feuerbach, Ludwig 2013, *Principles of the Philosophy of the Future*, New York: Prism Key Press.

Fisher, Eran and Christian Fuchs 2015, *Reconsidering Value And Labour In The Digital Age*. Palgrave.

Flaig, Joseph 2021, '"Monkey Mindpong" demonstration shows transformative power of brain-computer interfaces', *Institution of Mechanical Engineers*, 6 July, available at: https://www.imeche.org/news/news-article/'monkey-mindpong'-demonstration-shows-transformative-power-of-brain-computer-interfaces.

Flood, Linda 2020 'Utegångsförbud i Peru – men gruvarbetarna måste jobba' *Arbetet*, 25 March, available at: https://arbetet.se/2020/03/25/utegangsforbud-i-peru-men-gruvarbetarna-maste-jobba/.

Foorar, Rana 2019, 'How big tech is dragging us towards the next financial crash', *The Guardian*, 8 November, available at: https://www.theguardian.com/business/2019/nov/08/how-big-tech-is-dragging-us-towards-the-next-financial-crash.

Ford, Martin 2015, *The Rise of the Robots: Technology and the Threat of a Jobless Future*, New York: Basic Books.

Fortunati, Leopoldina 2018, 'Robotization and the domestic sphere', *New Media & Society*, 20, no. 8: 2673–90.

Foster, John Bellamy 1999, *The Vulnerable Planet: A Short Economic History of the Environment*, Revised Edition, New York: Monthly Review Press.

Foster, John Bellamy 2000, *Marx's Ecology: Materialism and Nature*, New York: Monthly Review Press.

Foster, John Bellamy and Brett Clark 2020, *The Robbery of Nature: Capitalism and the Ecological Rift*, New York: Monthly Review Press.

Foucault, Michel 1977, *Discipline and Punish: The Birth of a Prison*, Vintage: NY.

Fraiman, Nelson, Singh, Medini, Arrington, Linda and Carolyn Paris 2002, 'Zara', *Columbia Business School Case*, Columbia University.

Franck, Thomas 2019, 'Booming job market is leaving the retail industry behind', *CNBC*, 6 April.

Frase, Peter 2016, *Four Futures: Life after Capitalism*, London: Verso.

Fraser, Nancy 2016, 'Contradictions of capital and care', *New Left Review*, 100: 99–117.

Freeman, Richard B. 2008, 'The new global labour market', *Focus*, 26, no. 1: 1–6.

Frey, Carl Benedikt 2019, *The Technology Trap: Capital, Labor, and Power in the Age of Automation*, Princeton: Princeton University Press.

Frey, Carl. B. and Michael A. Osborne 2013, *The Future of Employment. How Susceptible are Jobs to Computerisation*, Resource Document, University of Oxford, available at: http://www.oxfordmartin.ox.ac.uk/downloads/academic/The_Future_of_Employment.pdf.

Fuchs, Christian 2002, 'Zu einigen Parallelen und Differenzen im Denken von Günther Anders und Herbert Marcuse', in *Geheimagent der Masseneremiten: Günther Anders*, edited by Dirk Röpcke and Raimund Bahr, St. Wolfgang: Edition Art & Science.

Fuchs, Christian 2014, *Digital Labour and Karl Marx*, New York: Routledge.

Fuscaldo, Donna 2019, 'Amazon's $23B R&D Budget Sets a Record: Recode', *Investopedia*, 25 June.

Galbraith, John Kenneth 1958, *The Affluent Society*, Boston: Houghton Mifflin.

García, Oseas, et al. 2015, 'Artisanal gold mining in Antioquia, Colombia: A successful case of mercury reduction', *Journal of Cleaner Production* 90: 244–52.

Gartner Inc. 2020, 'Gartner Supply Chain Top 25 Spotlights Leadership', available at: www.gartner.com/en/supply-chain/research/supply-chain-top-25.

Gehrke, Sean and Adrianna Kezar 2015, 'Unbundling the Faculty Role in Higher Education: Utilizing Historical, Theoretical, and Empirical Frameworks to Inform Future Research', in *Higher Education: Handbook of Theory and Research*, edited by M.B. Paulsen, 30, Switzerland: Springer: 93–150.

Gershuny, Jonathan 1978, *After Industrial Society?: The Emerging Self-Service Economy*, Palgrave.

Ghemawat, Pankaj and José Luis Nueno 2003, ZARA: *Fast fashion*, Vol. 1, Harvard Business School Case.

Giménez, Martha 2019, *Marx, Women, and Capitalist Reproduction: Marxist Feminist Essays*, Chicago: Haymarket.

Gligorevic, Nikola, Karol Schulz and Iryna Ivanochko 2021, 'Use of RFID Technology in Retail Supply Chain', in *Developments in Information & Knowledge Management for Business Applications*, edited by Natalia Kryvinska and Michal Greguš, Springer: 555–87.

Gnisa, Felix 2019, 'Das Maschinensystem des 21ten Jahrhunderts? Zur Subsumtion der Kommunikation durch digitale Plattformtechnologien', In *Marx Und Die Roboter. Vernetzte Produktion, Künstliche Intelligenz Und Lebendige Arbeit*, edited by Florian Butollo and Sabine Nuss, Berlin: Dietz, 276–92.

Goldthwaite, Richard 1982, *The Building of Renaissance Florence: An Economic and Social History*. Baltimore: Johns Hopkins.

Goodman, J. David 2019, 'Amazon Pulls Out of Planned New York City Headquarters', *The New York Times*, 14 February.

Gordon, Robert 2016, *The Rise and Fall of American Growth: The US Standard of Living Since the Civil War*, Princeton: Princeton University Press.

Gordon, Robert J. 2018, 'Why has economic growth slowed when innovation appears to be accelerating?', NBER Working Paper No. 24554.

Gordon, Whitson 2019, 'How much electricity does your PC consume?', *PC Magazine*, 8 February, available at: https://www.pcmag.com/how-to/how-much-electricity-does-your-pc-consume.

Gorz, Andre 1989, *Critique of Economic Reason*, London: Verso.

Graeber, David 2018, *Bullshit Jobs*, New York: Simon and Schuster.

Granter, Edward 2009, *Critical Social Theory and the End of Work*, Burlington: Ashgate.

Gratorp, Christina 2020, 'The Materiality of the Cloud: On the Hard Conditions of Soft Digitization', *Eurozine*, 24 September, available at: https://www.eurozine.com/the-materiality-of-the-cloud/.

Gray, Jim and Daniel P. Siewiorek 1991, 'High-Availability Computer Systems', *Computer*, 24, no. 9: 39–48.

Gray, Mary L. and Siddharth Suri 2019, *Ghost Work: How to Stop Silicon Valley from Building a New Global Underclass*, San Francisco, CA: HMH Books.

Griswold, Alison 2018, 'Amazon is hiring fewer workers this holiday season, a sign that robots are replacing them', *Quartz*, 2 November.

Griswold, Alison 2019, 'Americans are lining up to work for Amazon for $15 an hour', *Quartz*, 31 January.

Grundmann, Reiner 1988, *Marx and the Domination of Nature: Alienation, Technology, and Communism*, Florence: European University Institute Working Paper.

Grundmann, Reiner 1991a, 'The Ecological Challenge to Marxism', *New Left Review* 187: 103–20.

Grundmann, Reiner 1991b, *Marxism and Ecology*, Oxford: Clarendon Press.

Guendelsberger, Emily 2019, *On the Clock*, New York: Little, Brown and Company.

Guillot, Craig 2018, 'Are supply chain jobs at risk of automation?', *Supply Chain Dive*, 13 March.

Habermas, Jürgen 1971, 'Technology and Science as Ideology', in *Toward a Rational Society: Student Protest, Science, and Politics*, London: Heinemann Educational Books.

Hall, Gary 2018, *The Uberficaton of the University*, Incite Items for Educational Iconoclasm, Item 02, June, available at: https://inciteseminars.com/tag/education/page/2/.

Haraway, Donna 1991, *Simians, Cyborgs, and Women: The Reinvention of Nature*. New York: Routledge.

Hardt, Michael and Antonio Negri 2000, *Empire*, Cambridge: Harvard University Press.

Hardt, Michael and Antonio Negri 2017, *Assembly*, Oxford: Oxford University Press.

Harvey, David 2003a, 'The Fetish of Technology: Causes and Consequences', *Macalester International*: 13:7.

Harvey, David 2003b, *The New Imperialism*, New York: Oxford University Press.

Harvey, David 2008, 'The Right to the City', *New Left Review*, 53: 23–40.

Harvey, David 2014, *Seventeen Contradictions and the End of Capitalism*, Oxford, USA: Oxford University Press.
Harvey, David 2017, *The Condition of Postmodernity*, 35th ed., Oxford, UK: Blackwell Publishing.
Harvey, David 2019, *The Limits to Capital*, New York: Verso.
Harvie, David 1999. 'Alienation, Class and Enclosure in UK Universities', *Capital & Class*, 24, no. 2: 103–132, available at: https://www.researchgate.net/publication/242318502 _Alienation_Class_and_Enclosure_in_UK_Universities.
Harvie, David 2005, 'All Labour Produces Value for Capital and We All Struggle against Value', *The Commoner* 10: 132–171.
Harvie, David 2006, 'Value Production and Struggle in the Classroom: Teachers Within, Against and Beyond Capital', *Capital & Class*, 88, no. 1: 1–32, available at: https://www .researchgate.net/publication/27247410_Value_production_and_struggle_in_the_cl assroom_Teachers_within_against_and_beyond_capital.
Haug, Wolfgang Fritz 2003, *High-Tech-Kapitalismus: Analysen zu Produktionsweise, Arbeit, Sexualität, Krieg und Hegemonie*, Hamburg: Argument.
Haw, Mark 2019, 'Will AI Replace University Lecturers? Not if we make it clear why Humans Matter', *The Guardian*, available at: https://www.theguardian.com/education/2019/sep/06/will-ai-replace-university-lecturers-not-if-we-make-it-clear-why-humans-matter.
Heater, Brian 2022, 'Amazon debuts Sparrow, a new bin-picking robot arm', *TechCrunch*, 10 November.
Hecker, Rolf 2009, 'Vorwort', in Karl Marx, *Das Kapital 1.1: Resultate des unmittelbaren Produktionsprozesses*, edited by Rolf Hecker and Hildegard Scheibler, Berlin: Karl Dietz Verlag.
Hegel, Georg Wilhelm Friedrich 1969, *Jenaer Realphilosophie: Vorlesungsmanuskripte zur Philosophie der Natur and Giestes von 1805–6*. Herausgegeben von Johannes Hoffmeister. Hamburg: Felix Meiner.
Hegel, Georg Wilhelm Friedrich 1971, *Vorlesungen über die Äesthetik: Erster und zweiter Teil*. Stuttgart: Philipp Reclam Jun.
Hegel, Georg Wilhelm Friedrich 1975, *Aesthetics: Lectures on Fine Art*. Trans. T.M. Knox. Oxford: Clarendon Press.
Heidegger, Martin 1962, *Being and Time*, trans. John Macquarrie and Edward Robinson, Cambridge: Blackwell.
Heilbroner, Robert L. 1967, 'Do Machines Make History?', *Technology and Culture* 8.3: 335–45.
Heinrich, Michael 2004, *An Introduction to the Three Volumes of Karl Marx's Capital*, New York: Monthly Review.
Heinrich, Michael 2013a, 'The "Fragment on Machines": A Marxian Misconception in the *Grundrisse* and Its Overcoming in *Capital*', in *In Marx's Laboratory: Critical Inter-

pretations of the Grundrisse, edited by Riccardo Bellofiore, Guido Starosta, and Peter D. Thomas, Boston: Brill.

Heinrich, Michael 2013b, 'Crisis Theory, the Law of the Tendency of the Profit Rate to Fall, and Marx's Studies in the 1870s', *Monthly Review* 64.11, available at: https://monthlyreview.org/2013/04/01/Crisis-theory-the-law-of-the-tendency-of-the-profit-rate-to-fall-and-marxs-studies-in-the-1870s/.

Helle Panke 2021, 'The Pipeline Riots and the Riot Pipeline: Struggles in a Warming World, Joshua Clover, Andreas Malm', *Youtube*, available at: https://www.youtube.com/watch?v=vQocSOrwX9Y (last accessed 17 December 2021).

Hendrikse Reijer, David Bassens and Michiel Van Meeteren 2018, 'The Appleization of Finance: Charting incumbent finance's embrace of FinTech', *Finance and Society*, 4, no. 2: 159–180.

Hensel, Thomas and Jens Schröter 2012, 'Die Akteur-Netzwerk-Theorie als Herausforderung der Kunstwissenschaft', *Zeitschrift für Ästhetik und allgemeine Kunstwissenschaft*, 57, no. 1: 5–18.

Hern, Alex 2020, 'Microsoft productivity score feature criticised as workplace surveillance: Tool allows managers to use Microsoft 365 to track their employees' activity', *The Guardian, U.S. Edition*, 20 November, available at: https://www.theguardian.com/technology/2020/nov/26/microsoft-productivity-score-feature-criticised-workplace-surveillance.

Hildebrandt, Helmut 1990, *Weltzustand Technik: Ein Vergleich der Technikphilosophien von Günther Anders und Martin Heidegger*, Berlin: Metropol.

Hilferding, Rudolph 2019, *Finance capital: A study in the latest phase of capitalist development*, London: Routledge.

Hill, Phil 2019, 'Instructure: Plans to Expand beyond Canvas LMS into Machine Learning and AI', *eLiterate*, 2019, available at: https://eliterate.us/instructure-plans-to-expand-beyond-canvas-lms-into-machine-learning-and-ai/.

Hobsbawm, Eric 1952, 'The machine breakers', *Past and Present*, 1.

Hoff, Jan 2017, *Marx Worldwide: On the Development of the International Discourse on Marx Since 1965*. Leiden: Brill.

Hoffmann, Norbert 1992, 'Wegwerfwelt: Günther Anders' "Die Antiquiertheit des Menschen"', in *Günther Anders kontrovers*, edited by Konrad Paul Liessmann, München: C.H. Beck.

Horkheimer, Max and Theodor W. Adorno 2002, *Dialectic of Enlightenment: Philosophical Fragments*, Stanford: Stanford UP.

Hornborg, Alf 1992, 'Machine Fetishism, Value, and the Image of Unlimited Good: Towards a Thermodynamics of Imperialism', *Man: New Series*, 27, no. 1: 1–18.

Hornborg, Alf 2001, *The Power of the Machine, Global Inequalities of Economy, Technology, and Environment*, Walnut Creek: Alta Mira Press.

Hornborg, Alf 2008, 'Machine fetishism and the consumer's burden', *Anthropology Today*, 24, no. 5: 4–5.

Hornborg, Alf 2012, *The Myth of the Machine: Essays on Power, Modernity and Environment* (Myten om Maskinen: Essäer om makt, modernitet och miljö), 2nd ed., Göteborg: Daidalos.
Hornborg, Alf 2013, 'Technology As Fetish: Marx, Latour, And The Cultural Foundations Of Capitalism', *Theory, Culture & Society* 31, no. 4: 119–140.
Hornborg, Alf 2015, *Zero-Sum Game*, Göteborg: Daidalos.
Hornborg, Alf 2016, *Global Magic: Technologies of Appropriation from Ancient Rome to Wall Street*, New York: Palgrave Macmillan.
Hornborg, Alf 2019a, 'The Money – Energy – Technology Complex and Ecological Marxism: Rethinking the Concept of "Use-value" to Extend Our Understanding of Unequal Exchange, Part 2', *Capitalism Nature Socialism*, 30, no. 4: 71–86.
Hornborg, Alf 2019b, *Nature, Society, and Justice in the Anthropocene: Unraveling the Money-Energy-Technology Complex*, Cambridge: Cambridge University Press.
Huber, Matthew T. 2008, 'Energizing Historical Materialism: Fossil Fuels, Space and the Capitalist Mode of Production', *Geoforum* 40: 105–15.
Huber, Matt 2019, 'Climate and contradiction in Marx's theory of history', *Marxist Sociology Blog: Theory, Research, Politics*, 6 February, available at: https://marxistsociolo gy.org/2019/02/climate-and-contradiction-in-marxs-theory-of-history/.
Huck, Christian 2020, *Digitalschatten: Das Netz und die Dinge*, Hamburg: Textem.
Hudis, Peter 2013. *Marx's Concept of the Alternative to Capitalism*, Chicago: Haymarket Books.
Hugos, Michael 2011, *Essentials of Supply Chain Management*, Hoboken: John Wiley & Sons.
Huws, Ursula 2015, *Labour in the global digital economy the cybertariat comes of age*. [S.l.]: Aakar Books.
Huws, Ursula 2019. *Labour in Contemporary Capitalism: What Next?* London: Palgrave-MacMillan.
Ikeler, Peter 2016, 'Deskilling emotional labour: Evidence from department store retail', *Work, employment and society*, 30, no. 6: 966–983.
Inditex 2019, *Annual Report Inditex 2018*, available at: https://www.inditex.com/docu ments/10279/619384/Inditex+Annual+Report+2018.pdf/25145dd4-74db-2355-03f3-a 3b86bc980a7.
Inditex 2020, 'Inditex Logistics Practices', available at: www.inditex.com/how-we-do -business/our-model/logistics.
Ingold, Tim 2000, *The Perception of the Environment: Essays on Livelihood, Dwelling, and Skill*, New York: Routledge.
Irani, Lily 2015, 'Difference and Dependence among Digital Workers: The Case of Amazon Mechanical Turk', *South Atlantic Quarterly*, 114, no. 1: 225–234.
Isaacson, Walter 2018, *Leonardo da Vinci*. New York: Simon & Schuster.
Jamil, Jonna R. and John Bellamy Foster 2014, 'Beyond the Degradation of Labour:

Braverman and the Structure of the U.S. Working Class', *Employee Responsibilities and Rights Journal*, 26: 219–35.

Jay, Martin 2020, 'Irony and Dialectics: *One-Dimensional Man* and 1968', in Jay, *Splinters in Your Eye: Frankfurt School Provocations*, London: Verso.

Jee, Charlotte 2019, 'Amazon employees are going to strike over the firm's climate policies', MIT *Technology Review*, 10 September.

Johnson, Hans and Marisol Mejia 2014, 'Online Learning and Student Outcomes in California's Community Colleges', Public Policy Institute of California, available at: https://www.ppic.org/content/pubs/report/R_514HJR.pdf.

Jones, Janelle and Ben Zipperer 2018, 'Unfulfilled Promises', *Economic Policy Institute*, 1 February.

Jones, Sarah 2021, 'Amazon Defeats Union Drive in Alabama', *New York Magazine*, 9 April.

Jordan, Tim 2015, *Information Politics: Liberation and Exploitation in the Digital Society*, London: Pluto Press.

Kalleberg, Arne L. 2011, *Good Jobs, Bad Jobs: The Rise of Polarized and Precarious Employment Systems in the United States, 1970s–2000s*, New York: Russell Sage Foundation.

Kaminska, Izabella 2016, 'The robot revolution may be exaggerated, globalisation edition', *The Financial Times*, 24 October, available at: https://ftalphaville.ft.com/2016/10/24/2177921/the-robot-revolution-may-be-being-exaggerated-globalisation-edition/ (accessed 8 June 2021).

Kant, Immanuel, 1914, *Critique of Judgement. Second Edition*, translated by J.H. Bernard, London: MacMillan.

Kaori Gurley, Lauren 2019, 'Organised Amazon Warehouse Workers Just Got Two Fired Co-Workers Rehired', *Vice*, 4 October.

Kapferer, Norbert 1980, 'Commodity, Science and Technology: a Critique of Sohn-Rethel', in *Outlines of a Critique of Technology*, edited by Phil Slater, London: Ink Links.

Kaplan, Marisa n.d., 'Bridging the Chasm: Defining Success Beyond Traditional Academics', *EdSurge*, available at: https://www.edsurge.com/research/guides/bridging-the-chasm-defining-success-beyond-traditional-academics.

Kātz, Barry 1982, *Herbert Marcuse and the Art of Liberation: An Intellectual Biography*, London: Verso.

Kautsky, Karl 1927, *Die materialistische Geschichtsauffassung*, Berlin: Dietz.

Kelly, Kevin 2012, 'Better Than Human: Why Robots Will – And Must – Take Our Jobs', *Wired*, 24 December, available at: https://www.wired.com/2012/12/ff-robots-will-take-our-jobs/.

Kittler, Friedrich 1992, 'There Is No Software', *Stanford Literature Review*, 9, no. 1: 81–90.

Kjøsen, Atle Mikkola 2016, Capital's Media: The Physical Conditions of Circulation, PhD thesis, University of Western Ontario.

Kjøsen, Atle Mikkola 2019, 'Circulation', in *The Bloomsbury Companion to Marx*, edited by Jeff Diamanti, Andrew Pendakis and Imre Szeman, Bloomsbury Academic, 281–288.

Klein, Matthew C. 2015, 'Osbourne's Unorthodox Solution to the U.K. Productivity Puzzle', *The Financial Times*, 3 November, available at: https://ftalphaville.ft.com/2015/11/03/2143742/osbornes-unorthodox-solution-to-the-uk-productivity-puzzle/ (accessed 8 June 2021).

Klein, Matthew C. 2016, 'The Great American Make-Work Programme,' *The Financial Times*, 8 September, available at: https://ftalphaville.ft.com/2016/09/08/2173904/the-great-american-make-work-programme/ (accessed 8 June 2021).

Kolbert, Elizabeth 2016, 'Our Automated Future', *The New Yorker*.

Kosoff, Maya 2018, 'Forget Zuckerberg: Why Trump is Obsessed with Breaking Jeff Bezos', *Vanity Fair*, 28 March.

Krauss, Clifford 2017, 'Texas Oil Fields Rebound from Price Lull, but Jobs are Left Behind,' *The New York Times*, 19 February, available at: https://www.nytimes.com/2017/02/19/business/energy-environment/oil-jobs-technology.html (accessed 8 June 2021).

Krauss, Rosalind 1999, 'Reinventing the Medium', *Critical Inquiry*, 25, no. 2: 289–305.

Kühnen, Heiner and Stefanie Nutzenberger 2019, 'Digitalisierung im Handel – beteiligungsorientiert gestalten', in *Arbeitsschutz und Digitalisierung – Gute Arbeit Reader 2020*, edited by Lothar Schröder, 177–192, Frankfurt: Bund-Verlag.

Laboria Cuboniks 2018, *Xenofeminism: A Politics for Alienation*, London: Verso.

Lafargue, Paul 2011, *The Right to be Lazy*, Oakland: AK Press.

Lange, Elena Louisa 2019, 'Heißhunger Nach Mehrarbeit. Mit Marx Die Digitale Revolution Verstehen', In *Marx Und Die Roboter: Vernetzte Produktion, Künstliche Intelligenz Und Lebendige Arbeit*, edited by Florian Butollo and Sabine Nuss, Berlin: Dietz, 38–54.

Lecher, Colin 2019, 'How Amazon automation tracks and fires warehouse workers for "productivity"', *The Verge*, 25 April.

Lee, C.K.H., K.L. Choy, G.T. Ho and K.M.Y. Law 2013, 'A RFID-based resource allocation system for garment manufacturing', *Expert Systems with Applications*, 40, no. 2: 784–799.

Lee, Francis 2021, 'Enacting the pandemic: analyzing agency, opacity, and power in algorithmic assemblages', *Science & Technology Studies*, 34, no. 1: 65–90.

Lee, Min Kyung, Daniel Kusbit, Evan Metsky, and Laura Dabbish 2015, 'Working with machines: The impact of algorithmic and data-driven management on human workers', in *Proceedings of the 33rd Annual ACM Conference on Human Factors in Computing Systems*, 1603–1612.

Lee, MJ, Lydia DePillis and Gregory Krieg 2019, 'Elizabeth Warren's new plan: Break up Amazon, Google and Facebook', *CNN*, 8 March.

Lenin, Vladimir 1916, *Imperialism, the Highest Stage of Capitalism*, Marxists.org.
Lenin, V.I. 1960, *What is to be Done? Burning Questions of Our Movement*, in *Collected Works*, vol. 5, Moscow: Progress Publishers.
Lenin, V.I. 1964a, 'The Taylor System – Man's Enslavement by the Machine', in *Collected Works*, vol. 25, Moscow: Progress Publishers.
Lenin, V.I. 1964b, *The State and Revolution: The Marxist Theory of the State and the Tasks of the Proletariat in the Revolution*, in *Collected Works*, vol. 25, Moscow: Progress Publishers.
Lenin, V.I. 1966, 'Report on the Work of the Council of People's Commissars. December 1920', in *Collected Works*, vol. 31, Moscow: Progress Publishers.
Leonard, Matt 2019a, 'Amazon makes 7-year deal with French robotics company', *Supply Chain Dive*, 11 January.
Leonard, Matt 2019b, 'Amazon buys robotics company Canvas Technology', *Supply Chain Dive*, 11 April.
Levinson, Marc 2013, '"Hollowing Out" in U.S. Manufacturing: Analysis and Issues for Congress', *Congressional Research Service*, 15 April.
Levy, Steven 2018, 'Inside Amazon's Artificial Intelligence Flywheel', *Wired*, 1 February, available at: https://www.wired.com/story/amazon-artificial-intelligence-flywheel/.
Lichtenstein, Nelson 2012, 'The Return of Merchant Capitalism', *International Labor and Working-Class History*, 81: 8–27.
Licklider, Joseph C.R. 1960, 'Man-Computer Symbiosis', *IRE Transactions on Human Factors in Electronics*, HFE-1, no. 1: 4–11.
Linden, Marcel van der 1997, 'The Historical Limit of Workers' Protest: Moishe Postone, *Krisis* and the "Commodity Logic"', *International Review of Social History* 42.3. 447–58.
Liu, Wendy 2020, *Abolish Silicon Valley: How to Liberate Technology from Capitalism*, Repeater.
Lozano-Nieto, Albert 2010, *RFID design fundamentals and applications*, Boca Raton: CRC press.
Luhmann, Niklas 2000, *Art as a Social System*, Stanford: Stanford University Press.
Lukács, Georg 1966, 'Technology and Social Relations', in *New Left Review* I/39: 27–34.
Lukàcs, Georg 1975, *The Young Hegel: Studies in the Relations Between Dialectics and Economics*, Trans. Rodney Livingstone. Cambridge, MA: MIT Press.
McCowan, Tristan 2017, 'Higher Education, Unbundling, and the End of the University as we know it', *Oxford Review of Education*, 43, no. 6: 733–748.
Mcfarlane, Bruce 2011, 'The Morphing of Academic Practice: Unbundling and the Rise of the Para-Academic', *Higher Education Quarterly* 65, no. 1: 59–73.
MacKenzie, Donald 1984, 'Marx and the Machine', *Technology and Culture* 25.3: 473–502.
McKenzie, Lindsay 2019, 'Chatting with Chatbots', *Inside Higher Ed*, 6 September, avail-

able at: https://www.insidehighered.com/news/2019/09/06/expansion-chatbots-higher-ed.

McKinsey 2012, 'Seven Strategies to Beat the Retail Store Apocalypse', *Forbes*, 22 May.

McKinsey & Company 2022, 'Freelance, Side Hustles, and Gigs: Many More Americans have become Independent Workers', 23 August, available at: https://www.mckinsey.com/featured-insights/sustainable-inclusive-growth/future-of-america/freelance-side-hustles-and-gigs-many-more-americans-have-become-independent-workers.

McMurtry, John 2011a, 'Human Rights versus Corporate Rights: Life Value, the Civil Commons, and Social Justice', *Studies in Social Justice*, 5, no. 1: 11–61.

McMurtry, John 2011b, *Philosophy and World Problems Vol. 1: What is Good? What is Bad: The Value of all Values Across Tims, Places, and Theories*, Oxford: EOLSS Publishers.

McNeill, J.R. 2019, 'Cheap Energy and Ecological Teleconnections of the Industrial Revolution, 1780–1920', *Environmental History* 24: 492–503.

McPherson, Coco 2019, 'Prime Day for a union? Not yet at this Amazon warehouse', *Fast Company*, 20 July.

Maheshwari, Sapna, and Ben Casselman 2020, 'Pretty Catastrophic Month for Retailers, and Now a Race to Survive', *New York Times*, 15 April.

Maley, Terry, ed. 2017, One-Dimensional Man *Fifty Years On: The Struggle Continues*, Halifax: Fernwood Publishing.

Malm, Andreas 2016, *Fossil Capital: The Rise of Steam Power and the Roots of Global Warming*, New York: Verso.

Malm, Andreas 2018a, 'Marx on Steam: From the Optimism of Progress to the Pessimism of Power', *Rethinking Marxism*, 30, no. 2: 166–85.

Malm, Andreas 2018b, 'Long Waves of Fossil Development: Periodizing Energy and Capital', in *Materialism and the Critique of Energy*, edited by Brent Ryan Bellamy and Jeff Diamanti, Chicago: MCM'.

Malm, Andreas 2019, 'Against Hybridism: Why we Need to Distinguish between Nature and Society, Now More than Ever', *Historical Materialism* 27.2: 156–87.

Malm, Andreas 2020, *Corona, Climate, Chronic Emergency: War Communism in the Twenty-First Century*, London: Verso.

Mandel, Michael 2017, 'How E-commerce Creates Jobs and Reduces Income Inequality', *Progressive Policy Institute*.

Manzerolle, Vincent and Atle Mikkola Kjøsen 2012, 'The Communication of Capital: Digital Media and the Logic of Acceleration', *tripleC*, 10, no. 2: 214–229.

Manzerolle, Vincent and Atle Mikkola Kjøsen 2014, 'Dare et Capere: Virtuous Mesh and a Targeting Diagram', in *The Imaginary App*, edited by Paul D. Miller and Svitlana Matviyenko. MIT Press, 143–162.

Marachi, Roxana and Lawrence Quill 2020, 'The Case of Canvas: Longitudinal Datafication through Learning Management Systems', *Teaching in Higher Education*, 25, no. 4: 418–434.

Marachi, Roxana and Robert Carpenter 2020, 'Silicon Valley, Philanthro-Capitalism, and Policy Shifts from Teachers to Tech', in *Strike for the Common Good: Fighting for the Future of Public Education*, edited by Rebecca Kolins Givan and Amy Schrager Lang, Ann Arbor: University of Michigan Press, 217–233.

Marcuse, Herbert 1955, *Eros and Civilization: A Philosophical Inquiry into Freud*, Boston: Beacon Press.

Marcuse, Herbert 1960, *Reason and Revolution: Hegel and the Rise of Social Theory*, 2nd ed., Boston: Beacon Press.

Marcuse, Herbert 1964, *One-Dimensional Man: Studies in Advanced Industrial Society*, Boston: Beacon Press.

Marcuse, Herbert 1965, 'Industrialisation and Capitalism', *New Left Review*, I/30.

Marcuse, Herbert 1968, *Negations: Essays in Critical Theory*, Boston: Beacon Press.

Marcuse, Herbert 1969, *An Essay on Liberation*, Boston: Beacon Press.

Marcuse, Herbert 1972, *Counterrevolution and Revolt*, Boston: Beacon Press.

Marcuse, Herbert 1974, 'Marxism and Feminism', *Women's Studies*, 2: 279–288.

Marcuse, Herbert 1978, 'Some Social Implications of Modern Technology', in *The Essential Frankfurt School Reader*, edited by Andrew Arato and Eike Gebhardt, New York: Continuum.

Marcuse, Herbert 1978, *The Aesthetic Dimension: Toward a Critique of Marxist Aesthetics*, translated by Herbert Marcuse and Erica Sherover, Boston: Beacon Press.

Marcuse, Herbert 1979, 'The Reification of the Proletariat', *Canadian Journal of Political and Social Theory*, 3, no. 1: 20–23.

Marcuse, Herbert 1991, *One-Dimensional Man: Studies in the Ideology of Advanced Industrial Society*, London/New York: Routledge.

Marcuse, Herbert 2001, *The Collected Papers of Herbert Marcuse*, Volume Two: *Towards a Critical Theory of Society*, edited by Douglas Kellner, London: Routledge.

Marcuse, Herbert 2004, 'Some Social Implications of Modern Technology', in *Collected Papers*, volume 1, *Technology, War and Fascism*, edited by Douglas Kellner.

Marcuse, Herbert 2005, *Heideggerian Marxism*, edited by Richard Wolin and John Abromeit, Lincoln: University of Nebraska Press.

Marcuse, Herbert 2007, *The Collected Papers of Herbert Marcuse*, Volume Four: *Art and Liberation*, edited by Douglas Kellner, London: Routledge.

Martinez, Mario 2013, 'A conceptual foundation for understanding innovation in american community colleges', in *Disruptive Innovation and the Community College*, edited by Rufus Glasper and Gerardo de los Santos. League for Innovation in the Community College.

Marx, Karl 1953, *Grundrisse der Kritik der Politischen Ökonomie*, Berlin: Dietz, 1953.

Marx, Karl 1970, *The German Ideology: Part One*, Edited by CJ Arthur, International Publishers.

Marx, Karl 1975a, 'Critical Marginal Notes on the Article: "The King of Prussia and Social

Reform"', in Karl Marx and Friedrich Engels, *Collected Works, Vol. 3*, New York: International Publishers, 189–206.

Marx, Karl 1975b, 'Economic and Philosophical Manuscripts of 1844', in, Karl Marx and Friedrich Engels, *Collected Works, Vol. 3*, New York: International Publishers, 229–348.

Marx, Karl 1976, *The Poverty of Philosophy. Answer to the* Philosophy of Poverty *by M. Proudhon*, in *Karl Marx and Friedrich Engels: Collected Works*, vol. 6, New York: International Publishers.

Marx, Karl 1977, 'Wage Labour and Capital', in Karl Marx and Frederick Engels, *Collected Works*, vol. 9, London: Lawrence & Wishart.

Marx, Karl 1978a, 'Critique of the Gotha Program', in *The Marx-Engels Reader*. Ed. Robert C. Tucker. New York and London: W.W. Norton, 525–41.

Marx, Karl 1978b, *Economic and Philosophical Manuscripts of 1844* and *The German Ideology: Part 1* In *The Marx-Engels Reader*. Ed. Robert C. Tucker. New York and London: W.W. Norton. 66–125 and 146–202.

Marx, Karl 1979, 'To Vera Ivanova Zassulich', in, *The Letters of Karl Marx*, Saul K. Padover, ed, Englewood Cliffs, NJ: Prentice Hall, 335–336.

Marx, Karl 1980, 'Ökonomische Manuskripte und Schriften, 1858–1861', in Karl Marx Friedrich Gesamtausgabe (MEGA), II.2, Berlin: Dietz.

Marx, Karl 1982, 'Ökonomisch-philosophische Manuskripte', in Karl Marx Friedrich Gesamtausgabe (MEGA), I.2, Berlin: Dietz.

Marx, Karl 1986a, *Capital, Volume 1*, Moscow: Progress Publishers.

Marx, Karl 1986b, *Capital, Volume 3*, Moscow: Progress Publishers.

Marx, Karl 1986c, "Economic Manuscripts of 1857–58," in Karl Marx and Friedrich Engels, *Collected Works, Vol. 28*, Moscow: Progress Publishers, 5–541.

Marx, Karl 1987a, 'Outlines of the Critique of Political Economy (Rough Draft of 1857–58) [Second Installment]', in Karl Marx and Frederick Engels, *Collected Works*, vol. 29, New York: International Publishers.

Marx, Karl 1987b, *A Contribution to the Critique of Political Economy*, in Karl Marx and Frederick Engels, *Collected Works*, vol. 29, New York: International Publishers.

Marx, Karl 1989a, 'Economic Manuscripts of 1861–1863', in Karl Marx and Friedrich Engels, *Collected Works, Vol. 31*, Moscow: Progress Publishers, 6–583.

Marx, Karl 1989b [1883], *Das Kapital: Kritik der politischen Ökonomie*, vol. 1, in Karl Marx and Friedrich Engels Gesamtausgabe (MEGA), II.8, Berlin: Dietz.

Marx, Karl 1990a, *Capital: Critique of Political Economy*, vol. 1, trans. Ben Fowkes, London: Penguin.

Marx, Karl 1990b, *Capital: A Critical Analysis of Capitalist Production*, in Karl Marx Friedrich Gesamtausgabe (MEGA), II.9, Berlin: Dietz.

Marx, Karl 1991, *Capital: Critique of Political Economy*, vol. 3, trans. David Fernbach, London: Penguin.

Marx, Karl 1992, *Capital: Critique of Political Economy*, vol. 2, trans. David Fernbach, London: Penguin.

Marx, Karl 1993, *Grundrisse: Foundations of the Critique of Political Economy (Rough Draft)*, translated by Martin Nicolaus, London: Penguin.

Marx, Karl 1996, *Capital: A Critique of Political Economy*, vol. 1, in Karl Marx and Frederick Engels, *Collected Works*, vol. 35, New York: International Publishers.

Marx, Karl 1997, *Das Kapital: Kritik der Politischen Ökonomie Erster Band Hamburg 1872*, Karl Marx Friedrich Engels, *Gesamtausgabe* (MEGA), Zweite Abteilung, Band 6. Berlin: Dietz (Orig. pub. 1872).

Marx, Karl 2000, 'Ökonomisch-philosophiche Manuskripte aus dem Jahre 1844'. In Karl Marx and Friedrich Engels *Werke* (MEW), Ergänzungsband, 1 Teil. 465–588.

Marx, Karl 2013, *Capital: A Critical Analysis of Capitalist Production*. Volume 1 and 2, Hertfordshire: Wordsworth Classics of World Literature.

Marx, Karl and Friedrich Engels 1976a, *The German Ideology. Critique of Modern German Philosophy According to Its Representatives Feuerbach, B. Bauer and Stirner, and of German Socialism According to Its Various Prophets*, in *Karl Marx and Frederick Engels, Collected Works*, vol. 5, New York: International Publishers.

Marx, Karl and Friedrich Engels 1976b, *Manifesto of the Communist Party*, in *Karl Marx and Friedrich Engels: Collected Works*, vol. 6, New York: International Publishers.

Marx, Karl and Friedrich Engels 1978, *Manifesto of the Communist Party*, in *The Marx-Engels Reader*, 2nd ed., edited by Robert Tucker, New York: W.W. Norton & Company.

Mason, Paul 2016, *Postcapitalism: A Guide to Our Future*, n. p.: Penguin.

Mateescu, Alexandra and Aiha Nguyen 2019, 'Explainer: Algorithmic Management in the Workplace', *Data & Society*, available at: https://datasociety.net/wp-content/uploads/2019/02/DS_Algorithmic_Management_Explainer.pdf.

Mau, Søren 2021, *Stummer Zwang: Eine marxistische Analyse der ökonomischen Macht im Kapitalismus*, translated by Christian Frings, Berlin: Dietz.

Mazoué, James 2012, 'The Deconstructed Campus: A Reply to Critics', *Journal of Computing in Higher Education*, 24: 74–95.

Meiksins, Peter and Chris Smith 1996, *Engineering Labour: Technical Workers in Comparative Perspective*. London: Verso.

Melomics n.d., Website, available at: http://melomics.uma.es/.

Menegus, Bryan 2018, 'Amazon's Aggressive Anti-Union Tactics Revealed in Leaked 45-Minute Video', *Gizmodo*, 26 September.

Menegus, Bryan 2019, 'Amazon Workers Strike Outside Eagan, Minnesota, Delivery Station', *Gizmodo*, 8 August.

Meszaros, Istvan 1986. *Marx's Theory of Alienation*, London: Merlin Press.

Meyer, Christian 2019, 'Vorwärts und nicht vergessen. Ein Blick in materialistische Technologiediskussionen', In *Marx Und Die Roboter. Vernetzte Produktion, Künstliche Intelligenz und Lebendige Arbeit*, edited by Florian Butollo and Sabine Nuss, Berlin: Dietz, 113–129.

Meyer, Steve 2002, '"An Economic 'Frankenstein'": UAW Workers' Responses to Automation at the Ford Brook Park Plant in the 1950s', *Michigan Historical Review*, 28, no. 1: 63–89.

Miller, Arthur I. 2019, *The Artist in the Machine. The World of AI-Powered Creativity*, Cambridge/MA and London: MIT Press.

Moody, Kim 2017, *On new terrain: how capital is reshaping the battleground of class war*, Chicago: Haymarket Books.

Moody, Kim 2018, 'High Tech, Low Growth: Robots and the Future of Work', *Historical Materialism*, 26, no. 4: 3–34.

Moore, Jason W. 2015, *Capitalism in the Web of Life: Ecology and the Accumulation of Capital*, London: Verso.

Moraitis, Alexis and Jack Copley 2021, 'Capitalism in Decline: Automation in a Stagnant Economy', *Roar Magazine*, 14 May, available at: https://roarmag.org/essays/automation-benavav-smith-review/.

Morgan, Steve 2019, 'Global Cybersecurity Spending Predicted to Exceed $1 Trillion From 2017–2020', *Cybercrime Magazine*, 10 June, available at: https://cybersecurityventures.com/cybersecurity-market-report/. (last accessed 12 May 2021).

Morris-Suzuki, Tessa 1984, 'Robots and Capitalism', *New Left Review*, I/147.

Mouffe, Chantal 2020, 'Humans are not Resources: Coronavirus Shows Why We Must Democratise Work', *The Guardian*, 15 May, available at: https://www.theguardian.com/commentisfree/2020/may/15/humans-resources-coronavirus-democratise-work-health-lives-market (accessed 8 June 2020).

Muehlhauser, Luke 2013, 'What is AGI?', *Machine Intelligence Research Institute*, 11 August, available at: https://intelligence.org/2013/08/11/what-is-agi/.

Mueller, Gavin 2021, *Breaking Things at Work: The Luddites Were Right About Why You Hate Your Job*. London/New York: Verso.

Murray, Patrick 2017a, 'The Social and Material Transformation of Production by Capital: Formal and Real Subsumption in *Capital*, Volume 1', in *The Mismeasure of Wealth: Essays on Marx and Social Form*, Chicago: Haymarket.

Murray, Patrick 2017b, 'The Place of "The Results of the Immediate Production Process" in *Capital*', in *The Mismeasure of Wealth: Essays on Marx and Social Form*, Chicago: Haymarket.

Nakano, Mikihisa 2020, *Supply Chain Management*, Singapore: Springer.

National Center for Education Statistics 2011, 'Learning at a Distance: Undergraduate Enrollment in Distance Education Courses and Degree Programs', available at: https://nces.ed.gov/pubsearch/pubsinfo.asp?pubid=2012154.

National Center for Education Statistics 2019, 'Number and Percentage of Students Enrolled in Degree-Granting Postsecondary Institutions, by Distance Education Participation, and Level of Enrollment and Control of Institution, Fall 2017', available at: https://nces.ed.gov/fastfacts/display.asp?id=80.

Navlakha, Meera 2022, 'Amazon agrees to allow warehouse employees phone access permanently', *Mashable*, 28 April.

Nayak, Rajkishore 2019, *Radio Frequency Identification (RFID): Technology and Application in Garment Manufacturing and Supply Chain*, Boca Raton: CRC Press.

Neary, Lynn 2012, 'Publishers And Booksellers See A 'Predatory' Amazon', *NPR*, 23 January.

Negri, Antonio 1992, *Marx Beyond Marx: Lessons on the* Grundrisse, New York/London: Autonomedia/Pluto.

Negri, Antonio 2014, 'Some Reflections on the #Accelerate Manifesto', in *#Accelerate#: The Accelerationist Reader*, edited by Robin MacKay and Armen Avanessian, Falmouth: Urbanomic Media.

Ness, Immanuel 2015, *Southern Insurgency: The Coming of the Global Working Class*, London: Pluto.

Nicolaus, Martin 1968, 'The Unknown Marx', *New Left Review* 48: 41–61.

Nikola, Gligorevic, Karol Schulz and Iryna Ivanochko 2021, 'Use of RFID Technology in Retail Supply Chain', in *Developments in Information & Knowledge Management for Business Applications. Volume 1*, edited by Natalia Kryvinska and Michal Greguš, Cham: Springer, 555–87.

Noble, David F. 1978, 'Social Choice in Machine Design: The Case of Automatically Controlled Machine Tools, and a Challenge for Labor', *Politics & Society*, Vol. 8, Issue 3–4: 313–47. Best available copy in E. Bernard (ed.) Technological Change and Skills Development, Deakin University Press, 1991, 103–35.

Noble, David F. 1979, 'Social Choice in Machine Design: The Case of the Automatically Controlled Machine Tools', in *Case Studies on the Labor Process*, edited by Andrew Zimbalist, NY: Monthly Review: 18–50.

Noble, David F. 1993, *Progress without People: In Defense of Luddism*, Chicago: Charles H. Kerr.

Noble, David F. 2003, *Digital Diploma Mills: The Automation of Higher Education*, NY: Monthly Review Press.

Noble, David 2011, *Forces of Production: A Social History of Industrial Automation*, New Brunswick, NJ: Transaction Publishers.

Nolan, Mary 1994, *Visions of Modernity: American Business and the Modernization of Germany*, Oxford: Oxford University Press.

Noll, Michael 1967, 'The Computer as a Creative Medium', *IEEE Spectrum*, 4: 89–95.

Noonan, Jeff 2004, 'Between Egoism and Altruism: Outlines for a Materialist Conception of the Good', in *The Ethics of Altruism*, edited by Jonathan Seglow, London: Frank Cass, 68–86.

Noonan, Jeff 2012, *Materialist Ethics and Life-Value*, Montreal: McGill-Queen's University Press.

Noonan, Jeff 2017, 'Technology and the Place of Labour in a Future Socialist Society', in

Advances in Sociology Research, edited by Jared A. Jaworski, New York: Nova Science Publishers, 1–20.

Noonan, Jeff 2018, *Embodiment and the Meaning of Life*, Montreal: McGill-Queens University Press.

Noonan, Jeff 2020, 'All Work and No Play? The Role of Non-Alienated Labour in Marcuse's Emancipatory Vision', *Constellations*, Vol. 27, No. 2, (June): 300–312.

Noys, Benjamin 2014, *Malign Velocities: Accelerationism and Capitalism*, Winchester: Zero Books.

Nyberg, Mikael 2020, *The Automation of Capital – Human Robots & Systematic Stupidity* (Kapitalets automatik – Mänskliga robotar & systematisk dumhet), Stockholm: Verbal Förlag.

O'Connor, Brian 2018, *Idleness: A Philosophical Essay*, Princeton, NJ: Princeton University Press.

O'Connor, James, 1988, 'Capitalism, Nature, Socialism: A Theoretical Introduction'. *Capitalism Nature Socialism* 1, no. 1: 11–38.

O'Connor, Sarah 2017 "America's 'jobs for the boys' is just half the employment story," *The Financial Times*, 17 February, available at: https://www.ft.com/content/25a897bc-ec9a-11e6-ba01-119a44939bb6 (accessed 8 June 2021).

Online Learning Consortium 2020, 'OLC's historical timeline', available at: https://onlinelearningconsortium.org/about/pioneering-higher-educations-digital-future-timeline/.

Ovetz, Robert 1996, 'The Global Entrepreneurialization of the Universities', *Capital & Class*, 58: 113–151.

Ovetz, Robert 2015a, 'Migrant Mindworkers and the New Division of Academic Labour', *Working USA*, 18, no. 3: 331–347.

Ovetz, Robert 2015b, 'When Hephaestus Fell to Earth: Harry Braverman and the Division of Labour in Academia', *Labour Studies Journal*, 40, no. 3: 243–261.

Ovetz, Robert 2015c, 'The New Jim Crow in Higher Education: On-line Education and the Community Colleges', Pacific Sociological Association, paper presentation.

Ovetz, Robert 2017, 'Click to Save and Return to Course: On line Education, Adjunctification, and the Disciplining of Academic Labour in the Social Factory', *Work, Organisation, Labour and Globalisation*, 11, 1, Spring: 48–70.

Ovetz, Robert 2020a, 'The University is a Business: Interview with a Faculty Member on Strike in the UK', *Organizing Work*, March, available at: https://organizing.work/2020/03/the-university-is-a-business-interview-with-a-faculty-member-on-strike-in-the-uk/.

Ovetz, Robert 2020b, *Workers' Inquiry and Global Class Struggle: Strategies, Tactics, Objectives*, London: Pluto.

Ovetz, Robert 2021, 'The Algorithmic University: On-Line Education, Learning Management Systems, and the Struggle Over Academic Labor', *Critical Sociology*, 47: 7–8.

Palladino, Valentina 2019, 'Amazon to employees: Quit your job, we'll help you start a delivery business', *ArsTechnica*, 13 May.

Panzieri, Raniero 1980, 'The Capitalist Use of Machinery: Marx Versus the "Objectivists"', in *Outlines of a Critique of Technology*, edited by Phil Slater, London: Ink Links.

Patel, Raj and Jason W. Moore 2017, *A History of the World in Seven Cheap Things. A Guide to Capitalism, Nature, and the Future of the Planet*, Oakland, California: University of California Press.

Pathak, Ritesh 2021, 'How Apple Uses AI and Big Data', *Analytics Steps*, 21 January, available at: https://www.analyticssteps.com/blogs/how-apple-uses-ai-and-big-data.

Paulsen, Roland 2014, *Empty Labour: Idleness and Workplace Resistance*, Cambridge: Cambridge University Press.

Pellow, David Naguib and Lisa Sun-Hee Park 2002, *Silicon Valley of Dreams: Environmental Injustice, Immigrant Workers, and the High-Tech Global Economy*, New York and London: New York University Press.

Pfeiffer, Sabine 2021, *Digitalisierung als Produktivkraft*, Bielefeld: transcript Verlag

Phillips, Leigh 2015, *Austerity Ecology & The Collapse-Porn Addicts: A Defence of Growth, Progress, Industry and Stuff*, Winchester: Zero Books, ebook.

Pinker, Steven 2018, *Enlightenment Now*, New York: Penguin.

Pirani, Simon 2018, *Burning Up: A Global History of Fossil Fuel Consumption*, London: Pluto Press.

Pitts, Frederick Harry 2018, 'A Crisis of Measurability? Critiquing Post-Operaismo on Labour, Value and the Basic Income', *Capital & Class* 42:1, no. 3–21.

Pitts, Frederick Harry 2021, *Value*. Polity.

Pitts, Frederick Harry and Ana Dinerstein 2017, 'Corbynism's Conveyor Belt Of Ideas: Postcapitalism And The Politics Of Social Reproduction', *Capital & Class* 41, no. 3: 423–434.

Porter, Jon 2019, 'Amazon to spend $700 million retraining a third of its US workforce by 2025', *TheVerge*, 11 July.

Postone, Moishe 1978, 'Necessity, Labor, and Time: A Reinterpretation of the Marxian Critique of Capitalism', *Social Research*, 45, no. 4: 739–788.

Postone, Moishe 1993, *Time, Labor, and Social Domination: A Reinterpretation of Marx's Critical Theory*, Cambridge: Cambridge University Press.

Prendergast, Alan 2017, 'Did a Community College Plan to Pass More Students Fail its Teachers?', *Westword*, 1 August, available at: http://www.westword.com/news/community-college-of-aurora-may-pass-more-students-but-did-it-fail-teachers-9317325.

Pye, David 1979, *The Nature and Aesthetics of Design*, London: Barrie and Jenkins.

PYMTS 2019, 'Walmart-Amazon Battle In Mexico Forces Suppliers To Choose Side', 8 April.

Rabinbach, Anson 1990, *The Human Motor: Energy, Fatigue, and the Rise of Modernity*. New York: Basic Books.
Ramírez, J. Jesse 2012, 'Marcuse Among the Technocrats: America, Automation, and Postcapitalist Utopias, 1900–1941', *Amerikastudien/American Studies*, 57, no. 1: 31–50.
Ramírez, J. Jesse 2017, 'Marx vs. the Robots', *Amerikastudien/American Studies*, 62, no. 4: 619–32.
Ramtin, Ramin 1991, *Capitalism and Automation: Revolution in Technology and Capitalist Breakdown*, London: Pluto Press.
Ravenelle, Alexandrea J. 2019, *Hustle and Gig: Struggling and Surviving in the Sharing Economy*, Berkeley: Univ of California Press.
Rebentisch, Julia 2013, *Theorien der Gegenwartskunst. Zur Einführung*. Hamburg: Junius.
Reckwitz, Andreas 2012, *Die Erfindung der Kreativität. Zum Prozess gesellschaftlicher Ästhetisierung*, Berlin: Suhrkamp.
Reference for Business 2002, 'Amazon.com', available at: https://www.referenceforbusiness.com/history2/35/Amazon-com-Inc.html.
Reinfelder, Monika 1980, 'Introduction: Breaking the Spell of Technicism', in *Outlines of a Critique of Technology*, edited by Phil Slater, London: Ink Links.
Reitz, Charles 2014, *Celebrating Herbert Marcuse's* One-Dimensional Man: *Deprovincialization and the Recovery of Philosophy*, Limited Edition Publication.
Resnikoff, Jason 2021, *Labor's End: How the Promise of Automation Degraded Work*, Urbana: University of Illinois Press.
Rhoades, Gary 2013, 'Disruptive Innovations for Adjunct Faculty: Common Sense for the Common Good', *Thought & Action*, Fall, available at: https://www.nea.org/assets/docs/HE/k-pg71_TA2013Rhoades_SF.pdf.
Rhoades, Gary and Sheila Slaughter 2004, 'Academic Capitalism in the New Economy: Challenges and Choices', *American Academic* 1, June: 37–60.
Rifkin, Jeremy 1995, *The End of Work: The Decline of the Global Labor Force and the Dawn of the Post-Market Era*, New York: Putnam.
Roberts, William Clare 2017, *Marx's Inferno: The Political Theory of Capital*, Princeton: Princeton University Press.
Robinson, Joan 1962, *Economic Philosophy*. London: C.A. Watts.
Rockwell, Russell 2013, 'Marcuse's Hegelian Marxism, Marx's *Grundrisse*, Hegel's Dialectic', *Radical Philosophy Review* 16, 1: 289–306.
Rosdolsky, Roman 1957, 'Der esoterische und der exoterische Marx: Zur kritischen Würdigung der Marxschen Lohntheorie I', *Arbeit und Wirtschaft* 11.11: 348–51.
Rosenblat, Alex 2018, *Uberland: How Algorithms Are Rewriting the Rules of Work*, Berkeley: Univ of California Press.
Rowthorn, Robert and Ramana Ramaswamy 1994, 'Deindustrialization – Its Causes and Implications', IMF Working Paper, April.

Rubin, I.I. 2008, *Essays on Marx's Theory of Value*, Delhi: Aakar Books.

Saad-Filho, Alfredo 2020, *Value and Crisis: Essays On Labour, Money And Contemporary Capitalism*. Chicago: Haymarket Books.

Sainato, Michael 2022, 'US judge orders Amazon to 'cease and desist' anti-union retaliation', *Guardian*, 28 November.

Saito, Kohei 2022, *Marx in the Anthropocene: Towards the Idea of Degrowth Communism*, Cambridge: Cambridge University Press.

Samuelson, Paul 1948, 'Consumption theory in terms of revealed preference', *Economica* 15, no. 60:243–253.

San Jose State University Spring 2015 and Fall 2019, 'Canvas User Satisfaction', available at: https://www.sjsu.edu/ecampus/teaching-tools/canvas/canvas-statistics/canvas-user-satisfaction/.

Sandeen, Cathy 2014, 'Unbundling Versus Designing Faculty Roles', Presidential Innovation Lab White Paper Series, American Council on Education/Center for Education Attainment and Innovation, available at: https://www.acenet.edu/Documents/Signals-and-Shifts-in-the-Postsecondary-Landscape.pdf.

Sayers, Sean 2007, The Concept of Labor: Marx and His Critics, *Science and Society*, 71:4.

Sayers, Sean 1987. The Need for Work, *Radical Philosophy*, No. 46, (Summer, 1987): 17–26.

Schmidt, Alfred 2014, *The Concept of Nature in Marx*, London: Verso.

Schmidt, Dorothea 2019, '"Industrielle Revolution und Mechainisierung bei Marx: Ein Faktencheck"', in *Marx und Die Roboter: Vernetzte Produktion, Künstliche Intelligenz Und Lebendige Arbeit*, edited by Florian Butollo and Sabine Nuss, Berlin: Dietz, 55–73.

Schröter, Jens 2019, 'Digitale Medientechnologien und das Verschwinden der Arbeit', in *Mensch und Maschine: Freund oder Feind? Mensch und Technologie im digitalen Zeitalter*, edited by Thomas Bächle and Caja Thimm, Wiesbaden: Springer.

Schröter, Jens 2020, 'Imaginary Economies. The Case of the 3D Printer', *Review of Evolutionary Political Economy*, 1, no. 3: 357–370.

Schwartz, Andrew 2018, 'The Realities of Economic Development Subsidies', *Centre for American Progress*, 1 November.

Seccatore, Jacopo, et al. 2014, 'An estimation of the artisanal small-scale production of gold in the world', *Science of the Total Environment*, 496: 662–7.

Seidler, Paul, Paul Kolling and Max Hampshire 2016, 'terra0: Can an augmented forest own and utilise itself?' available at: https://www.terra0.org/assets/pdf/terra0_white_paper_2016.pdf.

Seifert, Dirk 2003, *Collaborative planning, forecasting, and replenishment: How to create a supply chain advantage*, New York: AMACOM.

Selyukh, Alina 2018, 'What Americans Told Us About Online Shopping Says A Lot About Amazon', *NPR*, 6 June.

Semuels, Alana 2018, 'I delivered packages for Amazon and it was a nightmare', *Atlantic*, 25 June.
Shaikh, Anwar 2016, *Capitalism*, New York: Oxford University Press.
Shapiro, Carl, and Hal R. Varian 1999, *Information Rules*, Cambridge: Harvard Business School Press.
Sigaki, Higor Y.D., Matjaž Perc and Haroldo V. Ribeiro 2018, 'History of Art Paintings through the Lens of Entropy and Complexity', *PNAS*, 115, no. 37: E8585–594, available at: www.pnas.org/cgi/doi/10.1073/pnas.1800083115.
Silver, Beverly 2003, *Forces of Labor: Workers' Movements and Globalization since 1870*, Cambridge: Cambridge University Press.
Simmel, Georg 1904, 'Fashion', *International Quarterly*, 10: 130–155.
Slaughter, Sheila and Gary Rhoades 2004, *Academic Capitalism and the New Economy*, Baltimore: The Johns Hopkins University Press.
Slaughter, Sheila and Larry Leslie 1999, *Academic Capitalism: Politics, Policies, and the Entrepreneurial University*, Baltimore: Johns Hopkins University Press.
Smith, Adam 1981, *An Inquiry into the Nature and Causes of the Wealth of Nations*, vol. 1, Indianapolis: Liberty Classics.
Smith, Jason E. 2020. *Smart Machines and Service Work: Automation in an Age of Stagnation*. London: Reaktion Books.
Smith, Molly 2019, 'The World's Biggest Corporate Borrowers Are Dominated by Tech', *Bloomberg*, 9 April, available at: https://www.bloomberg.com/news/articles/2019-04-09/the-world-s-biggest-corporate-borrowers-are-dominated-by-tech.
Smith, Neil 1984, *Uneven Development: Nature, Capital, and the Production of Space*, Oxford: Blackwell.
Smith, Ted and Phil Woodward (eds.) 1992, *The Legacy of High-Tech Development: The Toxic Lifecycle of Computer Manufacturing*, Silicon Valley Toxics Coalition, available at: https://icrt.co/wp-content/uploads/2019/07/1992_01_01_The_Legacy_of_H-T_development_toxic_lifecycle_of_computer_manufacturing_Ted_Smith.pdf.
Smith, Tony 2009, 'The Chapters on Machinery in the 1861–63 Manuscripts', in *Rereading Marx: New Perspectives After the Critical Edition*, edited by Riccardo Bellofiore and Roberto Fineschi, London: Palgrave Macmillan, 112–127.
Smith, Tony 2013, 'The "General Intellect" in the *Grundrisse* and Beyond'. In *In Marx's Laboratory: Critical Interpretations of the* Grundrisse. Peter Thomas, Guido Starosta, and Riccardo Bellafiore, Eds. Leiden and Boston: Brill. 213–231.
Smith, Vernon 2008, 'The Unbundling and Rebundling of the Faculty Role in E-Learning Community College Courses', Ph.D. dissertation, University of Arizona.
Sohn-Rethel, Alfred 1978, *Intellectual and Manual Labour: A Critique of Epistemology*, London: Macmillan.
Sohn-Rethel, Alfred 2018a, *Geistige und körperliche Arbeit*, in *Geistige und körperliche Arbeit: Theoretische Schriften 1947–1990*, vol. 1, edited by Carl Freytag, Oliver Schlaudt, and Françoise Willmann, Freiburg: ça ira.

Sohn-Rethel, Alfred 2018b, 'Technische Intelligenz zwischen Kapitalismus und Sozialismus', in *Geistige und körperliche Arbeit: Theoretische Schriften 1947–1990*, vol. 1, edited by Carl Freytag, Oliver Schlaudt, and Françoise Willmann, Freiburg: ça ira.

Sohn-Rethel, Alfred 2021, *Intellectual and Manual Labour: A Critique of Epistemology*, Leiden: Brill.

Solow, Robert 1956, 'A Contribution to the Theory of Economic Growth', *The Quarterly Journal of Economics* 70, no. 1: 65–94

Solow, Robert 1987, 'We'd Better Watch Out', *New York Times Book Review*, 12 July, www.standupeconomist.com/pdf/misc/solow-computer-productivity.pdf.

Soper, Kate 2020, *Post-Growth Living: For an Alternative Hedonism*, London: Verso.

Spencer, David and Slater, Gary 2020, 'No automation please, we're British: technology and the prospects for work', *Cambridge Journal of Regions, Economy and Society* 13, 1: 117–134.

Sprague, Shawn 2017, "Below trend: U.S. Productivity slowdown since the Great Recession," *Bureau of Labor Statistics* 6, no. 2, available at: https://www.bls.gov/opub/btn/volume-6/below-trend-the-us-productivity-slowdown-since-the-great-recession.htm (accessed 8 June 2021).

Srnicek, Nick 2017, *Platform Capitalism*. Cambridge: Polity.

Srnicek, Nick and Alex Williams 2015, *Inventing the Future: Postcapitalism and a World Without Work*, London: Verso.

Steinberg, Ronnie J. 1990, 'Social Construction of Skill: Gender, Power, and Comparable Worth', *Work and Occupations* 17, no. 4: 449–82.

Stross, Charles 2017, 'Dude, you broke the Future! Talk given at 34C3', *YouTube*, available at: https://www.youtube.com/watch?v=RmIgJ64z6Y4&t=3073s.

Svensson, Mikael 2019, 'How Class Makes a Difference. The Importance of Class for Racist and Negative Distinctive Practises', Ph.D. diss. Uppsala University.

Szabo, Nick 1997, 'The Idea of Smart Contracts', *Satoshi Nakamoto Institute*, available at: https://nakamotoinstitute.org/the-idea-of-smart-contracts/.

Taibbi, Matt 2018, 'Amazon's Long Game Is Clearer Than Ever', *Rolling Stone*, 14 November.

Taylor, Astra 2018, 'The Automation Charade', *Logic* 5, 1 August.

Taylor, Frederick Winslow 2004, *Scientific Management*, London: Routledge.

terrao 2018, 'Premna Daemon – An Introduction via a History of Autonomy in the Cryptosphere', *Medium*, 16 November, available at: https://terrao.medium.com/premna-daemon-an-introduction-via-a-history-of-autonomy-in-the-cryptosphere-3ce e15e92fe2.

The Analogue University 2019, 'Correlation in the Data University: Understanding and Challenging Targets-Based Performance-Management in Higher Education', *ACME: An International Journal of Critical Geographies*, 18, no. 6: 1184–1206.

The Economist 2015, 'The On-Demand Economy: Workers on Tap. The Rise of the On-

Demand Economy Poses Difficult Questions for Workers, Companies and Politicians', *The Economist*, 3 January, available at: http://www.economist.com/news/leaders/21637393-rise-demand-economy-poses-difficult-questions-workers-companies-and.

The Economist 2016, "Too Much of a Good Thing," *The Economist*, 26 March, available at: http://www.economist.com/news/briefing/21695385-profits-are-too-high-america-needs-giant-dose-competition-too-much-good-thing (accessed 8 June 2021).

Thompson, E.P. 1967, 'Time, Work-Discipline, and Industrial Capitalism'. *Past & Present* 38: 56–97.

Thornton, Jerry 2013, 'Community Colleges: Ready to Disrupt Again!', in *Disruptive Innovation and the Community College*, edited by Rufus Glasper and Gerardo de los Santos, League for Innovation in the Community College, 2013.

TIE Germany 2020, 'Who we are', available at: http://tie-germany.org/who_we_are/index.html.

Tomšič, Samo 2015, *The Capitalist Unconscious*. London: Verso.

Tronti, Mario 2008, 'Italy', in *Karl Marx's* Grundrisse: Foundations of the Critique of Political Economy *150 Years Later*, edited by Marcello Musto, New York: Routledge.

Troutt, William 1979, 'Unbundling Instruction: Opportunity for Community Colleges', *Peabody Journal of Education*, 56, no. 4: 253–9.

Tsogas, George 2012, 'The Commodity Form in Cognitive Capitalism', *Culture and Organization* 18, no. 5: 377–395.

Tung, Irene and Deborah Berkowitz 2020, 'Amazon's Disposable Workers: High Injury and Turnover Rates at Fulfilment Centres in California', *National Employment Law Project*, available at: https://www.nelp.org/publication/amazons-disposable-workers-high-injury-turnover-rates-fulfilment-centres-california/.

Uhl, Karsten 2019, 'Eine Lange Geschichte Der "Menschenleeren Fabrik". Automatisierungsvisionen Und Technologischer Wandel im 20ten Jahrhundert', In *Marx und die Roboter. Vernetzte Produktion, Künstliche Intelligenz und Lebendige Arbeit*, edited by Florian Butollo and Sabine Nuss, Berlin: Dietz, 74–90.

Ullrich, Peter 2019, 'In itself but not yet for itself – Organising the New Academic Precariat', In *The Radical Left in Europe – Rediscovering Hope*, edited by W. Baier, E. Canepa and H. Golemis. transform! europe, London: Merlin Press, 155–168.

UNI Global Union 2018, 'UNI shows Amazon the red card', 19 June, available at: https://www.uniglobalunion.org/news/uni-shows-amazon-red-card.

van Dijck, Jose 2014, 'Datafication, Dataism and Dataveillance: Big Data between Scientific Paradigm and Ideology', *Surveillance & Society*, 12, no. 2: 197–208.

van Dijk, Paul 2000, *Anthropology in the Age of Technology: The Philosophical Contribution of Günther Anders*, translated by Frans Kooymans, Atlanta: Rodopi.

van Parijs, Philippe, and Yannick Vanderborght 2017, *Basic Income: A Radical Proposal for a Free Society and a Sane Economy*, Cambridge: Harvard University Press.

Vertesi, Janet, Adam Goldstein, Diana Enriquez, Larry Liu and Katherine Miller 2020, 'Pre-Automation: Insourcing and Automating the Gig Economy', *Sociologica*, 14, no. 3: 167–193.

Virilio, Paul 1986, *Speed and Politics: an essay on dromology*, New York: Semiotext(e).

Virno, Paolo 2003, *Grammar of the Multitude*, Los Angeles: Semiotext(e).

von Greif, Bodo 1992, 'Produktion und Destruktion: Günther Anders' Theorie der industriellen Arbeit', in *Günther Anders kontrovers*, edited by Konrad Paul Liessmann, München: C.H. Beck, 181–99.

Wajcman, Judy 2015, *Pressed for Time: The Acceleration of Life in Digital Capitalism*, Chicago: The University of Chicago Press, 2015.

Wajcman, Judy 2019, 'Automatisierung: Ist es diesmal wirklich anders?', In *Marx und die Roboter. Vernetzte Produktion, Künstliche Intelligenz Und Lebendige Arbeit*, edited by Florian Butollo and Sabine Nuss, Berlin: Dietz, 22–35.

Walker, Richard 1985, "Is there a service economy? The changing capitalist division of labor," *Science & Society* 49, no. 1: 42–83.

Wang, William 1975, 'The Unbundling of Higher Education', *Duke L.J.* 53: 53–90, available at https://repository.uchastings.edu/cgi/viewcontent.cgi?article=1762&context=faculty_scholarship.

Wark, McKenzie 2019, *Capital is Dead: Is this something worse?* London: Verso.

Wasielewski, Amanda 2023. Computational Formalism: Art History and Machine Learning. Cambridge: The MIT Press.

Weber, Max 2002, *The Protestant ethic and the "spirit" of capitalism and other writings*, London: Penguin.

Weeks, Kathi. 2011. *The Problem with Work: Feminism, Marxism, Anti-Work Politics and Post-Work Imaginaries*, Durham, NC: Duke University Press.

Weil, David 2014, *The Fissured Workplace*, Cambridge: Harvard University Press.

Weise, Karen 2018, 'Amazon to Raise Minimum Wage to $15 for All U.S. Workers', *The New York Times*, 2 October.

Wellner, Galit 2016, *A Postphenomenological Inquiry of Cell Phones: Genealogies, Meanings and Becoming.* Lanham, MD: Lexington.

Wendling, Amy 2009, *Karl Marx on Technology and Alienation*, Springer.

Wendling, Amy 2013, 'Second Nature: Gender in Marx's *Grundrissse*'. In *In Marx's Laboratory: Critical Interpretations of the* Grundrisse. Peter Thomas, Guido Starosta, and Riccardo Bellafiore, Eds. Leiden and Boston: Brill, 213–231.

Wharton Business School 2019, 'Will Amazon's Plan to 'Upskill' Its Employees Pay Off?', 22 July.

Whittaker, Zack 2020 'Facebook to pay $52 million to content moderators suffering from PTSD', *Tech Crunch*, 12 May, available at: https://techcrunch.com/2020/05/12/facebook-moderators-ptsd-settlement/. (last accessed 1 June 2020)

Williams, Alex and Nick Srnicek 2014, '#Accelerate: Manifesto for an Accelerationist

Politics', in *#Accelerate#: The Accelerationist Reader*, edited by Robin MacKay and Armen Avanessian, Falmouth: Urbanomic Media.

Williamson, Ben, Sian Bayne and Suellen Shay 2020, 'The Datafication of Teaching in Higher Education: Critical Issues and Perspectives', *Teaching in Higher Education*, 25, no. 4: 351–365.

Wilmers, Nathan 2018, 'Wage Stagnation and Buyer Power: How Buyer-Supplier Relations Affect U.S. Workers' Wages, 1978 to 2014', *American Sociological Review*, 83, no. 2: 213–242.

Wingfield, Nick 2018, 'Amazon Chooses 20 Finalists for Second Headquarters', *The New York Times*, 18 January.

Woessner, Martin 2011, *Heidegger in America*, Cambridge: Cambridge University Press.

Woirol, Gregory 1996, *The Technological Unemployment and Structural Unemployment Databases*, Westport, CT: Greenwood.

Wolin, Richard 2001, *Heidegger's Children: Hannah Arendt, Karl Löwith, Hans Jonas, and Herbert Marcuse*, Princeton: Princeton University Press.

Wood, Ellen Meiksins 2016, *Democracy against Capitalism: Renewing Historical Materialism*. London/New York: Verso.

Woodcock, Jamie 2016, *Working the Phones: Control and Resistance in Call Centres*, London: Pluto Press.

Woodcock, Jamie and Sal Englert 2018, 'Looking Back in Anger: The UCU Strikes', *Notes From Below*, 30 August, republished from *The Worker and The Union* 3, available at: https://notesfrombelow.org/article/looking-back-anger-ucu-strikes.

Wright, Erik Olin 1997, *Class Counts: Comparative studies in class analysis*, Cambridge: Cambridge University Press.

Wright, Erik Olin 2015, 'How to be an Anticapitalist Today', *Jacobin Magazine*, 2 December.

Wright, Steve 2017, *Storming Heaven: Class Composition and Struggle in Italian Autonomist Marxism*, London: Pluto.

Wrigley, Neil, and Michelle Lowe 1996, *Retailing, consumption and capital: towards the new retail geography*, Harlow: Longman.

Wrigley, Neil, Michelle Lowe and Andrew Currah 2002, 'Retailing and E-Tailing' *Urban Geography* 23, no. 2: 180–197.

Wu, Jasmine and Katie Schoolov 2019, 'Amazon workers in Minnesota walk out as Prime Day orders roll in', *CNBC*, 16 July.

Yaffe-Bellany, David, and Corkery, Michael 2020. "Dumped Milk, Smashed Eggs, Plowed Vegetables: Wasted Food of the Pandemic", *The New York Times*, 4 April, available at: https://www.nytimes.com/2020/04/11/business/coronavirus-destroying-food.html (Accessed 1 June 2020).

Yates, Michael D. 1999, 'Braverman and the Class Struggle', *Monthly Review*, 1 January.

Yip, Amy and Minyi Huang 2016, 'Strategic values of technology-driven innovation in inventory management: a case study of Zara's RFID implementation', *International Journal of Inventory Research*, 3, no. 4: 318–336.

York, Richard 2012, 'Do Alternative Energy Sources Displace Fossil Fuels?', *Nature Climate Change* 2: 441–3.

Young, Jeffrey 2018, 'Do Online Courses Really Save Money? A New Study Explores ROI for Colleges and Students', *EdSurge*, 12 April, available at: https://www.edsurge.com/news/2018-04-12-do-online-courses-really-save-money-a-new-study-explores-roi-for-colleges-and-students.

Ziobro, Paul 2019, 'FedEx to End U.S. Express Business with Amazon', *The Wall Street Journal*, 7 June.

Zuboff, Shoshana 1988, *In the Age of the Smart Machine*, New York: Basic Books.

Index

accumulation
  Capital 2, 20, 42–43, 47, 77, 104, 114–15, 119, 136, 145, 148, 155, 222, 272, 294, 298
  Crisis of 94, 104
  By dispossession 183
  Over-accumulation 115
  Primitive 183
Adorno, Theodor W. 20n88, 71, 286–87
Algorithms 52–53, 88, 90, 111, 119–20, 124–25, 204, 206, 256, 264, 267, 269–71, 273, 274, 278–80, 286, 289, 294, 296, 302–04, 306
Althusser, Louis 4, 162, 182n14
Amazon 51, 53, 194, 203–24, 267, 293–96, 303, 308
  Mechanical Turk 124–125
Anders, Günther 48, 57–70, 73–74, 78–82
Aristotle 139, 159, 162, 173
Art 3, 51–53, 71, 80, 111, 124, 172, 274–87
Arthur, Christopher 31–32
Artificial intelligence (AI) 1, 173–174, 176, 49, 53, 90, 105, 123–24, 129, 135, 148, 151, 158, 165, 167, 186, 198, 216, 248, 258–59, 261, 265, 269, 271, 274–75, 283–84, 288–90, 292–97, 300, 302–04, 306n80, 308–10

Babagge, Charles 139, 141, 158
Bahr, Hans-Dieter 28–29
Bastani, Aaron 1, 38–39, 43, 59, 82, 135n2, 143, 179, 188, 190–91, 224n138
Benanav, Aaron 39, 135–36, 138n10, 146n37, 154, 157, 288–93, 295–96, 309–10
Benjamin, Walter 19–20, 71, 282n21
Bernes, Jasper 34–35, 47n248, 51, 92n16, 120n45, 196, 227
Boggs, James 22, 83–87, 89, 107–08, 220n120, 222n130
Bourgeoisie 17, 79–80, 164–65, 199, 223, 281; see also ruling class
Braverman, Harry 9, 26–27, 51–52, 92, 141, 152, 154, 164–166, 168–170, 172, 185n24, 203, 205, 222n130, 261–62, 281, 284–85, 299
Brenner, Robert 4n7, 39, 82n122, 91, 154, 290
Breuer, Stefan 29–32

Čapek, Karel 139
Capital; see also accumulation
  Circulating 223, 298n46
  Constant 92, 151, 156, 160, 203, 290
  Fixed 93, 107, 146, 169, 259, 298–302, 304n72
  Variable 53, 151–52, 154, 160, 203, 290, 298, 304n72
Caspary, Adolf 121, 131–132
Circulation 34, 52–53, 105n42, 115, 212, 223, 288–89, 292–96, 309; see also capital; circulating
Class
  Class-less society see socialism and communism
  Consciousness 26, 78, 183
  Struggle 6, 8, 22, 27, 33–34, 38, 40, 51, 58, 68, 84–85, 93, 106–07, 121, 126, 157, 175, 177, 180–81, 184, 188, 195, 200, 225, 226, 230n7, 246, 251, 258, 260, 270–72
  Ruling 18n83, 29, 61, 74, 179, 181, 191; see also bourgeoisie
  Working 6, 8–9, 11, 19, 22, 26, 30, 40, 58, 62–63, 66, 73–75, 78, 79, 80–82, 118, 139n15, 147, 164, 169, 172–73, 184, 188, 206, 218
Cohen, G.A. 2
Communism 3, 18, 33–36, 48, 70n64, 75, 86, 188, 190–91, 306n80
  "Fully Automated Luxury" 1, 35, 38, 190–91, 204; see also Bastani, Aaron; see also socialism
Communist, socialist, and left parties and movements 58, 74, 78
Computers 36, 49, 51, 85–86, 88, 90, 92–93, 95, 105, 108, 111, 118, 123, 130, 160, 165, 167, 172, 176, 184, 206, 228, 237, 239, 245, 247, 249, 263, 274–80, 290, 301–02, 308
Computer age 90, 113

Deindustrialization 97, 106, 154, 309
Deskilling see Labour; Deskilling
Digital technologies and infrastructure 49, 88, 112, 116–17, 123–24, 128, 132, 152, 155, 254, 256, 264–65, 271, 287n36, 294, 296, 304–05, 309

Dunayevskaya, Raya 22, 72–76
Dyer-Witheford, Nick 135, 139–40, 151, 153–54, 158, 165n23, 168n30, 171–74, 186n31, 274n3, 288–310

Ecology 36, 41–47, 80, 197, 309
Endnotes Collective 32–33; *see also* Benanav, Aaron; Bernes, Jasper
Engels, Friedrich 1, 4, 6, 15–16, 18, 45, 47, 59, 78, 158n3, 164, 183n17, 204n7, 223n134

Feminism 37–38, 80, 128, 188
Fetishism and fetishisation 19, 152, 177
  Commodity 28, 149
  Machine 46, 61, 70, 81, 130, 137–38, 140, 143–46, 148, 155
Fettering 1–2, 4, 6–7, 36–37, 41, 120, 132, 154
Feudalism 3–6, 45, 138–39, 150
Ford, Martin 89, 94, 138n11, 203n1, 222n130
Ford Motor Company 7–8, 148
Fordism 18, 48, 59, 63, 65, 119, 230–31, 289–90, 294
Fortunati, Leopoldina 127–129
Foster, John Bellamy 41–43, 113n8, 188n37, 205n9
Frankfurt School 14, 62, 70
Frase, Peter 36, 59, 93n19, 94n21, 98–99
Fraser, Nancy 113, 128n82
Freud, Sigmund 60, 72
Frey, Carl B. 99n32, 144n28, 164, 280n11
Fuchs, Christian 57n4, 118, 124

General intellect 1, 36, 50, 145, 156, 168–177
Gig work and economy 123, 135n1258, 270–71, 289, 296
Globalization 51, 95, 98, 198, 206, 293

Harvey, David 14, 35, 115, 118n34, 129–30, 143n26, 153, 157, 183, 213n75, 230n8
Hegel, G.W.F 71–72, 161–63, 166, 287
Heidegger, Martin 8, 57–58, 70
Heinrich, Michael 13–14, 146n38, 154–55
Hornborg, Alf 39n204, 45–47, 49, 114–117, 130, 143
Human Needs 70–71, 77, 179, 186, 200, 222–23

Kautsky, Karl 16–17
Kjøsen, Atle Mikkola *see* Dyer-Witheford, Nick
Keynes, John Maynard 35–36, 65, 86

Labour
  Academic 256–62, 264, 268–72
  Artistic 280–83
  Conditions of 22, 34, 63, 106, 219, 244, 249, 251–55
  Costs 129, 244–46, 248–49, 262, 296
  Deskilling 9, 26, 158–77, 204, 206, 213, 221–22, 256, 260–61, 263, 266, 271, 273, 299, 307, 309
  Division of 21–22, 101, 141, 107, 167–68, 271, 286
  Hidden 111–134
  Human vs. animal/machine 44, 158, 162, 174, 297–302, 308–
  Market 88, 93–97, 144, 263, 269
  Movement 78, 254
  Obsolescence of 57, 60
  Organization of 63; *see also* Communist, socialist, and left parties and movements
  Platform 123, 267, 270, 294
  Process 26, 59, 67, 71, 98–99, 103–6, 160–62, 203, 205, 298–99
  Separation of Conception from Execution 9–10, 170, 205, 261, 281, 284–87, 304
  Skill 9–10, 159, 163–177, 205, 214, 257, 269, 298, 307; *see also* Labour, Deskilling
  Socially necessary 13, 69, 149, 153, 179, 185, 197
  Strikes 5, 8, 72, 84–85, 107, 218–20, 243, 271; *see also* Communist, socialist, and left parties and movements
  Theory of value *see* Value
  Time 2, 13–15, 72, 92, 112–13, 133, 149–53, 185
Laboria Cuboniks 37–38
Laclau, Ernesto 59
Lenin, V.I. 17–18, 25, 38, 67, 144
Logistics 34, 203, 207, 211, 217, 221, 223, 227, 229–34, 238, 240–44, 250–51, 255, 288, 295–96, 303
Luddites and Luddism 40, 68, 139–40
Lukács, György 19, 62, 161

INDEX   349

Malm, Andreas   4n7, 19, 44–45, 82
Management   24–26, 79, 84–85, 126, 204–5, 219, 230, 232–33, 240–42, 244, 249, 250–51, 255, 260–61, 263–64, 269–71, 294, 296, 299
Marcuse, Herbert   22–24, 57–61, 222
Marginalism   50, 140–41, 143
Marx, Karl
   1844 Manuscripts (Paris Manuscripts)   162, 163, 173, 185
   Capital, Volume One   5, 8–11, 60, 158, 163, 166, 173, 203, 218, 221, 249, 260, 261, 271, 279, 282, 295, 297, 298, 300, 301
   Capital, Volume Two   212, 221, 293, 301
   Capital, Volume Three   15, 69, 167, 172, 205
   Communist Manifesto   6, 170, 204, 223
   Contribution to the Critique of Political Economy   6
   Grundrisse   1–2, 11–15, 71–72, 73, 76–77, 132, 163, 168–169, 173n7, 261, 298
   German Ideology, The   4, 297
   On Technology   7–15, 160, 204–05, 218, 221
   Poverty of Philosophy, The   5, 7
   post-Marxism   59, 62, 70
   Productive-force determinism   3–7
Mason, Paul   35–36, 59, 105
Mau, Søren   4n7
Mechanisation   9, 58–59, 62, 70, 73, 75, 77, 81–82, 154, 223, 288, 299
Monopoly Capitalism   80, 205, 209, 221, 224
Mouffe, Chantal   59, 178
Mueller, Gavin   40
Mumford, Lewis   23n112, 70n68, 71n70

Negri, Antonio   12, 37, 92, 156, 306n80
Noble, David   7, 126, 256, 260, 272

Panzieri, Raniero   20–21, 92
Phillips, Leigh   41n217, 43
Platform economy   *see* labour, platform
Plekhanov, Georgi   17
Political economy   6–7, 15, 26, 27–28, 31, 50, 83n1, 104, 105n42, 139, 141, 149, 154, 160, 180, 206–10, 224n137
Postone, Moishe   14, 27, 31, 60–61, 81–82, 92, 155, 175

Precarity   39, 97, 105, 108, 112, 123, 125, 155, 203–04, 206, 214–15, 221–23, 260, 269, 271, 273, 291, 296
Productivity   20, 22, 31, 39, 44, 70, 75, 79, 83, 85–86, 88–91, 93–94, 98–100, 102–05, 113, 114, 127, 146, 149, 152–53, 172, 176, 179, 185, 187, 204–05, 211–13, 224, 247, 249, 265–67, 269, 272, 289–91, 295, 297
Productive forces   1, 3–7, 16–19, 24, 25, 37, 39, 45–47, 69, 77, 86, 107, 120, 135, 148–51, 152–53, 187, 199; *see also* Marx, Karl, productive-force determinism
Proletariat *see* working class

Race and racialization   3, 38, 49, 78, 84–85, 112, 118, 133, 167
Reification   74, 81, 179, 181, 191, 199, 200
Retail   89, 94, 96, 97, 101, 107, 203, 206–10, 219, 221–24, 226, 230–33, 235, 239–45, 248, 250, 251–52, 254–55, 291, 293–96, 303
Revolution
   Proletarian   5–7, 14, 19, 21, 23, 29–30, 33, 42, 47, 58, 74, 77, 80–81, 82, 135–36, 196, 222
   Industrial and technological   8, 11, 18, 29–30, 34–35, 38, 44, 45, 58, 59, 60, 63, 65–67, 71, 75, 78, 88, 90, 92n16, 95, 115, 135, 139, 146, 148, 168, 169–70, 172, 227, 229–32, 240, 241–42, 290–91, 296, 308
Ricardo, David   139, 140
RFID   226–29, 241–42, 244–55, 269, 288–89
Robinson, Joan   143n24
Roberts, William Clare   11
Robots   66, 211, 216, 290–91, 303, 304n72, 306n80
Rubin, Isaak Illich   28

Sanders, Bernie   208
Service sector   40, 48–49, 83n1, 87–89, 94, 96, 98, 99–107, 129, 133n101, 290–92, 294, 296
Silicon Valley   113, 117, 118, 203, 217, 292
Smith, Adam   11, 139–41
Smith, Jason E.   39–40, 289–93, 295–96, 306, 309–10
Smith, Tony   169–71

Socialism   1, 3, 15, 17, 18, 21, 26, 33, 36, 50, 58–60, 74, 77, 78, 80–82, 107, 120, 133, 178–79, 181–82, 185–87, 188–200, 225; *see also* communism
  Actually existing   27, 31, 67, 77
  Ecosocialism   41–48
  National   19, 68–69, 71
Software   111–134, 143, 167, 175, 210, 239, 261, 301–02, 304, 306
Sohn-Rethel, Alfred   24–26, 141n19, 298–99
Soper, Kate   46–48, 128n82
Stagnation   39, 83n1, 90, 111, 129, 205, 221n128, 224, 288–310
Steam Engine   5, 8, 44–45, 166, 204
Steinhoff, James *see* Dyer-Witheford, Nick
Supply Chain   128, 146, 198, 226–27, 230–33, 240–45, 248, 250–55, 293
Surveillance   108, 135n1, 176, 203–04, 206, 212–15, 262, 265, 267, 289, 296, 303
Srnicek, Nick   1, 35–37, 59, 69, 92–93, 99, 179, 182, 189–91, 193–95
Surplus population   10, 22, 68, 87, 132, 154, 173, 309

Taylor, Astra   112, 128n82, 140
Taylor, Frederick Winslow   205, 265
Taylorism   17, 25–26, 59, 62–63, 65, 154, 165, 170–71, 205, 261, 294, 299
Technocracy   25, 62, 70–71
Thermodynamics   46, 113–114
Tronti, Mario   12
Trotsky, Leon   18–19, 72, 76

Universal Basic Income   35–37, 39, 135, 144, 225, 292, 309
Universal Basic Services   224
Unemployment   3, 36, 39, 61, 64–68, 70, 74, 78, 81, 84, 86–88, 91–94, 97, 100, 105, 139–40, 147, 222, 224, 296, 302
Unions   21–22, 58, 72, 78, 84–85, 106, 108, 205–06, 215, 218–20, 223–24, 226, 258, 270, 272, 309
Ure, Andrew   139, 141

Value (Marx)   2, 13, 27–31, 46, 76, 82, 83n1, 92, 124, 135–57, 272, 283, 286–87, 297–300, 302, 309
  Exchange   13, 143n24, 286
  Use   13, 15, 28, 30, 33, 103, 142, 143n24, 180–81, 286, 298n46
  Surplus   2, 9, 14, 104, 105n42, 131, 146–47, 150–52, 153–55, 186, 190, 203, 205, 223, 257, 293, 297, 300, 310

Wajcman, Judy   127n72, 73, 77, 128–29, 139n11
Williams, Alex   1, 35–37, 39, 59, 69n61, 92–93, 99, 135n2, 139n11, 179, 182, 189n40, 190–91, 193, 194–95
Wood, Ellen Meiksins   4n7
Work   *see* labour
Working Class   6, 8–9, 11, 17, 19, 26, 29–30, 40, 58, 61–63, 74, 79, 81, 118, 147, 155, 164, 184–85, 188, 204, 206, 218, 222, 227, 302n62

Xenofeminism   37–38

www.ingramcontent.com/pod-product-compliance
Lightning Source LLC
Chambersburg PA
CBHW070609030426
42337CB00020B/3725